FOURTH EUROPEAN ELECTRO-OPTICS CONFERENCE

Volume 164

Contents

Seminar Committee

FOURTH EUROPEAN ELECTRO-OPTICS CONFERENCE

Volume 164

Steering Committee
Dr. J. Paul Auton
Cambridge Consultants Ltd.
United Kingdon

Prof. Bertil N. Colding
Royal Institute of Technology
Sweden

Dr. Herbert A. Elion
Arthur D. Little Inc.
Cambridge, Massachusetts USA

Dr. Dirk J. Kroon
Philips Research Labs.
Eindhoven, The Netherlands

Jacques Ragot
SORO Electro-Optics SA
France

Dr. Hans Tiziani
Wild Heerbrugg Ltd.
Switzerland

Chairman Session 1—New Developments in Components and Electro-Optics
Dr. J. P. Krumme
Philips Research Laboratories
Eindhoven, The Netherlands

Co-Chairman
Dr. B. Hill
Philips Forschungslaboratorium GmbH
Hamburg, West Germany

Chairman Session 2—Electro-Optics in Communications and Imagery
Dean Brian J. Thompson
University of Rochester
Rochester, New York USA

Co-Chairman
Michel Treheux
C.N.E.T.
Lannion, France

Chairman Session 3—Electro-Optics in Chemistry
Dr. Dirk J. Kroon
Philips Research Laboratories
Eindhoven, The Netherlands

Chairman Session 4—Electro-Optics in Medicine
Dr. F. W. Hofmann
Siemens AG
Erlangen, West Germany

Co-Chairman
Dr. L. H. J. F. Beckman
NV Optische Industrie de Oude Delft
Delft, The Netherlands

Chairman Session 5—Electro-Optics in Measurement and Control
Dr. Lionel Baker
Sira Institute Ltd.
Kent, United Kingdom

Co-Chairman
Dr. Hans Tiziani
Wild Heerbrugg Ltd.
Heerbrugg, Switzerland

Chairman Session 6—Focus on Industry
J. Ragot
SORO Electro-Optics SA
France

Co-Chairman
Dr. J. P. Auton
Cambridge Consultants Ltd.
United Kingdom

Chairman Session 7—High Power Laser Metalworking
Dr. Helmut Walther
Fiat Research Center
Torino, Italy

Co-Chairman
Dr. J. Wright
J K Lasers Ltd.
United Kingdom

INTRODUCTION

This book was compiled from the proceedings of the Fourth European Electro-Optics Conference and Exhibition held in Utrecht, Netherlands. Organised by Sira Institute Ltd., EEO '78 was the fourth conference in a series which has presented recent major advances in electro-optics with emphasis on the applications of electro-optical devices and systems. Sessions were presented in the following areas: new developments in components and electro-optics, electro-optics in communications and imagery, electro-optics in chemistry, electro-optics in medicine, electro-optics in measurement and control, focus on industry and high power laser metalworking.

FOURTH EUROPEAN ELECTRO-OPTICS CONFERENCE

Volume 164

SESSION 1

NEW DEVELOPMENTS IN COMPONENTS AND ELECTRO-OPTICS

Session Chairman
Dr. J. P. Krumme
Philips Research Laboratories
Eindhoven, The Netherlands

Session Co-Chairman
Dr. B. Hill
Philips Forschungslaboratorium GmbH
Hamburg, West Germany

MINIATURE NEODYMIUM LASERS: PRINCIPLES AND ASPECTS FOR INTEGRATED OPTICS

Gunter Huber

Institut fur Angewandte Physik, Universitat Hamburg
Jungiusstr. 11, 2000 Hamburg 36

Abstract

The search for high gain Nd^{3+} laser materials has resulted in a new class of materials such as NdP_5O_{14}, $NdLiP_4O_{12}$, $NdAl_3(BO_3)_4$ and $Nd(Al,Cr)_3(BO_3)_4$. The basic properties and principles of the materials are described: limits of concentration, fundamental rare earth spectroscopy, fluorescence quenching and laser action. The lasers have typical active volumes of 10^{-6} to 10^{-8} cm^3 and operate at 1.06µm and 1.3µm with thresholds in the submilliwatt range. The connection between these lasers and integrated optics is shown in terms of fundamental considerations. Epitaxial growth as well as planar waveguide structures have been demonstrated. Various examples of lasers, e.g., bulk lasers, intracavity SHG, flashlamp pumping, and LED pumping are described, illustrating the materials and the design principles. The approach of cross pumping Nd^{3+} by Cr^{3+} allows to improve the spectral power density by an order of magnitude. A simple sun-pumped rare earth laser device should be realizable.

Introduction

The principle problem in miniaturization of lasers is how to achieve a sufficient high inversion density in a small volume. The maximum density of excited states N is always limited to

$$N \leqslant D_c^{-3} \tag{1}$$

where D_c is a critical distance depending on the ion type involved. In principal, rare earth inner 4f shell excitations can have $D_c \approx 5\text{Å}$. Unfortunately in the first Nd^{3+} laser Nd:YAG the critical distance is approximately 15Å. Due to a strong resonant quenching interaction (cross relaxation) the useful concentration od Nd^{3+} ions in Nd:YAG is limited to a few per cent.[1] Now the discovery of neodymium pentaphosphate NdP_5O_{14}[2] has opened the way to fabrication of a new class of miniature solid state lasers with low pump threshold. Neodymium is a stoichiometric constituent rather than a dopant. In spite of the high neodymium concentration of $4 \cdot 10^{21} cm^{-3}$, which is close to the limit of equation (1) the fluorescence properties are not degraded. The high concentration is possible because drastic resonant fluorescence quenching between Nd^{3+}-pairs is absent. In the meantime several compounds with similar properties have been discovered (see next section).

An inherent disadvantage of Nd^{3+} lasers is that they have to be pumped optically. Further improvements can be made by dividing up the absorption and emission processes in seperate channels. Two ways are possible:

(i) The compound is excited via a strong broad band absorber A, which transfers the energy efficiently to the acceptor Nd^{3+}. Thus the spatial and spectral power density is increased. Note, that fully concentrated stoichiometric materials are ideally suited for this case, because each sensitizer A has at least one nearest acceptor Nd^{3+} in the unit cell.

(ii) The rare earth ion is excited via inter band transitions or collisions in a semi-conductor, which opens the possibility of direct electrical pumping.

Up to now (i) has been succesfully demonstrated [3] and (ii) is in progress.

Basic Spectroscopic Properties

The Nd^{3+} ion is one of the most favourable laser ions of the periodic table for the near infrared region: narrow linewidths (<30Å), long lifetime of the metastable excited state (10^{-4}s), and low population of the lower laser level at room temperature ($\sim 10^{-5}$). From the last it follows that the required inversion density for laser action is low and is determined by the losses of the cavity in most cases. The left hand side of Figure 1 shows a pump cycle within the neodymium ion. LED pumping is efficient via the $^4F_{5/2}$ absorption at 0.8µm.

The $4f^n$ orbitals of the rare earths lie well inside the $5s^2 5p^6$ closed shells. Any

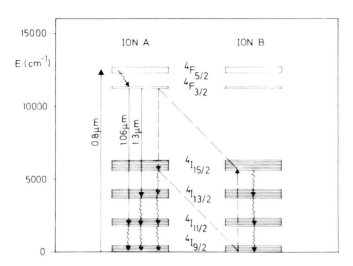

Fig. 1. Pump cycle and cross relaxation (dashed lines) in Nd^{3+} materials. Manifolds are roughly to scale, crystal field splittings are not. Involved phonons are indicated.

perturbations of the crystal field or ligand orbitals are small compared with the intraionic interaction. LS-coupling still holds and J remains a good quantum number. The 2J + 1 degenerate energy levels $^{2S+1}L_J$ are split by the crystal field depending on symmetry. The electric dipole transitions between 4f states are parity forbidden and only become allowed if the rare earth site is noncentrosymmetric (forced electric dipole transitions). The lack of inversion symmetry produces admixtures of opposite parity wavefunctions. Both intraionic 5d-4f admixtures as well as admixtures of ligand wavefunctions (covalency) are used to explain the observed transition probabilities and selection rules[4]. However, pure intraionic 4f-5d orbital mixing does not represent the full truth, because the unscreened, relatively expanded 5d wavefunctions cannot be described by an unaffected local ionic state. This is confirmed by the observed strong

Stokes shifts of 4f-5d transitions[5] due to 5d-lattice interaction. Further details are reported elsewhere.[4]

As mentioned before, cross relaxation is the basic quenching process in Nd^{3+} materials. The dashed lines in Figure 1 show a resonant cross relaxation which is the dominant quenching interaction in Nd:YAG. Although the coupling mechanism between Nd^{3+} pairs is not fully understood (dipole-dipole, dipole-quadrupole or exchange interaction), it turns out that the position of the $^4I_{15/2}$ level determines in general whether the material is suitable for high Nd^{3+} concentrations or not. If cross relaxation is non resonant (e.g. in NdP_5O_{14}) drastic quenching is prevented even in the case of 5 Å Nd-Nd spacings.

Stoichiometric Laser Materials

The first publications on room temperature laser action in NdP_5O_{14} have caused specific search for similar compounds and resulted so far in several succesful materials:

NdP_5O_{14}[6-8], $NdLiP_4O_{12}$[9,10], $NdKP_4O_{12}$[11], $NdAl_3(BO_3)_4$[11], $Nd(Al,Cr)_3(BO_3)_4$[3], $NdK_3(PO_4)_2$[12,13], $NdNa_3(PO_4)_2$[13], $NdK_5Li_2F_{10}$[14].

Table 1 compares three typical and important examples of stoichiometric materials.

Table 1. Comparison of Three Important Stoichiometric Lasermaterials

Material	Space Group	Nd Site Symmetry	Minimum Nd Spacing [Å]	Nd Concentration N [cm^{-3}]	Fluorescence Lifetime τ [μsec]	Largest Cross Section σ [10^{-19}cm^2]	Lowest Exp. Threshold A_T [mW]
NdP_5O_{14}	$P2_1/c$	1	5,194	$3,9\cdot10^{21}$	120	2	0,45
$NdLiP_4O_{12}$	C2/c	2		$4,4\cdot10^{21}$	120	3,2	0,20
$NdAl_3(BO_3)_4$	(R32)	(32)	5,917	$5,4\cdot10^{21}$	20	10	0,55

In table 1 τ is the fluorescence lifetime of the upper laser level $^4F_{3/2}$ and σ is the largest cross section of the transitions $^4F_{3/2} - ^4I_{11/2}$. The parenthesis in table 1 means that R32 is not exactly valid. This was currently demonstrated by He temperature absorption and emission measurements[15] as well as X-ray spectra[16]. The exact symmetry in the borate system is an important question because from the space group R32 non zero electro-optic coefficients r_{11} and r_{41} are expected. This would open the possibility of integrating the active laser material and the modulator in a single device. Contradictory published results on second harmonic generation[17] and electrooptic coefficients[18] may be

HUBER

caused by different crystal symmetries realized in the investigated samples.

NdK$_5$Li$_2$F$_{10}$ represents an exception among the materials. In spite of energetic resonance for cross relaxation, drastic fluorescence quenching is absent[19].

Bulk Lasers

Miniature lasers do not require high pump power but high pump power density according to the formula for the threshold pump power density P_T/V

$$P_T/V = \frac{h\gamma}{\sigma_E\tau} \cdot \left(\sigma_A N + \frac{L + T}{2l} \right) \qquad (2)$$

The symbols have the following meaning : P_T = threshold pump power, V = active volume, h = Planck's constant, γ = frequency of pump light, σ_E = effective emission cross section, τ = fluorescent lifetime, σ_A = effective absorption cross section (including the Boltzmann factor for the population of the lower laser level), N = density of Nd ions, L = losses of the cavity, T = transmission of mirrors, and l = active length in the crystal. For longitudinal pumping, l is of the order of the absorption length (50µm ....1mm) and the reabsorption losses $\sigma_A N$ can be neglected, if L+T is of the order of a few per cent. For transverse pumping, l can be increased so that $\sigma_A N$ dominates. The threshold pump power density can be calculated from Equation 2 and the parameters listed in Table 1.

A typical set-up for investigating bulk lasers is shown in Figure 2. When pumping with a focused laser beam active volumes of 10^{-6}cm^3 to 10^{-7}cm^3 are typical.

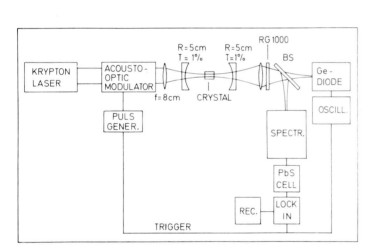

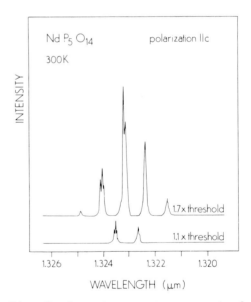

Fig. 2.Set-up for investigation of bulk lasers; cw and pulsed operation mode can be selected by an accoustooptic modulator.

Fig. 3. Room temperature spectral output of a cw NdP$_5$O$_{14}$ laser at 1.3µm wavelength.

Equation 2 can be checked exactly. Figure 3 shows the spectral output at 1.3µm of a NdP$_5$O$_{14}$ platelet 600µm thick. The output power of a NdLiP$_4$O$_{12}$ laser as a function of the absorbed pump power at 5145 Å is shown in Figure 4[20]. The pump threshold is below 1mW and slope efficiency is 43 per cent.

Intracavity second-harmonic generation was demonstrated with a Ba$_2$NaNb$_5$O$_{15}$ frequency doubling crystal in a NdP$_5$O$_{14}$ laser [21].One per cent optical power conversion of the low power dye laser pump to the second-harmonic was achieved. The results show the feasibility of obtaining visible laser radiation from miniature Nd lasers pumped by near infrared semiconductor diodes. A pocket size room temperature NdP$_5$O$_{14}$ laser has been excited by a small pulsed Xe flash-lamp[22]. The device works uncooled with 2 Hz repetition rate and 1.5mJ output at less than 1J electrical input.

The first step towards miniaturazation of the lasers is to have a cavity which is defined by the active material itself (see Figure 5). In the case of cube lasers mirrors are directly applied to the endfaces of the laser material. The cube lasers are fabricated by cementing the crystals in a glass capillary or a milled slit in a glass platelet. The

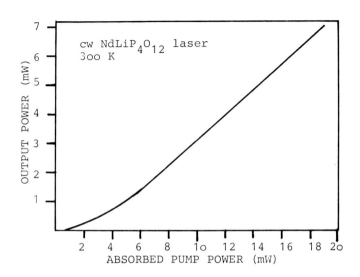

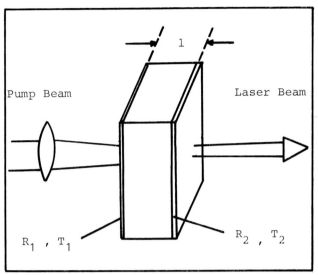

Fig. 4. Output power versus absorbed pump power of a cw $NdLiP_4O_{12}$ laser at 1.06 µ . Taken from Otsuka and Yamada[20].

Fig. 5. Typical set-up of a cube laser. R_i and T_i denote mirror reflectivity and transmittance respectively.

endfaces of the crystals are then polished plane parallel (together with the holder) and coated. Typical lengths are 100 µm up to millimeters. $LiNdP_4O_{12}$[23], NdP_5O_{14}[24] and $NdAl_3(BO_3)_4$[25] cube lasers have been fabricated.

Waveguide Lasers

The basic concept of integrated optics is to avoid the natural diffraction spreading of light by confining it in a narrow dielectric optical waveguide. Waveguiding lasers (thin film and stripe geometry lasers) can be realized by epitaxial growth of a laser on a transparent substrate. For stripe-geometry lasers an additional etching process is required to reduce the stripe width. In order to achieve perfect epitaxial growth the lattice constants of the substrate have to be matched accurately to those of the layer. The planar guides permit two-dimensional propagation of light whereas the stripe lines permit only one-dimensional propagation like optical fibers. In general, one has to distinguish between monomode guides (~ 1µm dimension) and multimode guides (dimensions up to 100µm). However, the confinement of light is limited by scattering losses due to the roughness of the surface. The scattering losses increase with decreasing thickness of the guide and increasing mode order of the guided light.

Epitaxial layers of NdP_5O_{14} were grown on $(Gd,La)P_5O_{14}$[26] as well as layers of $(Nd,Y)P_5O_{14}$ on $(Gd,Y)P_5O_{14}$[27]. The latter system was chosen to avoid ferroelastic twinning domains. The refractive index of $(Gd,Y)P_5O_{14}$ is $5 \cdot 10^{-3}$ lower than that of $(Nd,Y)P_5O_{14}$. Thus waveguiding is possible. Lasing of the layers has been observed with thresholds in the milliwatt range[27].

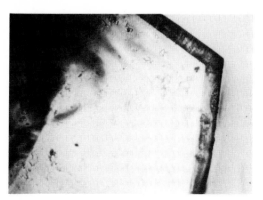

Fig.6. Epitaxial layer of $NdGa_{2.97}Cr_{0.03}(BO_3)_4$ grown on a $Gd_{0.59}La_{0.41}Ga_3(BO_3)_4$ substrate. The layer is 15 µm thick.

Very recently leaky waveguiding (refractive index of the layer lower than that of the substrate) was demonstrated for $NdGa_{2.97}Cr_{0.03}(BO_3)_4$ layers epitaxially grown on $Gd_{0.59}$ $La_{0.41}Ga_3(BO_3)_4$ [28,29] Unfortunately the layers do not yet have laser quality. Fig. 6 shows a layer 15 µm thick grown perpendicular to the crystallographic c-axis.

The use of LED's for pumping such layers is of basic interest. Depending on the type of diode the waveguide laser has to be pumped longitudinally or transversally. The required LED intensity at 0.8 µm wavelength was estimated by several workers and ranges from 10 W/cm^2 up to several 100 W/cm^2. For transverse pumping lowest intensity is required. A $NdLiP_4O_{12}$ bulk laser was longitudinally and transversally pumped by spontaneous LED's at-30°C [30]. (Nd,La)P_5O_{14} with bonded mirrors has been side pumped with two arrays of LED's in a double cylindrical-elliptical cavity [31]. Pulsed and cw lasing was obtained at-13,5°C and -49°C respectively. The threshold was 17 W/cm^2 LED intensity.

Cross Pumping of Nd^{3+} via Cr^{3+}.

$Nd(Al,Cr)_3(BO_3)_4$ [3] and $Nd(Ga,Cr)_3(BO_3)_4$ [29] are attractive materials for waveguide lasers as well as surface lasers. Figure 7 shows the transmittance and excitation spectrum for a 5 % Cr sample of $Nd(Al,Cr)_3(BO_3)_4$. The excitation

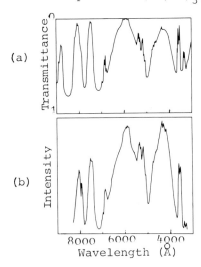

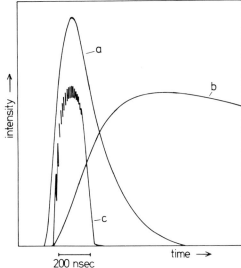

Fig.7.(a) Room temperature unpolarized transmittance of $Nd(Al,Cr)_3(BO_3)_4$.
(b) Excitation spectrum .

Fig.8.(a)Exciting dye laser pulse
(b)Response of Nd^{3+} fluorescence
(c)Superradiance. All curves are measured pumping Cr^{3+} at 6120 Å wavelength.

spectrum reproduces the absorption spectrum which means full energy transfer from Cr^{3+} to Nd^{3+}. This is confirmed by the observed fast transfer time, which is less than 100 ns in Figure 8. The reason for the efficient transfer is that each excited Cr^{3+} has a nearest Nd^{3+} ion which is 3,5 Å apart. Note that this is a consequence of rare earth stoichiometry. Very recently cw laser operation by cross pumping has been observed in $Nd(Al,Cr)_3(BO_3)_4$ [32]. Due to the strong broad band absorption of Cr^{3+} the material offers a order of magnitude reduction of the spectral density of pump light. Sun-pumping should be possible using a simple arrangement of focusing lens and cavity.Unfortunately powerful LED's in the visible region are not yet available but the combination of sun-pumping, waveguide structures and possible electrooptic coefficients encourages the work on this material.

Acknowledgements

The author wishes to thank Prof. Hans-Günter Danielmeyer for encouragement and helpful discussions. He wishes to thank Thomas Härig, Horst-Dieter Hattendorff, Melchior Leiß, Wilfried Lenth, Dr. Friedrich Lutz, Johannes Müller, Dieter Rüppel and Gert Weber for support.

References

1. Geusic, J.E., Solid State Maser Research (optical), Final Report AD-482-511, 1965.
2. Danielmeyer, H. G. and Weber, H. P., "Fluorescence in Neodymium Ultraphosphate", IEEE J. Quantum Elect., QE-8, pp. 805-808. 1972.
3. Hattendorff, H.-D., Huber, G. and Danielmeyer, H. G., "Efficient Cross Pumping of Nd^{3+} by Cr^{3+} in $Nd(Al,Cr)_3(BO_3)_4$ Lasers, "J. Phys. C, Vol. 11, pp. 2399-2404. 1978.
4. Huber, G., Current Topics in Materials Science, North Holland, to be published.
5. Lenth, W., Ph. D. Dissertation, Hamburg Univ., to be published.

6. Weber, H.P., Damen, T.C., Danielmeyer, H.G., and Tofield, B.C., "Nd-Ultraphosphate Laser", Appl. Phys. Lett., Vol.22, pp. 534-536. 1973.

7. Krühler, W.W., Jeser, J.P., and Danielmeyer, H.G., "Properties and Laser Oscillation of the (Nd,Y) Pentaphosphate System", Appl. Phys., Vol. 2, pp. 329-333. 1973

8. Danielmeyer, H.G., Huber, G., Krühler, W.W., and Jeser, J.P., "Continuous Oscillation of a (Sc,Nd) Pentaphosphate Laser with 4 Milliwatts Pump Threshold," Appl.Phys., Vol. 2, pp. 335-338. 1973.

9. Yamada,T., Otsuka, K., Nakano, J., "Fluorescence in Lithium Neodymium Ultraphosphate Single Crystals," J. Appl. Phys., Vol. 45, pp. 5096-5097. 1974.

10. Otsuka, K., Yamada, T., Saruwatari, M., and Kimura, T., "Spectroscopy and Laser Oscillation Properties of Lithium Neodymium Tetraphosphate," IEEE J. Quantum Electr., Vol. QE-11, pp. 330-335. 1975.

11. Chinn, S.R. and Hong, H.Y.-P., "Cw Laser Action in Acentric $NdAl_3(BO_3)_4$ and $KNdP_4O_{12}$, "Opt. Comm., pp. 345-350. 1975.

12. Hong, H.Y.-P. and Chinn, S.R., "Crystal Structure and Fluorescence Lifetime of Potassium Neodymium Orthophosphate, $K_3Nd(PO_4)_2$, a New Laser Material," Mat. Res. Bull., Vol. 11, pp. 421-428. 1976.

13. Chinn, S.R., Hong, H.Y.-P., "Fluorescence and Lasing Properties of $NdNa_5(WO_4)_4$, $K_3Nd(PO_4)_2$," Opt. Comm., Vol. 18, pp. 87-88. 1976.

14. Lempicki, A., McCollum, B., Chinn, S.R., and Hong, H.Y.-P., "Lasing and Fluorescence in $K_5NdLi_2F_{10}$, "Tenth IQEC, Atlanta, May 1978.

15. Härig, T., Masters Thesis, Hamburg Univ. 1978.

16. Lutz, F., private communication.

17. Filimonov, A.A., Leonyuk, N.I., Meissner, L.B., Timchenko, T.I., and Rez., I.S., "Nonlinear Optical Properties of Isomorphic Family of Crystals with Yttrium-Aluminium Borate (YAB) Structure", Kristall und Technik, Vol. 9, pp. 63-66, 1974.

18. Winzer, G., Möckel,P.G., and Krühler, W.W., "Laser Emission from Miniaturized $NdAl_3$ $(BO_3)_4$ Crystals with Directly Applied Mirrors," IEEE, J. Quantum Electr., to be published.

19. Lempicki, A., and Chinn, S.R., private communication.

20. Otsuka, K., and Yamada, T., "Continuous Oscillation of a Lithium Neodymium Tetraphosphate Laser with 200/uW Pump Threshold", IEEE, J. Quantum Electr., QE-11, pp. 845-846. 1975.

21. Chinn, S.R., "Intracavity Second-Harmonic Generation in a Nd Pentaphosphate Laser", Appl. Phys. Lett., to be published.

22. Chinn, S.R., and Zwicker, W.K.,"Flashlamp Excited NdP_5O_{14} Laser", Appl. Phys. Lett., to be published.

23. Kimura, T., Uehara, S., Saruwatari, M., and Yamada, T., "800 Mbit/s Optical Fiber Transmission Experiments at 1.05/um", Int. Conf. Integrated Optics and Optical Communication Paper C 7-2, Tokyo, July 1977.

24. Winzer, G., Möckel, P.G., Oberbacher, R., and Vité, L.,"Laser Emission from Polished NdP_5O_{14} Crystals with Directly Applied Mirrors," Appl. Phys., Vol. 11, p. 121. 1976.

25. Winzer, G., Möckel, P., and Vité, L., "Experimental Investigation of an $NdAl_3(BO_3)_4$ Miniature Laser with 215/um Resonator Length", Int. Conf. Integrated Optics and Optical Fiber Communication, Paper B 3-3, Tokyo, July 1977.

26. Huber, G., Jeser, J.P., Krühler, W.W., and Danielmeyer, H.G., "Laser Action in Pentaphosphate Crystals," IEEE, J. Quantum Electr., Vol. QE-10, p. 766. 1974

27. Krühler, W.W., Plättner, R.D., Fabian, W. Möckel, P., and Grabmaier, J.G.,"Laser Oscillation of $Nd_{0.14}Y_{0.86}P_5O_{14}$ Layers Epitaxially Grown on $Gd_{0.33}Y_{0.67}P_5O_{14}$ Substrates, Optics Comm., Vol. 20, pp. 354-355. 1977.

28. Rüppel, D., and Ulrich, R., to be published.

29. Rüppel, D., and Lutz, F., to be published.

30. Saruwatari, M., and Kimura, T., "LED Pumped Lithium Neodymium Tetraphosphate Lasers", IEEE, J. Quantum Electr., Vol. QE-12, pp. 584-591. 1976.

31. Budin, J.-P., Neubauer, M., and Rondot, M., "Miniature Nd-Pentaphosphate Laser with Bonded Mirrors Side Pumped with Low-Current-Density LED's, " Appl.Phys. Lett., Vol.33, pp. 309-311. 1978.

32. Hattendorff, H.-D., and Huber, G., to be published.

OCTAL, A NEW NEODYMIUM GLASS LASER BUILT IN THE
CENTRE d'ETUDES de LIMEIL   FOR LASER FUSION EXPERIMENTS

M. Andre, J. C. Courteille, J. Y. Le Gall, P. Rovati

Commissariat a l'Energie Atomique — Centre d'Etudes de Limeil
B. P. No. 27-94190, Villeneuve-Saint-Georges, France

OCTAL (Figure 1) is an eight beams Neodymium Silicate glass laser working in the 2 TW and 1 kJ region with pulse durations ranging from 30 ps to a few ns.

It has been built for implosion experiments with D.T. filled microspheres targets of typically 100 µm in diameter.

High flux spatial filters and new technology rod amplifiers insure a good beam quality. Routine performances are 1.6 TW in 250 ps at the exit of the 90 mm diameter rod amplifiers, which corresponds to 1.3 TW focusable on target.

Figure 1. Laser Assembly

Figure 2 shows a schematic diagram of the laser assembly:

> . the two Yag oscillators (one is single mode and Q-switched, the second mode-locked) followed by preamplifiers and pulse shaping;

> . the common part of the laser ending by a 64 mm diameter rod amplifier and a spatial filter (60 GW level) and beginning by a photographic film soft aperture

(apodiser);

. four beams, each of them including: soft aperture (quadratic profile lens and
dye liquid), 64 and 90 mm diameter rod amplifiers, spatial filter and Faraday
Rotator;

. eight beams composed of 90 mm diameter rod amplifiers and dye cell which is
necessary to minimize the amplified spontaneous emission on the target at a
level lower than 1 mJ per beam.

The structure of the laser has been optimized using a theoretical code.

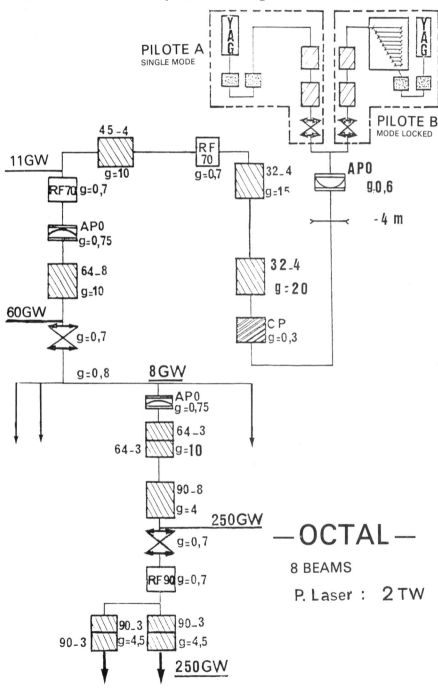

Figure 2.   Schematic Diagram of the laser assembly

Figure 3 shows the result of such calculations which agree very well with the experiment, except for the beginning of the laser chain where the radial distribution of energy is strongly concerned by diffraction.

Centre $\tau = 200\,pS$

	J/cm^2	GW/cm^2	B	ΣB	α cm^{-1}	Gain in Energy	J	GW à t = 180 pS	$\overline{W}p$ KJ/F
Ø90	0,62	3,1	1,54	2,64	0,042	4,4	54	271	8
						0,450			
Beam splitter									
Fa	0,43	2,15	1,1				27,3	137	
Spatial filter						0,75			
Ø90	0,86	4,3	3	4,56	0,029	3,4	52	262	7
V									
Ø64	0,58	2,9	1,56		0,047	6,9	15,2	77	5,7
	0,095	0,48							
Apodiser / Beam splitter / Spatial filter						0,155	2,2	11	8
Ø 64	0,93	4,6	3,45	6,72	0,035	7,7	14,2	71,5	
Fa			0,69			0,7	1,84	9,26	
V									
Ø45	0,47		1,41		0,047	7,4	2,79	14	
Fa			0,32			0,7	0,376		
V									
Ø32	0,31		0,74		0,06	12,6	0,57		
V									
Ø32			0,1			5,6	0,045		

Figure 3. Calculation of the main performances

Figure 4 presents the position of the laser elements on iron girders and shows a 9th laser beam which is used for diagnostic purpose of the plasma (interferometry and X-ray shadowgraphy).

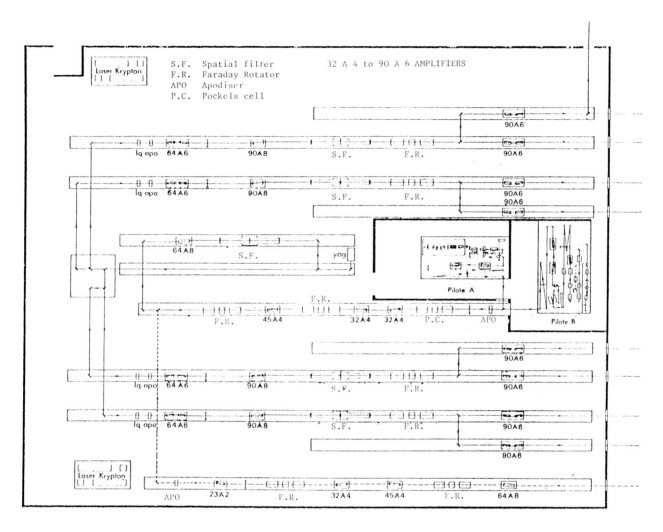

Figure 4. Laser elements positioning

On figure 5 one can see the CILAS new technology 90 mm rod amplifier which presents a small signal gain of 6.0 with Silicate glass 0.7 % Nd doped. The cooling liquid is now water but index-matching liquid may be used which should provide a 10 % increase in the extracted energy.

Figure 5. CILAS 90 mm rod amplifier

The laser is driven by a microcomputer. Figure 6 shows the operating and control desk with the color T.V. visualisation monitor used by the operator to give necessary orders. The microcomputer is a Micral M from REE in a configuration using three Central Units (UC1, UC2, UC3) to insure the following functions:

. sequence of orders for laser elements;

. control of the "ready-to-shot" state of those elements;

. checking of all securities (doors, positioning of alignment mirrors,..);

. connection with the calculator assembly used to collect the diagnostics of the interaction phenomena;

. collection of calormetric measurements of the laser itself.

Figure 6. Operating and control desk

As previously said, two pulse generators can be used which have been manufactured by Quantel. The first one is presented on figure 7. It is made of a single mode Q-switched Yag oscillator delivering a spatially and temporally smooth pulse followed by a fast double Pockels-cell device cuting out a pulse of 120 ps rise time and 250 ps to a few ns duration.

This pulse of 0.1 mJ energy is then amplified by two 16 mm diameter glass rod amplifiers, spatially filtered and driven to the beginning of the laser itself at a level of 20 mJ.

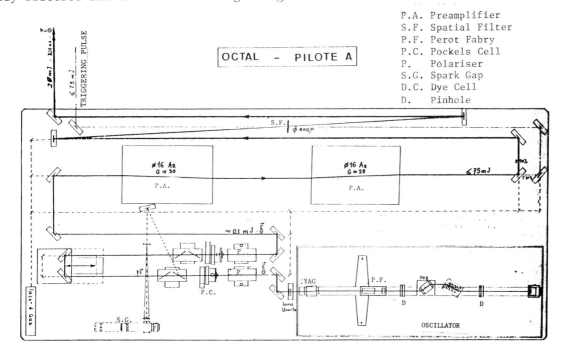

P.A. Preamplifier
S.F. Spatial Filter
P.F. Perot Fabry
P.C. Pockels Cell
P. Polariser
S.G. Spark Gap
D.C. Dye Cell
D. Pinhole

Figure 7. Schematic diagram of the first pilote

This generator assembly is shown on figure 8. An energy stability of ± 5% on 95% of the shots and the accurate single mode operation is obtained by the use of "Zero dur" ceramic glass support and molecular adhesion of the Fabry-Perot interferometer.

Figure 8. Single-mode oscillator with fast Pockels-cells device and preamplifiers

The second generator presented in figure 9 uses a passive mode locked oscillator.

One pulse of the emitted train is selected by the same kind of Pockels-cell assembly used in the previously described generator. This pulse is then amplified and temporally shaped to obtain:

. a prepulse of adjustable amplitude from 0 to 1 ns before the main pulse;

. the main pulse composed of 10 pulses of adjustable amplitude, 80 ps time separated.

This complex pulse is then preamplified, spatially filtered, and introduced at a level of 5 to 10 mJ in the laser itself.

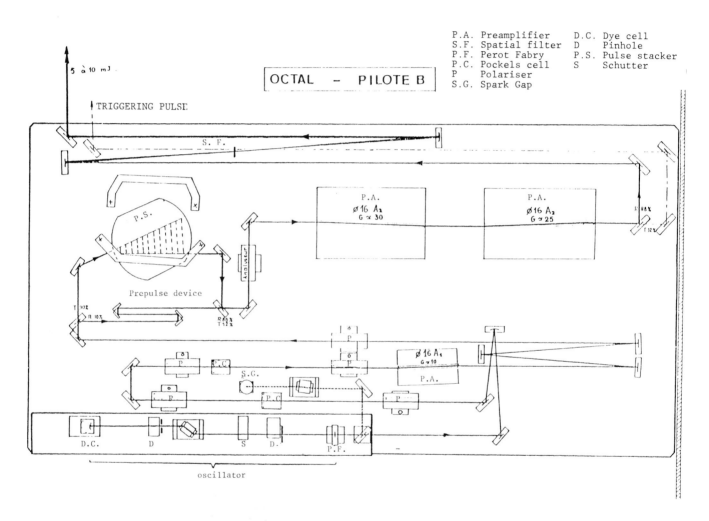

Figure 9. Schematic diagram of the second pilote

Figure 10 shows the generator which energy stability is about = 10% for 80% of the shots.

The arrangement of the two generators has been conceived so that it is very easy to use either one or the other.

Figure 10. Passive mode locked oscillator with pulse selection, pulse stacker and preamplifiers

Figure 11, concerning the beam quality presents the values of the divergence between the real wavefront of the laser and the nearest perfect sphere. This measurement made by the Hartmann method shows that this value is less than $\lambda/2$.

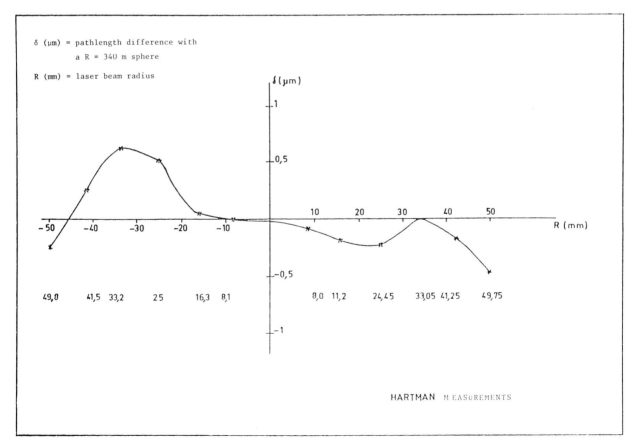

Figure 11. Wavefront distortion of the output laser beam

Figure 12, concerning the spatial repartition of the energy in the beam, shows burn patterns at different levels of the laser. At the exit of the last amplifier, for a power of 250 GW, the noise is in the range of ± 30%.

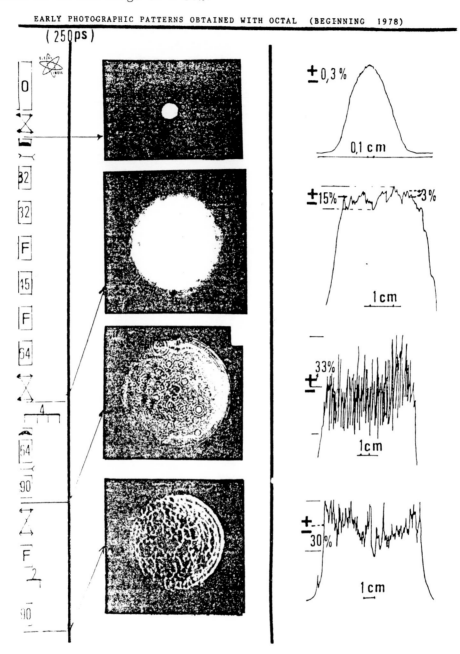

Figure 12. Early photographic patterns obtained with Octal (Beginning 1978)

OCTAL is now operating since the beginning of 1978. The synchronization of the beams has been obtained with an accuracy of ± 6 ps by a method using solid-state devices together with a new Thomson-CSF 4 GHz oscilloscope.

The laser itself is very reliable and the mechanical supports are stable in the ± 1% C temperature range of the room.

Alignments controls are achieved using Infrared T.V. detectors which can measure about 5.10^{-6} rd angle variations.

Recent measurements have shown that the power contrast ratio of the main pulse is better than 10^6 when using the mode locked oscillator.

NARROW-BANDWIDTH CW DYE LASER WITH HIGH OUTPUT POWER

W. Jitschin [+] **and G. Meisel**
Institut fur Angewandte Physik der Universitat
D-5300 Bonn, West Germany

Abstract

For investigations of fine details of atomic energy spectra light sources are required which emit light with very well defined frequency. A powerful tunable light source is the cw dye laser. Since the frequency of a free-running laser jitters, it has to be stabilized.

In order to avoid the high internal losses of a stabilization system with an intracavity electrooptic crystal, a very fast piezoelectric mirror translator has been developed. It was driven by properly designed electronics. Thus a small laser frequency bandwidth of 270 kHz rms with an output power of up to 200 mW was obtained. The laser system was tested in a sub-natural linewidth two-photon experiment with Na atoms.

Introduction

Within the different types of light sources the continuous-wave dye laser exhibits several unmatched features: well collimated light beam, high output power, easy tunability and very narrow frequency bandwidth if operated in the single-frequency mode. Thus it is almost an ideal tool for investigating the energy spectra of atoms and molecules. Several models are nowadays commercially available with a typical frequency stability of the free-running laser of 5 MHz rms, corresponding to a relative stability of $1:10^8$, which is sufficient for many applications. New techniques of laser spectroscopy now permit investigations of very fine details of atomic spectra, e.g. small hyperfine splittings of energy levels, which are caused by the electromagnetic moments of the atomic nuclei. For such investigations lasers with even higher frequency stability are required. Therefore the frequency has to be stabilized by means of a servo loop.

As the laser frequency is determined by the optical length of the laser resonator, the length has to be kept constant to a very high degree. Unfortunately the resonator is sensitive to perturbations: firstly to mechanical vibrations and acoustical noise of the environment and secondly to small irregularities in the flowing dye-liquid. A frequency stabilization servo loop compensates the disturbances by inducing opposite changes in length. If a commercial piezoelectric mirror translator which reacts up to a few kHz is used as servo element most of the mechanical and acoustical noise can be compensated resulting in a smaller laser bandwidth of about 1 MHz. A further reduction of the bandwidth requires compensation of the fast disturbances caused by the flowing dye. The upper limit of the speed of disturbance is roughly given by the inverse of the time, an irregularity in the dye jet needs to cross the laser beam. With a beam radius of 14 μm and a flow velocity of 10 m/sec one obtains $(2\pi \cdot 1.4 \ \mu sec)^{-1} \sim 100$ kHz as upper limit. A servo system in which a fast reacting electrooptic crystal is used as a servo element, can also compensate these fast disturbances, resulting in an extremely small laser bandwidth [1,2]. However the use of a crystal in the laser resonator inhibits disadvantages [3]: The resonator has to be modified and the insertion of the crystal causes a reduction of output power. Therefore fast reacting piezoelectric transducers were developed which were driven by properly designed electronic equipment. Thus an effective frequency stabilization which avoids the disadvantages mentioned above was obtained.

Stabilization of Intensity and Frequency

The laser used was a commercially available single-frequency dye laser model 580 A by Spectra-Physics. To reduce the frequency jitter of the free-running laser it was carefully shielded against noise and vibrations. The dye fluid was filtered by a membrane filter of 0.8 μm pore size to retain small particles. Air bubbels in the fluid were removed by applying a vacuum of about 1 mbar to one of the dye fluid storage tanks.

The output power of the laser was stabilized by a simple servo system: the laser output beam was sent through an electrooptic crystal (ADP) and a polarizer. A small part of the transmitted light entered a light detector and the light induced electric current was subtracted from a fixed DC current. The resulting small difference current provided the error signal which was amplified by a simple oscilloscope. The y-plate voltage of the scope was directly fed to the electrooptic crystal. Thus the intensity of the laser beam was kept constant to better than 10^{-4}.

[+] Now at the Fakultät für Physik der Universität, D-4800 Bielefeld, West Germany

For the stabilization of the laser frequency a commonly employed method was used, see e.g. [2,3]: the frequency was fixed to a reference etalon by a servo loop using the slope of an etalon transmission peak as frequency discriminator. As servo element for the frequency stabilization a special piezoelectric mirror transducer has been developed. Since a servo system cannot work stable beyond a mechanical resonance of the transducer, its lowest eigenfrequency defines the cutoff frequency of the servo system. In commonly used designs, the translator consists of a laser mirror, a supporting plate and a tube or disk of piezoelectric material between mirror and plate. The lowest resonance frequency of such a device is the eigenoscillation of the supporting plate. As the size of the plate has to be relatively high to ensure a good mount, the lowest resonance frequency is just a few kHz.

This undesired oscillation of the supporting plate has been suppressed by using a symmetric translator design with two piezos and two mirrors. The disks are electrically connected in such a way that they are driven in opposite directions. As the resulting force on the supporting plate is zero, no resonance oscillation of the plate is excited. The resonances of the symmetric transducer occur at higher frequencies. From the speed of sound data of the materials used the lowest eigenfrequencies are calculated to be 340 kHz for the longitudinal and 394 kHz for the radial mode. These values are in good agreement with the measurements. A reduction of the geometric dimensions of the translator should easily be possible, so that the first resonance would be shifted to still higher frequencies. This would permit an even higher efficiency for the laser frequency stabilization.

The small piezoelectric disks used in the fast translator only permit a small mechanical motion, which is too small to compensate large cavity variations or to tune the laser frequency. Therefore an additional piezoelectric mirror translator was constructed with almost conventional asymmetric design, but with its lowest resonance at 30 kHz. Both mirror translators were inserted into the laser resonator and driven via a frequency dividing network, so that the slower translator compensates for large, slow length changes and the fast translator compensates for fast changes which are small. Figure 1 shows a schematic of the

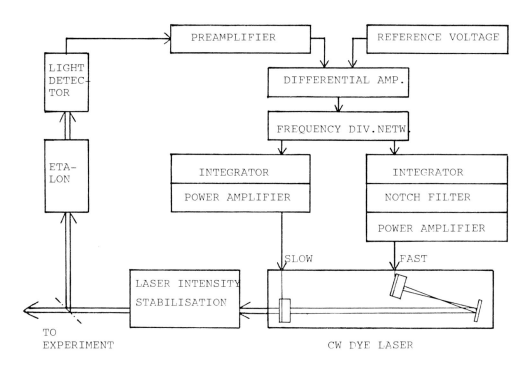

Fig. 1. Schematic of the stabilized cw dye laser system.

stabilized dye laser. The frequency stabilization servo loop is of the simple integrator type to ensure stable operation. The electronic equipment had to work from DC up to 1 MHz with negligible phase lack, so its upper frequency limit as specified by the common 3 dB bandwidth is several MHz. As the output amplifiers have to drive the capacitive load of the piezoelectric disks (1 nF for the slow translator) at high frequencies and also at high voltages, power circuit design techniques were applied.

Conclusions

The laser frequency stabilization described above compensates the frequency jitter to a high degree. Since the gain of the servo loop is proportional to the inverse frequency, slow changes are almost completely compensated, whereas some fast frequency jitter remains. Thus the frequency bandwidth of the stabilized dye laser is 270 kHz, whereas the stability of the time averaged frequency is much higher, e.g. better than 1 kHz for 1 msec averaging time.

The tuning of the laser frequency can easily be accomplished by tuning the reference etalon to which the laser is locked. By employing a novel tuning technique [4] it was possible to tune the laser with rf-accuracy. Long term stability of the reference etalon was achieved by locking the etalon to an Iodine stabilized He-Ne laser. The details of the complete laser system are given elsewhere [5].

The laser system was checked in a spectroscopic investigation. With the high output power a two photon experiment on Na atoms could be performed without any laser beam colli- mation. Thus high resolution was obtained. It was found that the laser frequency could be tuned over an interval of 1 GHz with an absolute accuracy of $\pm$ 15 kHz. Thus the small dif- ferences in the separation of the hyperfine lines in the Na 3s $\rightarrow$ 4d transition could be accurately measured resulting in a determination of the extremely small hyperfine splitting in the 4d level [6].

References

1. R.L. Barger, J.B. West a. T.C. English
   Appl. Phys. Lett. 27, 31 (1975)
2. F.Y. Wu a. S. Ezekiel
   Laser Focus, March 1977, p. 78
3. D. Anafi, R. Goldstein a. J. Machwirth
   Laser Focus, Aug. 1977, p. 72
4. W. Jitschin a. G. Meisel
   Proc. of Laser 77 Conference, Munich 1977, p. 88
5. B. Burghardt, W. Jitschin a. G. Meisel
   to be published
6. B. Burghardt, M. Dubke, W. Jitschin a. G. Meisel
   to be published

SIGNAL PROCESSING IN INTEGRATED OPTICS EMPLOYING GEODESIC LENSES

Giancarlo C. Righini, Vera Russo and Stefano Sottini
Istituto di Ricerca sulle Onde Elettromagnetiche of C. N. R.
Firenze, Italy

Abstract

The implementation of integrated optics signal processors for one dimensional signals, as for instance radar signals, is a problem attracting increasing attention. The devices in mind are miniaturized analogic processors involving Fourier transformation properties of waveguide lenses. Among different types of lenses already suggested, geodesic lenses are the one currently feasible solution when, as usual, crystal substrates with high refractive index must be used. A method for designing generalized perfect geodesic lenses has been developed by the authors. The advantage of this method is represented by the easy theoretical approach and the large flexibility in the design without requiring long computations. The employment of such lenses in integrated spectrum analyzers and two lenses correlators is discussed taking into account the characteristics of the other components of these devices, as array detectors and Bragg acousto-optic modulators. A different geodesic correlator constituted by a portion of spherical surface is also briefly described.

Introduction

The word "integrated optics" indicates a wide area of phenomena involving light guided in thin films of transparent materials [1]. Such films are laid on substrates made of glass or special crystals having a refractive index less than that of the guiding thin layer.

By using the physical properties of crystal substrates and/or suitably shaping the film guide, it is possible to realize a variety of components which can perform several elementary operations on optical waves. They are the counterparts either of bulk optics components or of microwave components. More complex devices are obtained assembling some of these elementary components.

Several fields of application of integrated optics have been already studied. Among them, the implementation of optical waveguide signal processors for one dimensional signals, as, for instance radar or communication signals, is a problem attracting a continuously increasing attention. The devices in mind are miniaturized analogic processors involving the Fourier transformation properties of optical waveguide lenses. They are expected to be compact, insensitive to vibrations and low power consuming.

In theory, integrated optical data processors can be achieved by extending to guided optics the working principles of bulk optical systems. Many configurations can be realized with a suitable assembling of integrated optical modulators, lenses and detectors. Such elementary components are among the most studied and a variety of prototypes have been tested in the last few years. However an important factor limiting so far the implementation of integrated optical processors has been the realization of high quality waveguide lenses, that is perfect or corrected lenses, compatible with the other elements of the optical circuit.

Following different approaches related either with bulk optics or with microwave optics, geodesic lenses, Luneberg lenses and Fresnel lenses have been tested in integrated optics showing good performances. However geodesic lenses are the one currently feasible solution with substrates having high refractive index as for instance $LiNbO_3$ or GaAs, which are usually used because of their convenient acousto-optic or electro-optic coefficients.

Processors Employing Geodesic Lenses

Geodesic lenses [2] are based on the possibility of distorting the 2D guide into the third dimension with a suitable shaping of the surface of the substrate. The rays follow the geodesics of the curved surface in a two-dimensional Riemann space. Therefore they are the only lenses that have performances independent of the wavelength and the mode of the guided waves.

The simplest geodesic lens consists of a quarter of a spherical surface, it perfectly focuses a collimated beam (Figure 1), but it cannot be easily inserted in a planar circuit. On the other hand every portion of spherical surface focuses with strong spherical aberrations. This drawback can be overcome giving to the lens surface an aspherical shape.

Moreover, in order to avoid serious difficulties of fabrication and high losses, a good geodesic lens has to be constituted by a surface of revolution, coupled with the planar optical circuit without discontinuities for the tangent plane.

A method for designing lenses satisfying all the above described requirements has been recently developed by the authors [3]. It has been found that perfect "generalized" geodesic lenses with revolution symmetry exist (Figure 2), they are able to image perfectly a point source P at any predetermined distance from the lens axis. The image point P' is also chosen without limitations, on the plane surface external to the lens depression.

The theoretical approach is a generalization of that followed by Toraldo di Francia to find only a particular family of lenses for microwave applications, able to focus a collimated beam at the lens edge [4]. Starting from a general theorem on the revolution surfaces, our procedure allows an analitical evaluation of the meridional curve characteristic of the desired lens without approximation.

Chosen the actual lens diameter 2a and the distance d and c of P and P' from the lens axis, the lens depression, limited by the circle B of radius b > a, must be considered as constituted by two parts. The outer part is a smooth toroidal junction to the plane, while the inner part, limited by the circle A, constitutes the true lens because only the rays crossing A can be perfectly focused in the image point.

The junction surface must satisfy well defined conditions to avoid discontinuities of the lens profile, however many different solutions exist. The most convenient profile of the junction has to be chosen among them, then the equation of the inner depression can be uniquely derived through well known theorems on the geodesics, by requiring that the condition of perfect imaging be satisfied. In general it is clear that the more b is greater than a, the more the junction between the lens and the planar surface becomes smooth.

The advantage of this general method with respect to other generalized lenses suggested by other authors, is represented by the perfect operation and the easy theoretical approach that does allow a large flexibility in the design of such lenses without requiring long and expensive computations. In fact all the lenses are described by relatively simple equations, where one has to insert the values of the four parameters a,b,d,c. The characteristics of the lens depend on their values: focal length $F = dc/(d+c)$, aperture $D/2F = a/F$, magnification $X = c/d$ while, as above mentioned, b-a gives an idea of the smoothness of the junction.

A particular case of large interest is that of a lens which focuses a collimated beam ($d \rightarrow \infty$). Moreover, if also the condition b = c is satisfied, we obtain the particular case of the family of the Toraldo lenses, having the focus at the depression edge. Figure 3 shows a lens of this type focusing a 5mm wide laser beam. The waveguide is an epoxy film doped to make visible the ray paths. The lens has F = 4.5mm and aperture 0.78.

This lens, realized with a simple technique, is a good test to valuate how a defective fabrication can deteriorate the lens performances. Concerning with spherical aberration, only a rough test has been done, reducing the beam width entering the lens (from 5mm to paraxial rays only) and looking at the focus through a microscope. No displacement of the focus itself was observed within a precision of 50μm. The resolution capability was tested by placing a grating in front of the coupling prism and looking if the undiffracted beam and the first order diffracted beams appear to be separated in the focal region. An angular separation of 3.2 mrad has clearly resolved, corresponding to an arc length separation of 14.3μm.

Spectrum Analyzer

Let us now discuss what performance could offer such a lens if actually employed in a simple signal processor. Therefore, let us consider the integrated spectrum analyzer device sketched in Figure 4 where light, coupled in from an injection laser, is collimated by a geodesic lens and then is deflected via a Bragg interaction with an acoustic surface wave modulated by the signal to be analyzed. A second geodesic lens focuses the diffracted waves on to a detector array. For instance, with a resolution of 1-5 MHz and a bandwidth of 0.5-1 GHz, such a device could be of great interest to process radar signals.

As above mentioned, in order to obtain such performances, a Bragg acousto-optic modulator seems the one possible solution, using a tilted or a phased array of transducers [5]. The substrate of the device must be made of a crystal having high piezoelectric coupling constant. The best solution seems to be a $LiNbO_3$ crystal with a Ti-diffused guiding film. An efficiency of about 5% is expected with $\lambda_{optic} = 0.9$ μm and ∿ 100 mW of input electrical power [6].

The main factors which can limit the resolution capability of the spectrum analyzer are the width of the light beam and the spacing of the detector array. It can be easily shown [6-7] that the width D of the optical beam (equal to the effective lens diameter 2a) and the resolution Δf are related by the equation: $D = 1.41$ Q v/Δf where the beam intensity distribution is supposed gaussian and truncated at e^{-2}, v is the acoustical velocity and Q is a factor describing lens quality and is equal to the actual spot size divided by the diffraction limited spot size. At the same time the spacing S of the detector array is related to the focal length F by: $F = n_{eff} v S/\lambda\Delta f$ with n_{eff} effective refractive index of the guided wave. From this equation it turns out that, if the lens above described were used with S = 14μm, the resolution available would be only Δf = 26 MHz. Therefore it is necessary to increase the focal length but, on the other hand, the substrate size must be as small as possible. For instance, in order to obtain a resolution of 2 MHz with S = 14 μm and a Q factor of 1.5 (which seems to be available with special fabrication technique of the lenses) [8-9], a focal length F = 58.8 mm and an effective lens diameter D = 2a = 4.4 mm are requested. Supposing that, in the near future, a spacing S = 7 μm become available, the focal length can be reduced to F = 29.4 mm. In practice it means that a crystal substrate of 60 x 12 mm² and 3-4 mm thick would be sufficient.

Another possibility is that of an optical magnification of the output distribution using a third geodesic lens. In this way S could be increased. It can reduce the resolution losses due to a possible displacement of the output signal line as a consequence of defective fabrication of the lenses. For instance, with magnification X = 4 and supposing S = 14μm, the same resolution of 2 MHz could be obtained with a chip of 56 x 12 mm.^{2}

Geodesic Correlators

Geodesic lenses are affected by field curvature, as well as the other lenses with revolution symmetry. It makes difficult to assemble two lenses in order to constitute an integrated correlator analogous to a bulk double diffraction system. A disposition which can reduce this effect is shown in Figure 5. The input signal must be introduced on the I line behind the first lens via a light modulator of the acousto-optic or electro--optic type. The Fourier Transform is obtained on the FT line where a suitable filter is placed. The second lens reimages the signal on the O curved line. In this way the field curvature of the lenses has no consequence at least when a pulse response is expected, while the optical system presents the advantage of a good compactness.

A different geodesic correlator having some attractive characteristics, can be realized by using a portion of spherical surface [10-11]. It is well known that a hemispherical surface images without aberration and its symmetry properties allow us to expect a FT locus on the great circle that divides in half the surface. In Figure 6 such a geodesic system is compared with its bulk analogous.

The aberrations on the FT line have been evaluated. Only a quite small amount of coma was found. As an example, if we consider, according to Rayleigh, a maximum wave aberration of $\lambda/4$, an aperture of 10° and a field angle of 9° could be used. It is to be noted, however, that the aberrations are completely corrected by the second half of the correlator.

Numerical evaluations have been made to clarify the effects of wave aberrations on the auto and cross--correlations of binary signals. The practical realization of the filter to be put on the FT line constitutes a difficult problem, therefore approximated filters with only two or three transmission levels have been also tested numerically.

For example, in Figure 7 the autocorrelation of a five pulse input signal is shown, taking into account the aberrations of the system. Both the cases of a perfect filter and of a binary filter are considered. It seems that, for pattern recognition applications, binary filters could often be used with acceptable performances.

In the communication or radar fields, several applications of the geodesic correlator can be suggested. Figure 8 shows, for instance, the sketch of the geodesic analog of a time integrating acoustooptic correlator. The comparison with the bulk system shows that even rather complex optical systems can be realized with a spherical surface. This application could be particularly interesting for the possibility of introducing the input signal directly via an optical fiber.

Conclusions

The realization of integrated signal processors is in progress even if there are many problems to solve before demonstrating the operation of a completely integrated device. The performances of such devices, obtained by assembling high quality elementary components recently developed, are expected to be much better than their bulk counterparts.

In particular, geodesic lenses have been designed suitable for a variety of integrated processors. Moreover it seems possible that, by using recent techniques tested to build aspherical surfaces, mass production of high quality geodesic lenses can be available.

References

1. Tamir,T. (ed), Integrated Optics, Springer Verlag, NY 1975

2. Righini,G.C., Russo,V., Sottini,S., and G.Toraldo di Francia, "Geodesic Lenses for Guided Optical Waves", Appl.Opt., vol.12, pp.1477-1481, 1973.

3. Righini,G.C., Russo,V., and Sottini,S., "Geodesic Optical Systems for Signal Processing", Fourth European Conference on Optical Communication, Genova, 1978.

4. G.Toraldo di Francia, "Un problema sulle geodetiche delle superfici di rotazione che si presenta nella tecnica delle microonde", Atti Fond.Ronchi, vol.12, pp.151-172, 1957.

5. Tsai,C.S., Nguyen,L.T., Kim,B., and Yao,I.W., "Guided-Wave Acousto-Optic Signal Processors for Wideband Radar Systems", SPIE vol. 128, Effective Utilization of Optics in Radar Systems, Huntsville (USA) 1977.

6. Hamilton,M.C., Wille,D.A., Miceli,W.J., "An Integrated Optical R.F. Spectrum Analyzer", Optical Engineering, vol.16, pp.475-478, 1977.

7. Anderson,D.B., Boyd,J.T., Hamilton,M.C., and August,R.R, "An Integrated Optical Approach to the Fourier Transform", IEEE J. of Quantum Electronics, vol. QE-13, pp.268-274, 1977.

8. Bor-Uei Chen, Marom,E., and Lee,A., "Geodesic Lenses in Single-Mode LiNbO$_3$ Waveguides", Appl.Phys. Lett., vol.31, pp.263-265, 1977.

9. Bajuk,D.J., "Computer Controlled Generation of Rotationally Symmetric Aspheric Surfaces", Opt.Engin. vol.15, pp.401-406, 1976.

10. Righini,G.C., Russo,V., Sottini,S., "An Unusual Correlator for Guided Optical Waves", Colloque sur l'Optique des Ondes Guidées, Paris, 1975, paper II.7

11. Righini,G.C., Russo,V., Sottini,S., "Thin Film Integrated Signal Processors", in AGARD Conference Proceedings n.219, London pp.25.1-25.14, 1977.

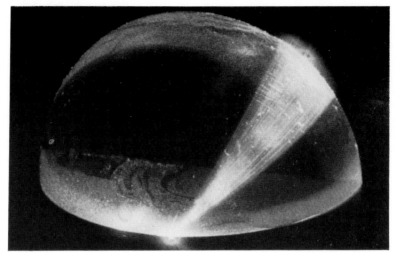

Fig.1. A geodesic lens constituted by a quarter of spherical surface focusing a collimated laser beam.

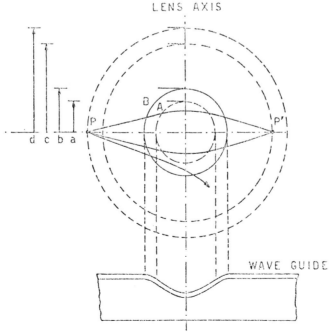

Fig.2. Top view and profile of an aspheric geodesic lens giving on the circle of radius c the perfect image P' of a point P on the circle of radius d.

Fig.3. Aspherical Toraldo lens (F = 4.5 mm) focusing a collimated laser beam.

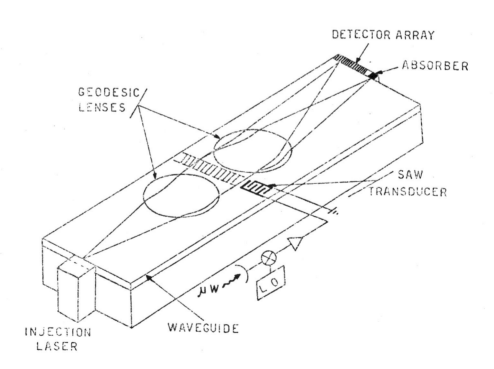

Fig.4. Sketch of an integrated optical spectrum analyzer.

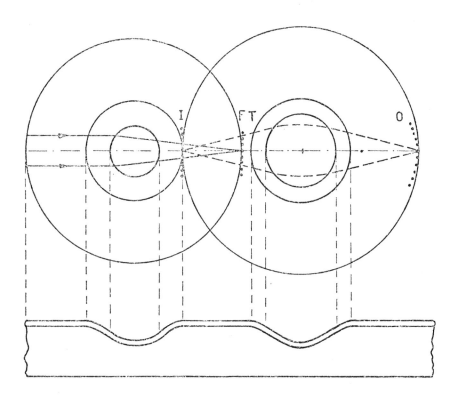

Fig.5. Top view and profile of a two-lens correlator. The input signal is placed on the I line and the filter on the FT line. O indicates the output line.

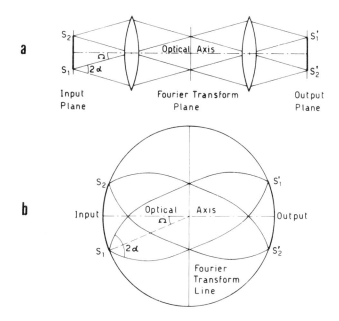

Fig.6. Top view of a hemispherical double-diffraction system (b) compared with the analogous bulk double-diffraction system (a).

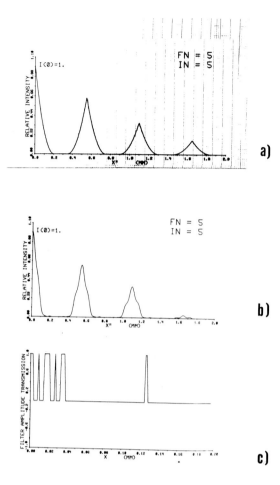

Fig.7. Autocorrelation of a five-pulse input signal: (a) with a perfect filter; (b) with a binary filter having amplitude transmission (c).

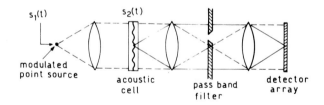

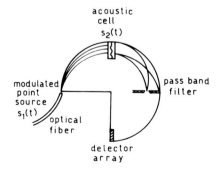

Fig.8. Sketch of the geodesic analogous of a time-integrating acoustooptic correlator already suggested in bulk optics.

RAPID SWITCHING OF PHOTOTHERMOPLASTIC DEVICES

U. Killat, G. Rabe and D. R. Terrell

Philips GmbH Forschungslaboratorium Hamburg,
2000 Hamburg 54, West Germany

Abstract

Information recording in conventional photothermoplastic devices (PTD's) entails charging, illumination and heat-development of the device. These three steps are examined with respect to the ultimate speed of photothermoplastic recording. It is shown that to a first order approximation the sensitivity of a PTD is proportional to the electrophotographic sensitivity ζ_e of the photoconductor layer. Several 'fast organic photoconductors' for use at 515 nm and 633 nm have been developed.

Rapid heat development, which is largely limited by the viscoelastic properties of the thermoplastic used, has been found to be limited to the µs-range. However, such short development times lead to overheating and therefore development times of about 100 µs are recommended for prolonged cycling.

Some recent investigations are reported on a new class of PTD's developed for point storage applications. These devices operate in the 50-200 µs range and utilize photo-induced development.

Introduction

In photothermoplastic recording optical information, generally in the form of holograms, is stored as a surface relief pattern in the thermoplastic layer. Current research in photothermoplastic recording[1],[2],[3], concerns the exploitation of PTD's to their very limits. Our research has focused upon the ultimate speed of the recording process. Before discussing this theme in detail, it is helpful to recount the basic principles of photothermoplastic recording. Figure 1 shows a conventional two layer PTD. It consists of (from top to bottom)

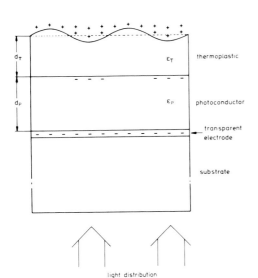

a thermoplastic layer, a photoconductor layer and a transparent electrode sputtered onto a glass substrate. Holograms are recorded by depositing a surface charge on the initially flat surface of the thermoplastic layer and then modulating this homogeneous charge distribution by exposing the photoconductor to the desired interference pattern. Upon heating the thermoplastic layer to its softening point, the layer deforms under the influence of the electrostatic stress distribution thereby producing a thin phase hologram. Rapid cooling renders the deformation permanent and heating to a still higher temperature results in the restoration of the initially flat surface.

Fig. 1. Photothermoplastic device exposed to light.

Charging

Rapid switching in PTD's will depend upon the respective speeds of the constituent processes i.e. charging, exposure and heat development and erasure. In the case of charging, various attempts have been made at replacing the slow corona charging with a faster process[2],[4]. The most successful approach has been that of Colburn et al.[4], who proposed the parallel plane charging system. In this configuration areas of 9 cm² have been charged in less than 1 ms. However, such systems are not without their drawbacks, since construction and maintenance are difficult. Moreover, the ions produced in the gas discharge apparently slowly sputter away the thermoplastic layer. In our opinion, therefore, the problem of rapid charging has not yet been satisfactorily solved. Perhaps the simplest solution is to use the corona in the precharging system of a photothermoplastic tape recorder as proposed by Moraw[3] and Lee et al.[1].

Exposure

The required exposure time, t, can be described in terms of the device sensitivity ζ and the light source intensity I_0 i.e.

$$t \sim (I_0 \zeta)^{-1} \qquad (1)$$

The device sensitivity is defined as the reciprocal of that exposure required to achieve a certain diffraction efficiency under simple two beam interference conditions. However, the quantity that is frequently available, when a particular system is being designed, is the electrophotographic sensitivity ζ_e. (This is here defined as the reciprocal of the exposure required to reduce the surface potential on the photoconductor to one half its initial value.) Since the photoconductor is the only photosensitive element in a PTD, one would expect this to determine the device sensitivity. Moreover, the electrophotographic sensitivity ζ_e is much more easily measurable than the device sensitivity ζ. Indeed, photoconductors for use in PTD's have been chosen on the basis of their ζ_e values, but in the opinion of the authors without any justification.

In order to establish a relationship between ζ_e and ζ one must consider the physical processes involved in relating the input signal (in the form of an interference pattern) to the final output signal (which can be characterized by the diffraction efficiency). Four separate steps can be identified:

  i) the conversion of the light intensity distribution into a charge and potential distribution, using the photoconductive properties of the photoconductor,

 ii) the relationship of this potential distribution to a corresponding stress distribution,

iii) the deformation resulting from the stress distribution, thus producing a phase grating,

 iv) the scattering of the reconstructing beam by the phase grating.

This somewhat complicated picture can be simplified by considering only small deformations. Under such conditions most of these relationships can be linearized. In fact the only critical relation turns out to be that between exposure and potential distribution. This can be described in terms of the equivalent circuit model shown in Figure 2. Here the

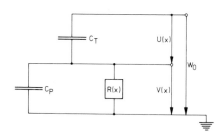

thermoplastic is represented by the capacitance C_T and the photoconductor is given by the capacitance C_P in parallel with the photoresistance R. R has been chosen such as to comply with the experimental observation that the following relationship holds for a wide range of organic photoconductors in the field range appropriate to PTD's:

$$J_P = K(1 - e^{-\alpha d_e}) I_0 E^m \qquad (2)$$

Fig. 2. Equivalent circuit model for a photothermoplastic device.

where J_P is the photocurrent; E, the electric field; m is a constant greater than unity; α and d_e are the extinction coefficient and layer thickness of the photoconductor layer; and K is a photoconductor constant related to the electrophotographic gain. This is demonstrated in Figures 3a and 3b, in which log J_P is plotted against log E and log I_0 respectively for poly(N-vinylcarbazole) (PVK) doped with 2 wt% malachite green. Similar dependences have been observed for PVK doped with a large variety of triphenylmethane and xanthene dyes (5).

One can now construct a differential equation for U, the potential across the thermoplastic layer, and calculate the initial stress exerted upon the thermoplastic's surface for a two beam interference pattern i.e. $I(x) = I_0(1 + \cos kx)$. This stress will then be proportional to the diffracted light amplitude, according to the assumptions made above, and the following relationship for the diffraction efficiency η is obtained if transmission losses are neglected:

$$\sqrt{\eta} \sim G \frac{d}{\epsilon} E^{m+1} K\alpha \cdot I_0 t \qquad (3)$$

where G represents a geometrical factor, which for two layer PTD's is given by the expression:

$$G = \frac{\epsilon^2}{d_T} \frac{1}{C_P + C_T} \quad ,$$

where ϵ is the dielectric constant of the photoconductor and d_T is the thermoplastic layer thickness; E is the initial electric field across the photoconductor layer; d is the photoconductor thickness in the PTD.

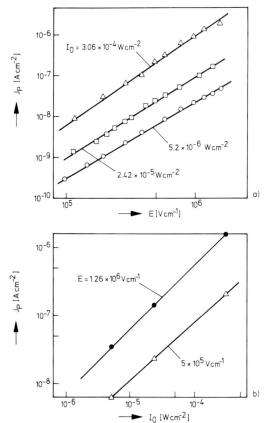

Fig. 3. Dependences of a) photocurrent, J_p, upon electric field, E, and b) photocurrent, J_p, upon incident light intensity, I_0, for PVK doped with 2 wt% malachite green.

The term $K\alpha$ is a material constant and can be determined independently by electrophotographic sensitivity measurements on the photoconductor layer itself. $K\alpha$ is related to ζ_e by the expression:

$$K\alpha = f(m) \frac{\epsilon}{d_e} \frac{\zeta_e}{E_e^{m-1}} \qquad (4)$$

where $f(m) = (2^{m-1}-1)/m-1$ and E_e is the electrical field appertaining to the electrophotographic sensitivity measurements.

The object of the derivation of (3) was a relationship between device sensitivity and the electrophotographic sensitivity of photoconductors contained in the PTD. This can be achieved by substituting (4) in (3). This produces a rather complicated expression, which can be further simplified by equating E and E_e since both electrophotographic and device sensitivity measurements are usually carried out at the maximum attainable photoconductor field E_a (i.e. $E = E_e = E_a$). This involved expression then reduces to:

$$\sqrt{\eta} \sim G \ f(m) \ d \ E_a^2 (\zeta_e(E_a)/d_e) I_0 t \qquad (5)$$

where $\zeta_e(E_a)$ is the electrophotographic sensitivity determined at an initial field E_a. Since $f(m)$ varies very little for the values of m encountered with dye-doped PVK photoconductors, the quantity $(E_a^2 \zeta_e(E_a)/d_e)$ can be taken as a figure of merit for such photoconductors in PTD's. The device sensitivity for a two beam interference pattern (see (1)) is than given by:

$$\zeta \sim G \ f(m)d \ (E_a^2 \zeta_e(E_a)/d_e) \qquad (6)$$

In the case of almost identical E_a-values, a thickness correction was sufficient to extrapolate device sensitivities from electrophotographic measurements. This is demonstrated in Table I for several photoconductors used with Staybelite Ester 10 as the thermoplastic.

	Photoconductor, d = 0.6 μm	λ(nm)	ζ_e^{-1} [μJ cm^{-2}] *	ζ^{-1} [μJ cm^{-2}] for η=0.15%	t [ms] ** for η=0.15%
Two Layer PTD's	PVK + 1 wt% crystal violet	633	71	70	7.0
	PVK + 2 wt% malachite green	633	33	33	3.3
	PVK + 1.5 wt% rhodamine 6G	515	32	30	0.03
Single Layer PTD	Copolymer of 20 mol% n-dodecyl acrylate and 80 mol% VK + 1.5 wt% rhodamine 6G	515	75	325	0.33

* normalized to d = 0.6 μm

** I_0 = 10 mW cm^{-2} at 633 nm and 1 W cm^{-2} at 515 nm

Table I: Electrophotographic and PTD-Sensitivities for Several Photoconductors.

The bottom row of numbers refer to a recently developed single layer PTD and are not directly comparable with the other values, because the thermoplastic properties are different from those of Staybelite Ester 10. In a single layer PTD the thermoplastic and photoconductor layers in Figure 2 have been replaced by a single photoconductive thermoplastic layer. An equation identical to (3) can be derived for such a device with $G = \epsilon_{PT}$, where ϵ_{PT} is the dielectric constant of the photoconductive thermoplastic.

Two points emerge from the above discussion:

1) m sec exposure times are feasible with both single and two layer PTD's in conjunction with an Ar-laser (see Table I),

2) the selection of photoconductors in PTD's should not be based on ζ_e alone, but should also include the field dependence.
The quantity $E_a^2 \zeta_e(E_a)/d_e$ is proposed as a suitable figure of merit.

Development and Erasure

Let us assume that the photoconductor fulfils the system requirements. The question then remains of how fast the thermoplastic can deform. In the absence of theoretical predictions we have tried to explore the limits experimentally. The experimental set up is shown in Figure 4. The PTD is exposed to a two beam interference pattern and the development process is monitored by a photomultiplier positioned to collect the second order light scattered by the deforming thermoplastic layer. In the top right hand corner of this Figure typical oscilloscope traces are shown, corresponding to the heat pulse - upper trace - and the photomultiplier signal - lower trace. The multiplier signal will rise to a maximum, corresponding to maximum diffraction efficiency, and then decrease due to erasure, unless the heat supply is stopped. Figure 5 shows an actual oscilloscope trace. The upper trace again corresponds to the heat pulse, which in this case is inverted, and the lower trace to the multiplier signal. It can be seen that the time taken from the beginning of the heat pulse to the maximum diffraction efficiency was 10 µs, of which 4 µs was required for the deformation process itself.

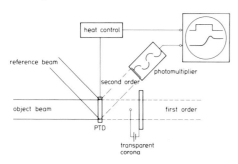

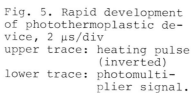

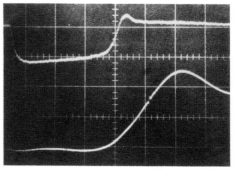

Fig. 4. Rapid cycling of photothermoplastic devices: experimental set up.

Fig. 5. Rapid development of photothermoplastic device, 2 µs/div
upper trace: heating pulse (inverted)
lower trace: photomultiplier signal.

An essential parameter for the interpretation of these results is the variation of temperature with time, before and during the development process. It was unfortunately not possible to measure this variation experimentally, due to the nonavailability of a temperature measuring device with a sufficiently fast response. A model of linear heat transfer has therefore been used to calculate this variation for several experimentally observed developments. In Figures 6a and 6b the calculated dependences of temperature (ΔT) upon time are shown together with the observed dependences of multiplier signal (L) upon time. These figures represent the two opposite extremes that can be observed in the development process. Two heat pulses of the same height, but of different lengths were used as indicated by the dotted rectangles. In Figure 6a the multiplier signal (L) passes through a maximum and is then erased. This rapid erasure is due to the steadily increasing temperature (ΔT). In Figure 6b, on the other hand, just enough heat is supplied for development to take place. It is also worthy of comment that in Figure 6b, the diffraction efficiency (multiplier signal) is seen to increase while the temperature at the surface is already decreasing. This seemingly curious phenomenon can be explained as follows: the electrostatic pressure is the deforming force and as long as it is greater than the restoring forces (i.e. surface tension and viscoelastic forces), deformation can proceed. In the process of deformation, the field distribution across the thermoplastic will change and this will give rise to a reinforcing effect upon electrostatic pressure. Thus one can arrive at a situation in which deformation will be favoured despite an increase in the restoring viscous forces due to the temperature decrease. This increase in electrostatic forces upon deformation has been theoretically described by various authors[6]-[10].

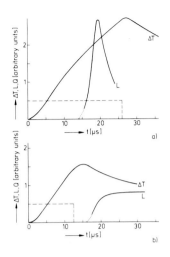

At development times less than 15 μs our results were frequently not very reproducible. Indeed the samples often showed signs of overheating such as cracks and milkiness. As a result it was often impossible to erase the image. However, in the 100 μs range such effects were not observed and hence this time range is recommended for prolonged cycling.

Fig. 6. The dependences of calculated surface temperature ΔT and scattered light intensity L upon time during the development process, for two extreme cases. The corresponding heat pulses are shown as dashed rectangles.

Self-Developing Photothermoplastic Devices

Up to now we have only been concerned with aspects of the conventional photothermoplastic device. Recently we have developed a new photothermoplastic device which we have called self-developing, because it combines the development and illumination steps. In Figure 7

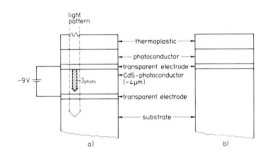

this new device, shown on the left hand side, is compared with the conventional two layer device on the right hand side. The upper three layers comprising: a thermoplastic layer, a photoconductor layer and a transparent electrode, are identical in each case. A Cu-doped cadmium sulphide layer has been inserted between two transparent electrodes in this new device, and placed between the organic photoconductor layer and the glass substrate. Upon illumination the organic photoconductor produces an electric field modulation as in the conventional device. However, at the same time illumination of the cadmium sulphide layer results in such a high photocurrent, that sufficient heat is produced to enable local development to take place.

Fig. 7. Two types of Photo-thermoplastic Devices:
a) self-developing type exposed to light;
b) conventional type.

The diffraction efficiences that we have thus far obtained with this new device have been much poorer than expected. In addition there is an unexplainably strong frost contribution, which spoiled the signal to noise ratio of the reconstructed image. The term frost denotes a random type of surface deformation, which is observed upon heating a charged photothermoplastic or thermoplastic device in the presence or absence of homogeneous illumination. This phenomenon has been previously discussed by one of the authors [6]. The reasons for the low diffraction efficiencies and strong frost contributions observed are unknown.

However, the scattering induced by the frost was sufficiently large to make the component interesting for point storage applications. In Figure 8 the exposed areas exhibit a typical frost-like appearance, each bit having a pitch diameter of the order of 10 μm. The scattering power from these spots is about 12% of the incoming light and is largely confined to a cone defined by the predominant spatial frequency of the frost pattern. Figure 9 shows the voltage applied to the CdS layer (lower trace) and the optically detected output signal (upper trace). In this mode of operation the organic photoconductor layer can, of course, be dispensed with thereby simplifying the device considerably. The performance of the self-developing PTD used in the frost-mode is summarized in Table II. The observation of scattering efficiencies of 12% at exposures of 150 μJ cm^{-2} compares very favourably with point storage in other optical memory systems.

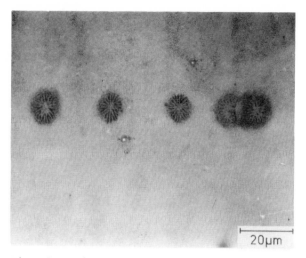

Fig. 8. Point storage in self-developings PTD's in the frost-mode.

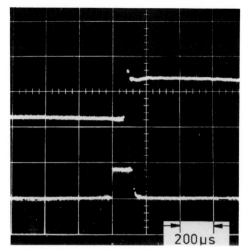

Fig. 9. Rapid development of self-developing PTD's in the frost-mode
upper-trace: optically detected output signal
lower-trace: voltage applied to CdS layer.

Table II: Point Storage in Self-developing Photothermoplastic Devices in the Frost-Mode

spot size	=	$10 \ \mu m$
laser intensity	=	$2 \ W \ cm^{-2}$
exposure time	=	$75 \ \mu s$
sensitivity	=	$150 \ \mu J \ cm^{-2}$
scattering efficiency	=	12%

Conclusion

We have shown the feasibility of millisecond switching for conventional single and two layer photothermoplastic devices. This also holds for a new self-developing photothermoplastic device designed for point storage applications.

References

1. Lee, T.C., Marzwell, N.I., Schmit, F.M. and Tufte, O.N., Appl. Opt. 17, 2802 (1978).
2. Killat, U., and Terrell, D.R., Opt. Acta 24, 441 (1977).
3. Moraw, R., Bundesministerium für Forschung und Technologie, Bericht, T 77-37 (1977).
4. Colburn, W.S., and Tompkins, E.N., Appl. Optics 13, 2934 (1974).
5. Terrell, D.R., unpublished.
6. Killat, U., Proc. Soc. Photo-optical Instr. Eng. 99, 144 (1977).
7. Budd, H.F., J. Appl. Phys. 36, 1613 (1965).
8. Killat, U., J. Appl. Phys. 46, 5169 (1975).
9. Kermisch, D., Appl. Optics 15, 1775 (1976).
10. Handojo, A., J. Appl. Phys., to be published.

OPTICAL INFORMATION STORAGE IN LiTaO₃:Fe-CRYSTALS

E. Kratzig, R. Orlowski, V. Doormann and M. Rosenkranz

Philips GmbH Forschungslaboratorium Hamburg

D 2000 Hamburg 54

Abstract

The holographic storage process is investigated in LiTaO₃:Fe-crystals. We discuss the photorefractive principles, the physical effects determining the light-induced charge transport and the recording of volume phase holograms. The experimental results clearly demonstrate that LiTaO₃:Fe is a very attractive material for optical information storage.

Introduction

Compared to conventional technologies optical methods offer many attractive advantages for data storage. Extremely large capacities are obtained, if the information is stored in superimposed volume phase holograms. This can be achieved in materials showing light-induced refractive index changes. Large effects of this kind have been observed in pyro-electric crystals like LiNbO₃ and LiTaO₃.[3] Upon exposure to light interference patterns electrons are excited and transferred to different sites. The resulting space charge fields modulate the refractive index via electro-optic effect. Uniform illumination erases the space charge fields and brings the crystals back to its original state.

Of special interest for the photorefractive process is the light induced charge transport. In most cases the main contribution to the charge transport is determined by a photovoltaic effect [2] characteristic for pyroelectric crystals. In weakly Fe-doped LiTaO₃-crystals, however, it has been found [3], that a considerable enhancement of photorefractive sensitivity may be obtained by externally applied fields utilizing the photoconductivity.

In the present contribution we discuss the photorefractive principles, and the physical processes determining the photoconductivity in LiTaO₃:Fe. The storage of volume phase holograms is experimentally demonstrated.

Photorefractive Effect

We consider two coherent light beams interfering in a crystal as shown in Figure 1. The c-axis is chosen along the z-coordinate perpendicular to the interference pattern. Then the intensity distribution in the crystal can be written:

$$I(z) = I_0 \ (1 + m \cos K \cdot z). \tag{1}$$

The modulation index m is determined by the intensities of the two incident beams:

$$m = 2 \cdot (I_1 \cdot I_2)^{\frac{1}{2}} \cdot (I_1 + I_2)^{-1}. \tag{2}$$

In the case of symmetrical incidence (interference angle 2 θ) the spatial frequency K is given by:

$$K = 4\pi \cdot \sin \theta \cdot \lambda^{-1}, \tag{3}$$

λ is the light wavelength.

An external field is applied along the c-axis, the crystal is completely illuminated and electrically short-circuited. The light-induced charge transport in the direction of the c-axis generates an internal space charge field $E_i(z,t)$ in the crystal according to the continuity and Poisson equations [4]:

$$\frac{\partial}{\partial z} \ [j(z,t) + \varepsilon\varepsilon_0 \cdot \frac{\partial}{\partial t} E_i \ (z,t)] = 0, \tag{4}$$

j is the photocurrent density along the c-axis and ε the static dielectric constant.

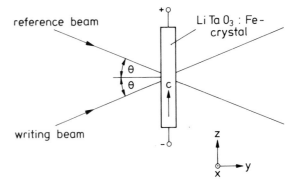

Fig. 1. Experimental set-up for the photorefractive measurements (schematically)

The drift length of the electrons is small compared to the grating parameter $l = 2\pi \cdot K^{-1}$ of the interference pattern. The electric fields modulate the refractive index via electro-optic effect:

$$\Delta n = 0.5 \cdot n^3 r \cdot E_i \ (z,t), \tag{5}$$

r is the corresponding electro-optic coefficient.

The following contributions determine the photocurrent density j(z): photovoltaic effect, drift in an electric field and diffusion [2]:

$$j(z) = \varkappa_0 \ \alpha \ I(z) + [\varkappa_1 \ \alpha \ I(z) + \sigma_d] \cdot E(z,t) + e \ D \frac{\partial}{\partial z} N(z), \tag{6}$$

$\varkappa_0$ is the photovoltaic constant, α the absorption coefficient, $\varkappa_1$ the specific photoconductivity, σ_d the dark conductivity, e the elementary charge, D the diffusion constant and N the concentration of optically excited charge carriers.

The photovoltaic contribution is proportional to the absorbed intensity. The electric drift field E(z,t) is a linear combination of the externally applied field E_{ex} and the internal space charge field $E_i(z,t)$. The contribution of diffusion according to the inhomogeneous charge concentration N(z) is small compared to the photovoltaic and drift term and can be neglected. [2]

The refractive index distribution is slightly changed compared to the intensity distribution I(z). The non-linearities grow with increasing modulation index [4,5]. The first order diffraction, however, is determined by the fundamental wave of the refractive index grating. For small modulation index ($m \leq 0.3$) the time dependence of the fundamental component of refractive index change during the recording process can be written:

$$\Delta n_f \ (z,t) = \Delta n_{fs} \cdot (1 - \exp - \gamma t),$$

with

$$\gamma = (\varkappa_1 \ \alpha \ I_0 + \sigma_d)/\varepsilon \varepsilon_0. \tag{7}$$

In the case of subsequent homogeneous illumination with light of intensity I_0 charge carriers are excited in the whole crystal volume thus compensating the space charge fields (optical erasure):

$$\Delta n_f \ (z,t) = \Delta n_{fs} \ (z) \cdot \exp - \gamma t. \tag{8}$$

The refractive index grating is in phase with the light intensity distribution because of the small electronic drift length:

$$\Delta n_{fs} \ (z) = \Delta n_{fs} \cdot \cos K \cdot z \tag{9}$$

with

$$\Delta n_{fs} = 0.5 \cdot n^3 r \ m' \ (\varkappa_0/\varkappa_1 + E_{ex}),$$

$$m' = m \cdot (1 + \sigma_d/\varkappa_1 \ \alpha \ I_0)^{-1}.$$

The saturation value Δn_{fs} of the amplitude of refractive index change is determined by the electro-optic effect and the saturation value of the internal space charge field containing contributions of the photovoltaic field $\varkappa_0/\varkappa_1$ and the external field E_{ex}.

The photorefractive recording sensitivity is significantly characterized by the figure of merit $\mathcal{E}(1\%)$ representing the energy density necessary for recording an elementary volume phase hologram of 1% read-out efficiency. From eqs. (7) and (9) we obtain [6]:

$$\mathcal{E}(1\%) = \frac{2\lambda \ \varepsilon \varepsilon_0 \ \cos \theta}{\pi \ n^3 r \ m} \cdot \frac{1}{(\varkappa_0 + \varkappa_1 \ E_{ex})} \cdot \frac{\text{arc sin} \ (0.01 \cdot \exp \ \alpha \ d/\cos \theta)^{\frac{1}{2}}}{1 - \exp - \alpha \ d} . \tag{10}$$

The ratio $\varepsilon \cdot \varepsilon_0/r$ has been found to change not very much for a wide variety of electro-optic oxides with a structure similar to that of petrovskites. [7] Therefore $\mathcal{E}(1\%)$ is essentially a function of the term $\varkappa_0 + \varkappa_1 \ E_{ex}$.

Photoconductivity of LiTaO$_3$:Fe-crystals

In the case of LiTaO$_3$:Fe the smallest energy densities $\mathcal{E}(1\%)$ are obtained with externally applied fields utilizing the large photoconductivity $\sigma_1 = \varkappa_1 \ \alpha \ I$. [3] In the near uv-region at 351 nm $\mathcal{E}(1\%) = 11$ mJ/cm^2 has been measured and in the visible region at 488 nm ε(1%) = 20 mJ/cm^2 with an external field of 15 kV/cm. In addition large dark storage times have been observed. An optimization of the storage properties requires detailed knowledge of the photoconductivity σ_1. In this section we investigate the physical processes involved in the light induced charge transport.

For the measurements single domain LiTaO₃:Fe crystals (Union Carbide Corporation) containing between 0.002 and 0.05 wt% Fe have been used. The crystals comprise Fe^{2+}- and Fe^{3+}-ions and the concentration ratio $C_{Fe^{2+}}/C_{Fe^{3+}}$ can be varied by chemical annealing treatments in oxygen or argon atmosphere. (3)

The photoconductivity measurements have been performed with incoherent light of photon energies between 1.5 and 4.0 eV. The crystals were uniformly illuminated and edge electrodes of silver paste were used. The deviation of the photocurrent density j under the influence of an external field has been found to be proportional to the field strength up to the highest applied fields of 15 kV/cm. A linear relation between σ_1 and the light intensity has also been verified. No dependence of σ_1 on electrode configuration or on illumination of the electrodes has been observed.

While photovoltaic effects are induced in the direction of the c-axis only, the photoconductivity has been found to be independent of the crystal orientation within the measuring accuracy of ± 10%. Because of the small influence of light polarization, the results are presented for unpolarized light.

The photoconductivity σ_1 can be written as $\sigma_1 = g\tau e\mu$, where g is the optical generation rate (free carriers $\cdot$ cm^{-3} $\cdot$ s^{-1}), τ the lifetime, e the charge and μ the mobility of excited carriers. Informations on electronic centers and transitions involved in photoconductive processes are obtained by correlating the quantities σ_1, g and τ with the concentrations $C_{Fe^{2+}}$ and $C_{Fe^{3+}}$.

The ratio $C_{Fe^{2+}}/C_{Fe^{3+}}$ can be determined from optical measurements. (3) A measure for $C_{Fe^{2+}}$ is an absorption band at about 3 eV resulting from intervalence transfers (8) Fe^{2+} → Ta^{5+}. The concentration $C_{Fe^{3+}}$ is correlated to an absorption band at 4 eV due to charge transfer processes from oxygen π-orbitals of the valence band to Fe^{3+}-ions.

The photoconductivity of LiTaO₃:Fe-crystals in the visible and near uv spectral region is expected to result from electronic excitations of Fe^{2+}-ions and subsequent migration of the electrons in the conduction band under the influence of the external field until the electrons are trapped by Fe^{3+}-ions. Then in analogy to LiNbO₃:Fe the following relations should be valid (9): g $\sim$ $C_{Fe^{2+}}$, τ $\sim$ $(C_{Fe^{3+}})^{-1}$ and consequently σ_1 $\sim$ $C_{Fe^{2+}}/C_{Fe^{3+}}$. This proportionality is confirmed by the experimental results shown in Figure 2. The photoconductivity σ_1/I is plotted versus $C_{Fe^{2+}}/C_{Fe^{3+}}$ for various Fe-concentrations. Within the experimental error σ_1 does not depend on the entire Fe-content C_{Fe} but only on the ratio $C_{Fe^{2+}}/C_{Fe^{3+}}$.

For $C_{Fe^{2+}}/C_{Fe^{3+}} < 0.1$ additional transitions due to charge transfer processes from oxygen orbitals to Fe^{3+}-ions contribute considerably to σ_1 in the near uv-region. The holes migrate in the valence band until they are trapped by Fe^{2+}-ions. In this case the following relations

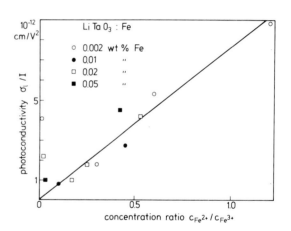

Fig. 2. Concentration dependence of photoconductivity σ_1/I at 3.5 eV.

are valid (9): g $\sim$ $C_{Fe^{3+}}$, τ $\sim$ $(C_{Fe^{2+}})^{-1}$ and consequently σ_1 $\sim$ $C_{Fe^{3+}}/C_{Fe^{2+}}$. For this reason σ_1/I increases again for small values of $C_{Fe^{2+}}/C_{Fe^{3+}}$ in Figure 2. A direct confirmation of hole contributions to the light induced charge transport has been given by a holographic method based on the generation of space charge fields by diffusion. (10)

In the case of LiNbO₃:Fe hole currents are very attractive for photorefractive storage because the dark resistivity and hence the dark storage time are very large in oxidized crystals, whereas in slightly reduced samples ($C_{Fe^{2+}}/C_{Fe^{3+}} \geq 0.1$) these quantities decrease considerably. This argument does not hold for LiTaO₃:Fe in the same manner. Dark storage times of several years have been measured for $C_{Fe^{2+}}/C_{Fe^{3+}}$-ratios of about 1.

In comparison to photoconductivity σ_1 the short-circuited photocurrent density j induced by illumination without external electric field shows a different behaviour. As in the case of LiNbO₃ j can be written (2): $j/I = \kappa_0$ $\alpha = \alpha e\mu\tau' \cdot E_p/\hbar\omega$, where τ' is the time, in which the excited carriers contribute to the anisotropic charge transfer and E_p the photovoltaic field depending on the dopant and on wavelength only.

To a first approximation κ_0 in LiTaO₃ is a constant for Fe-dopants and a certain wavelength: The current density j is mainly determined by α proportional to $C_{Fe^{2+}}$ in the visible spectral region. Beyond that κ_0 slightly increases with increasing Fe-concentration C_{Fe}. However, this dependence cannot be correlated to the ratio $C_{Fe^{2+}}/C_{Fe^{3+}}$ indicating that a relation τ' $\sim$ $(C_{Fe^{3+}})^{-1}$ is not valid. The electrons contributing to the anisotropic charge transfer loose their directional properties before they are finally trapped by Fe^{3+}-centers.

Holographic measurements

The holographic measurements have been performed using an experimental set-up similar to that schematically outlined in Figure 1. The results of recording and optical erasure of an elementary volume phase hologram in $LiTaO_3$:Fe for light polarized parallely to the c-axis are shown in Figure 3. The experimental values represented by the dots completely coincide with the solid lines calculated from eqs. (7), (8) and (9) using $\varkappa_0$ and $\varkappa_1$ values obtained from measurements with incoherent light.

The saturation value Δn_{fs} of the fundamental component of refractive index change depends linearly on the applied field strength thus confirming eq. (6) and the linear combination $E(z,t) = E_i(z,t) + E_{ex}$. The time constants γ for recording and erasure are equal and not influenced by the external field.

The holographic storage of pictures containing many details requires crystals of excellent optical quality to avoid the influence of unwanted stray light. As an experimental demonstration of the optical quality of $LiTaO_3$:Fe-samples a page of a telephone book containing more than 10^6 bit has been stored in a crystal using a hologram area of about 10 mm². In Figure 4 a section with an area of about 1% of the whole image is chosen. Light of 514 nm wavelength of an argon layer was used and the crystal thickness was 5 mm. Figure 4a shows the image transmitted through the crystal and Figure 4b the holographic reconstruction, which is not stable against light.

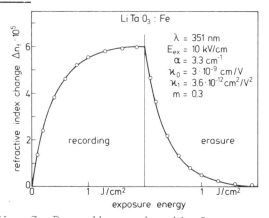

Fig. 3. Recording and optical erasure of an elementary hologram in $LiTaO_3$:Fe (0.002 wt% Fe, $C_{Fe}2+/C_{Fe}3+ = 1$):
○ ○ ○   experimental data
——   calculated from the eqs. (7), (8) and (9) using $\varkappa_0$ and $\varkappa_1$-values obtained from measurements with incoherent light.

Fig. 4. Optical storage of a volume phase hologram in $LiTaO_3$:Fe (section with an area of about 1% of the whole image)

a
transmitted image

b
reconstruction

c
reconstruction after thermal fixing

For the resolution no restriction due to the photorefractive process has been found. Two beam interference storage experiments with an interference angle $2\theta = 180°$ yield phase gratings with more than 4000 lines/mm representing the limitations of optical systems.

Holograms can be fixed by a thermal treatment as discovered by Amodei and Staebler in the case of $LiNbO_3$:Fe. [11] This stabilization process works in the case of $LiTaO_3$:Fe too: After the recording procedure the crystal containing the hologram is heated to about 150 °C. At this temperature ions in the lattice become relatively mobile, while the trapped electronic charges remain still thermally stable. Hence the ions drift in the electric field compensating the electronic charge. Then the crystal is cooled down and the ionic pattern is frozen in. When the crystal is again exposed to light, the trapped electronic charges redistribute leaving an ionic electric field pattern which mirrors that of the original hologram (Fig. 4c). The new pattern is stable against light and can only be erased by heating the crystal again.

Furthermore, the thermal fixing process is of importance for the superposition of holograms in the same volume under different angles. [12] In this case the crystal is kept at 150 °C during the whole recording process. Then each hologram relaxes completely after recording has been finished and a latent hologram is formed. By these means recording of an individual hologram is not disturbed by beam coupling effects caused by already written holograms.

The realization of large storage capacities requires the superposition of many holograms in a certain angular region. In the case of a 5 mm thick $LiTaO_3$:Fe-crystal an angular selectivity of $0.07°$ has been measured for a half interference angle $\theta = 15°$, a read-out efficiency of a few percent and a signal-to-noise ratio larger than 10:1.

Conclusions

The photorefractive process in $LiTaO_3$:Fe can be well described by the theory summarized in the second chapter. The photo-induced charge transport is also understood on a macroscopic scale: The main contribution results from electrons of Fe^{2+}-ions, which are directionally excited along the c-axis without an externally applied field (photovoltaic effect). Before the electrons are trapped by Fe^{3+}-ions they loose their anisotropic properties and can be influenced by an external electric field.

Holographic experiments demonstrate that $LiTaO_3$:Fe is a very attractive material for optical storage. Compared to $LiNbO_3$:Fe relatively small recording energies $\mathcal{E}(1\%) = 11$ mJ/cm² are measured using crystals with $C_{Fe}^{2+}/C_{Fe}3+ = 1$ and external fields. Furthermore, large dark storage times of several years are observed and crystals of excellent optical quality can be grown.

Utilizing the thermal fixing process and the experimental values of angular sensitivity storage densities larger than 10^{10} bit/cm³ are derived.

References

1. Chen, F.S., J. Appl. Phys. 10, 3389 (1969).
2. Glass, A.M., D. von der Linde, T.J. Negran, Appl. Phys. Letters 24, 4 (1974).
3. Krätzig, E., R. Orlowski, Appl. Phys. 15, 133 (1978).
4. Alphonse, G.A., R.C. Alig, D.L. Staebler, and W. Philips, RCA Review 36, 213 (1975).
5. Kurz, H., E. Krätzig, W. Keune, H Engelmann, U. Gonser, B Dischler, and A. Räuber, Appl. Phys. 12, 355 (1977).
6. Kurz, H., V. Doormann, and R. Kobs: Applications of Holography and Optical Data Processing, ed. by E. Marom and A.A. Friesem (Pergamon Press, Oxford and New York 1977), p. 361.
7. Wemple, S.M., M. DiDomenico, and J. Camlibel, Appl. Phys. Letters 12, 209 (1968).
8. Clark, M.G., F.J. DiSalvo, A.M. Glass, and G.E. Peterson, J. Chem. Phys. 59, 6209 (1973).
9. Krätzig, E., Ferroelectrics to be published 1978.
10. Orlowski, R., and E. Krätzig, Solid State Communications to be published 1978.
11. Amodei, J.J., and D.L. Staebler, Appl. Phys. Letters 18, 540 (1971).
12. Staebler, D.L., W.J. Burke, W. Philips, and J.J. Amodei, Appl. Phys. Letters 26, 182 (1975).

SESSION 2

ELECTRO-OPTICS IN COMMUNICATIONS AND IMAGERY

Session Chairman
Dean Brian J. Thompson
University of Rochester
Rochester, New York

Session Co-Chairman
Michel Treheux
C.N.E.T.
Lannion, France

COST EFFECTIVE FIBER COMMUNICATIONS IN THE MIDDLE EAST

Herbert A. Elion
Managing Director, Electro-Optics
Arthur D. Little, Inc.
Cambridge, Massachusetts 02140

Abstract

Fiber optic communications is a "must" in the Middle East for economic, sociological and military defense reasons. This paper outlines the reasons and the applications.

Introduction

The rationale and economics for cost effective fiber communications and electro-optic information systems in the Middle East differs from that of developing countries elsewhere, and differs from other industrial or industrializing nations. The concept of Value-In-Use (V-I-U), introduced for fiber optics in 1975, points to a number of needs that can be economically satisfied. Indeed, it appears that the nature of the needs are really of the "must" variety, if certain societal and economic structures are to grow or be maintained in this part of the world.

The Middle East

This paper addresses itself to regions encompassing Morocco, Algeria, Tunisia, Libya, Egypt, Sudan, Ethiopia, Somalia, Yemen, South Yemen, Oman, United Arab Emirates, Saudi Arabia, Jordan, Israel, Lebanon, Syria, Iraq, Cyprus, Turkey, and other Persian Gulf countries. (see Figure 1.) One of the regions currently undergoing dramatic changes is that bordered by the Red Sea, Arabian Sea, and Persian Gulf. (see Figure 2.) For example

Figure 1. THE MIDDLE EAST

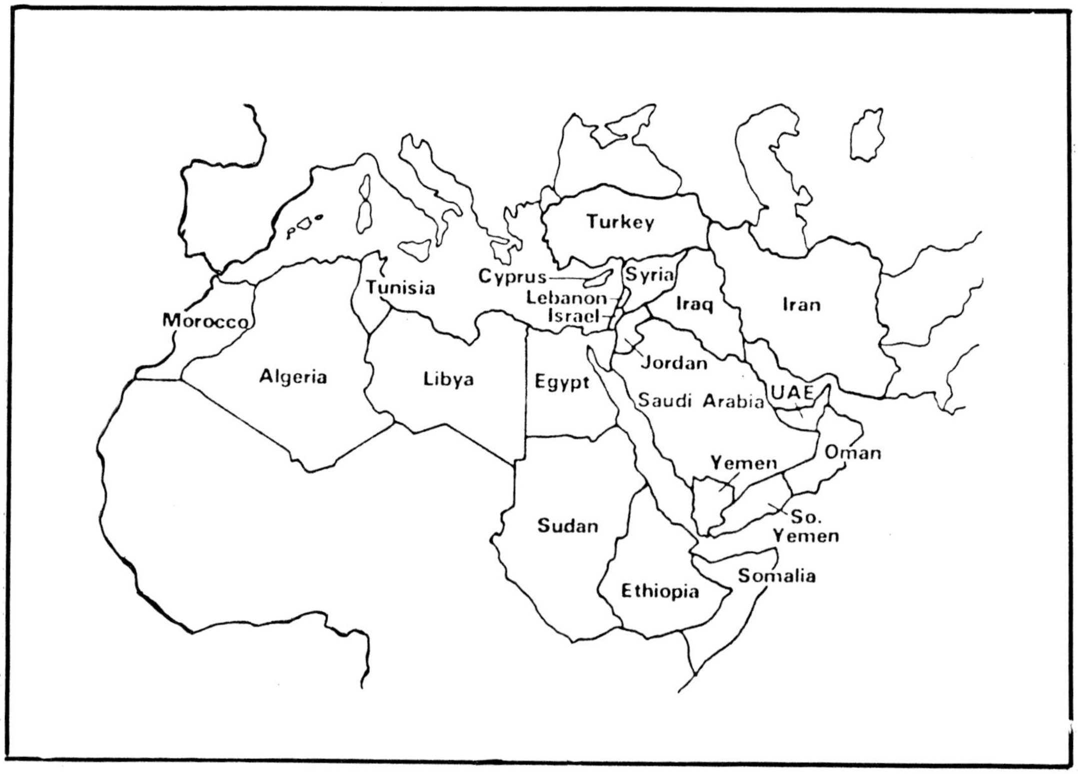

Jubail, once a small fishing village on the Persian Gulf is the site of a major new port that will rival Rotterdam in size. It will serve a 56 square mile industrial and petrochemical complex with 282 square miles available for expansion. From a once sleepy fishing village to a population of 170,000 by 1987 with $20 billion budgeted just for the construction of port roads, communications and other basic facilities. Another $50 billion for plants, industries, housing, and commercial and government facilities. At Yanbu, on the Red Sea, is a twin industrial city expected to have 150,000 population by the year 2000. It will be linked to the Eastern Province by a 48" diameter crude-oil pipeline.

Communications is a big problem, here, as in Iran. Saudis talking to Americans who are talking to Koreans who work alongside Filipinos and Maylasians on jobs designed by Germans and Dutch with British surveyors. Some of these people don't even like each other. The existing communications system is not only inadequate, but unsuitable. The Saudis traditional way of life is straining and the social fabric is in jeopardy. Something other than traditional communications is clearly necessary to handle the morals, traditions, religions, social, cultural and entertainment gaps. Military security is a problem here as it is in Israel, Egypt, Iran and elsewhere in the Middle East. In Saudi Arabia a key question to be answered is how to create and maintain an Arab-Islamic society in a developed, industrial world.

The crux of the matter is that the limits on cooperation are set by the limits on communication. Without communication, there can be no common shared values. Without fast, voluminous two-way communication, the only cohesive force between disparate cultures is raw force. Broadband intermixed voice, video, and data is a must, particularly with high visual content. Differences in different cultures, differences in dialect and languages, are proof of lack of interaction and communication in the past. If those differences were real in the past, they remain as barriers now. History teaches us that the story of all empire is the story of forced cooperation that weakened and dissipated with prosperity. To produce an enduring society the broadband communications software will have to emphasize that the values of each subgroup must be consistent with and subordinate to the values of the larger society. Timing is critical, as is the need.

Figure 2. ARABIAN SEA REGION

It is another lesson of history that all societies, all civilizations, all nations, all tribes, all social entities have failed and disintegrated when this is no longer true. Here indeed is the demand for "leap-frog" technology.

Present Telecommunications In The Middle East

Present telecommunications in the Middle East is basically telephones, although Saudi Arabia is expected to have cable T.V. in Jubail and Jidda in 1979. Even the cable T.V., however, will be the conventional one way type, and not the new interactive 2-way fiber optic T.V.

The growth rates of telephone communications differ greatly between countries ranging from 5% in Morocco to 50% in the United Arab Emirates. If we consider two categories of rate expansion, some countries will have 5-12% expansion and have a supply that will never satisfy demand. The second is a telephone growth rate of about 25% which will catch up with demand, but not meet the true overall communications needs, such as previously cited. The systems installed in the Middle East represent a wide range of manufacturers and a number of technological generations of equipment. Upgrading such a system means basically completely replacing the system with modern-day technology. If such replacement takes place, it would appear economical to make the replacements the leap-frog replacements already known, as they will be cheaper within the very near future as demand continues, and as the newest technologies can begin to deal with the societal requirements. Because of the great distances involved and the large number of small towns and villages that must be served, conventional telephone standards are not useful in estimating the per-main-station costs involved. The present state of optical fibers has been demonstrated to permit 32Mb/sec over a 53 km fiber at 1.27 microns wavelength and up to 100 km indicated, with no repeaters. Such fiber lengths have been achieved by two methods, namely fusion splicing or continuous manufacture of fibers by vapor phase deposition (VAD). The later process eliminates the need for splices every several kilometers, and reduces the manufacturing cost. The advantages of transmission around 1.3 μ have been generally well known for almost two years. For satellite earth stations as another option, fiber optics can carry studio-quality TV signals from a satellite earth station antenna to the video operating center that distributes the signals to individual receivers. Fiber optics can outperform ordinary coaxial cable now, under the same set of conditions in such fields as transmission methods and picture-quality factors, signal-to-noise ratio, insertion loss, linearity requirements, and practical transmission distances. Some wave division multiplexing or simultaneous up-stream and down stream communications has been achieved. These fibers are the same as used for conveying telephone conversations. Thus the new population centers being built for petrochemicals and natural gas raw materials for added value products are prime candidates, since these systems are "starting from scratch". Such systems can also be made compatible with the rest of the nation concerned and interconnected with older technology installations.

50 Major Applications Of Fiber Optic Communications

At the 1976 meeting, the third EEO meeting, I listed forty-four uses of fiber optic communications that appeared to meet value-in-use criteria that in many cases were higher than those of metallic wire systems. Today we list fifty. (See Table 1.)

Table 1. APPLICATIONS

1. TELEPHONE LOOPS
2. TELEPHONE TRUNKS
3. TELEPHONE TERMINALS AND EXCHANGE USES
4. INTERNAL COMPUTER LINKS
5. INTER-COMPUTER LINKS
6. HARD-WIRED COMPUTER TERMINALS
7. LONG-HAUL DATA TRANSMISSION
8. CATV TRUNKS
9. CATV DISTRIBUTION
10. CCTV
11. HIGH CAPACITY LINES (VS. MICROWAVE CABLE)
12. PRIVATE BROADBAND NETWORKS
13. MUNICIPAL SERVICES
14. AIRCRAFT-MILITARY
15. AIRCRAFT-CIVILIAN
16. SHIPS-MILITARY
17. TOWED ARRAYS-NAVAL
18. SPECIAL TETHERS
19. SUBMARINE CABLE
20. SHIPS AND TANKERS-CIVILIAN
21. ARMY CABLE CONNECTORS AND SYSTEMS
22. SECURE COMMUNICATIONS
23. CABLE CONGESTION RELIEF
24. NUCLEAR PLANTS-FUSION DEVELOPMENT
25. OIL AND CHEMICAL-STATIONARY AND MOBILE
26. ELECTRONIC AND ELECTRICAL INSTRUMENT SYSTEMS
27. SPACE PROGRAM
28. MEDICAL
29. SUPERVISORY CONTROL (LOW DATA RATE)
30. SONOBUOYS
31. ELECTRIC POWER CONTROL-LINE TYPE
32. SUBSTATION CONTROL
33. COMPUTER-VIDEO ELECTRIC POWER SYSTEMS
34. HAZARDOUS AREAS
35. LONG LINE ISOLATORS
36. MICROWAVE-OPTICAL INTERCONNECTIONS-ENTRANCE LINKS
37. LASER POWER GUIDES
38. SIGNAL SWITCHING (PROXIMITY COUPLING)
39. SATELLITE GROUND LINKS
40. FACSIMILE SYSTEMS
41. MEMORY DELAY LINES
42. LIGHTNING INSTRUMENTATION
43. HOSPITAL PATIENT MONITORING
44. TELEMETERING
45. RURAL BROADBAND SYSTEMS (VOICE, VIDEO, AND DATA)
46. BORDER AND PERIMETER SECURITY SYSTEMS
47. MWD SYSTEMS (MEASURE WHILE DRILLING--OIL AND GAS)
48. COMPLEX MULTI-NATIONAL LANGUAGE & EDUCATIONAL SYSTEMS
49. MICROPROCESSOR SYSTEMS (E.G., PROCESS CONTROL HIERARCHIES)
50. BIO-MASS MONITORING & CONTROL SYSTEMS

Value - In - Use

What is value-in-use? Value-in-use is the value of an electro-optics or other system to the user. I listed the twenty-one key values in use in the 1976 paper as follows. (See Table 2.) These give you more information than cost/effectiveness and often show why fiber optic systems or electro-optic dependent systems already outrate many other types of cable communications without showing the absolute value. They also show why fiber communications, on a world basis, will begin to have a major installed base in the 1980's and will overtake satellite communications in dollar volume by 1990.

Table 2.   ELECTRO-OPTIC VALUE-IN-USE CHARACTERISTICS

- OPERATING

    - Data Rate
    - Data Form
    - Power Requirements
    - Transmission Quality
    - Ground Independence

- SERVICE/RELIABILITY

    - Reliability
    - Ease of Service/Repair
    - Maintenance Level
    - Service Life

- ENVIRONMENTAL

    - Size/Weight
    - Stress/Vibration (Mechanical)
    - Temperature/Humidity
    - EMI/EMP
    - Detectability
    - Privacy
    - Safety

- COSTS

    - Cost Per Unit
    - Installation Cost
    - Operating Cost (OEM)

- OTHER

    - Upgradability
    - Non-Radio

Diversity In The Middle East

Before reviewing ten value-in-use applications of fiber communications that makes a prime solution to certain difficulties in the Middle East, and an attractive growth and profit factor, let us examine the problems related to "DIVERSITY".

The largest producers of revenue are oil and gas. These resources exist in several Middle East countries in considerable quantity. By way of example, Table 3 shows the eight largest oil producers. Table 4 shows the populations of the Middle East, the areas involved, and the per capita GNP.

TABLE 3.

OIL RESERVES AND PRODUCTION

	1976 PROVEN RESERVES ('000 MM BBLS.)	1975 PRODUCTION ('000 BBLS. DAILY)
SAUDI ARABIA	148.6	7,075
KUWAIT	68.0	2,084
IRAN	64.5	5,350
IRAQ	34.3	2,262
ABU DHABI	29.5	1,663*
LIBYA	26.1	1,480
ALGERIA	7.4	1,020
QATAR	5.9	438
TOTAL MIDDLE EAST OPEC	384.3	21,372
TOTAL ALL OPEC	447.3	27,192
WORLD TOTAL	658.6	55,095

*UNITED ARAB EMIRATES.

Table 4. <u>DIVERSITY IN THE MIDDLE EAST</u>
(SOME INDICES 1976)

	POPULATION ('000)	AREA (KM)	GNP (PER CAP.$)
ALGERIA	17,330	2,381,741	$ 650
BAHRAIN	2,727	598	2,250
CYPRUS	675	9,252	1,380
EGYPT	38,110	1,001,000*	280
IRAN	34,100	1,645,000	1,060
IRAQ	11,487	438,446	970
ISRAEL	3,600	20,700	3,380
JORDAN	2,794	97,740	400
KUWAIT	1,050	16,918	11,640
LEBANON	2,950	10,400	1,080
LIBYA	2,500	1,759,400	3,360
MOROCCO	17,873	458,730†	430
OMAN	790	300,000	1,250
QATAR	190	10,360	5,830
SAUDI ARABIA	9,150	2,149,000	2,080
SUDAN	18,201	2,506,000	150
SYRIA	7,555	185,180	490
TUNISIA	5,915	163,610	550
TURKEY	40,385	780,576	690
UAE	229	85,800	13,500
YEMEN AR	7,050	195,000	120
YEMEN PDR	1,729	288,000	120

*Including Sinai.
†635,000 KM with Northern Spanish Sahara.

The language and religion diversity is substantial, and the growth of wealth in such countries as Iran with large diverse populations can perhaps indicate some causes of recent difficulties. Table 5. shows some of the language diversity in Iran, while Table 6. shows their religious diversity.

Table 5.

EXAMPLE OF LANGUAGE DIVERSITY PROBLEM

IRANIAN LANGUAGES:	NUMBERS
PERSIAN (FARSI)	17,515,000
GILAKI	1,800,000
LORI	2,500,000
KURDI	2,000,000
MAZANDERANI	1,500,000
BALUCHI	600,000
PUSHTOO, TAJIK & TALESHI	25,000

TURKISH LANGUAGES:	
AZARBAIJANI TURKISH	5,000,000
TURKOMANI	450,000

OTHERS:	
ARABIC	450,000
ARMENIAN	260,000
SYRIAN	100,000
TOTAL (1973 ESTIMATES)	32,200,000

Table 6.

RELIGION (1976 Estimates)

TOTAL POPULATION	34,000,000
MUSLIMS (90% SHIITE)	33,300,000
ARMENIANS	380,000
ASSYRIANS	35,000
OTHER CHRISTIANS	45,000
JEWS	90,000
ZOROASTRIANS	40,000

The military expenditures show the necessity of force to control differences. These obviously must be augmented in several ways by communications, both military and social. Figure 3. shows the extent of military expenditures in the Middle East as compared with other developing regions.

Figure 4. shows a housing design being applied in East Asia that appears better suited to the societal conditions and future of the Middle East than some of the high-rise apartment buildings currently being installed or already in operation. These new designs are created for communications, ecology, societal needs, and energy conservation (including solar energy geometries).

Value-In-Use In The Middle East

Table 7. lists ten values-in-use of fiber communications in the Middle East that meet needs of the "must" variety or other criteria such as pure immediate positive economics.

Table 7. SELECT PROJECTED V-I-U NEEDS OF FIBER COMMUNICATIONS IN THE MIDDLE EAST

1. M-W-D Drilling and Logging of Natural Resources.

2. Process Control in Energy Production Plants and Telemetering.

3. Border and Perimeter Security Systems.

4. Secure Communications.

5. Defense Servicing.

6. Electric Power Production and Power-Line Control and Sub-Stations.

7. Rural Broadband Systems - Long Haul (Voice, Video and Data).

8. Multi-National Language and Educational Communications Systems.

9. Bio-Mass Monitoring and Control Systems.

10. Fiber Optic Components, Sub-Systems and Systems Production.

1. Measure-while-drilling is one of the newest cost effective means of oil well exploration. It supplements logging for natural resources and permits instant analysis as drilling proceeds. It permits directional changes to be made at appropriate times and is best done by broad-band communications even under the severe environments of drilling. Fiber optics is a natural candidate for severe V-I-U reasons.

2. Process control in energy production plants and telemetering is recognized for safety, S/N, and other reasons of V-I-U, although it is often up to the user to insist on fiber optics as process control systems evolution is always slow. Some one-for-one replace-

Figure 3.

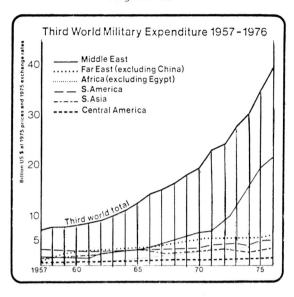

SOURCE: Middle East Yearbook-1978, IC Magazines Ltd; England

Figure 4.

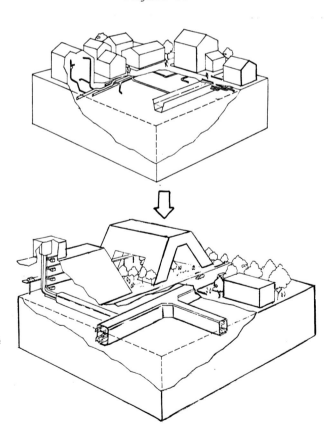

ment is more rapid in occurring.

3., 4. Border and perimeter security systems are vital in the Middle East and the V-I-U characteristics of what is possible with fiber optics almost immeasurably large. By way of one example, there is the combined fiber optic and barbed wire perimeter fence that warns as to both break and location. The system also has an "all-quiet" mode in cases where attempts may be made to bridge the wire before cutting. The optical fiber used has a low breaking strain. A far more sensitive fiber optic sensor and perimeter system is possible that would identify friend-or-foe and can not be tunnelled under without alarm. It is more sensitive than fiber optic hydrophones. Similarly fiber optic C^3I_3 systems permit secure transmission of messages, as well as permitting effective C^3I of all types particularly associated with potential or actual crises.

5. Defense servicing is of vital interest to the Middle East as most defense material is purchased outside of the region. A great deal of the equipments have electro-optic devices associated with them and servicing within a country is another "must" as a deterrent or in active use.

6. Electric power production is increasing markedly and power-line control and sub-station control is becoming vital with industrialization. e.g., The General Electric Corporation of Saudi Arabia is charged with the responsibility of electrifying the Kingdom as of 29 June 1976. It must build and operate electrical utilities (e.g., in effect a national utility company). The customer base will approximate just short of about 546,000 by 1980 or about double the present base. New plants will be required and the latest, most economical controls tested in several Japanese and American locations point to the increasing use of fiber optic electric power transmission control.

7. Rural broadband systems are being tested in a number of countries, and in some cases fiber optics V-I-U is clearly indicated although often hampered by old regulatory restrictions. The Middle East is in an excellent position to leap-frog such regulations that are barriers to other geographic regions.

8. As discussed earlier multi-national languages and the need for educational systems customized to the society are a "must" for overcoming diversity situations in communications, and a nation that doesn't heed the need may face an historical answer that is sure and unpleasant.

9. Bio-mass monitoring and control systems are a defensive and investment opportunity for much of the Middle East which borders warm waters. Combined bio-mass systems are economical and competitive with gasoline if Ocean Thermal Energy, gaseous conversion, and food & fodder applications are jointly designed into a system. The production unit cost is high but the yield cost is economical. A defensive and offensive strategy appears to be called for. Fiber optic control communications can be usefully employed in such systems.

10. How can a nation using fiber optic components, sub-systems and systems be totally dependent on the outside for defense, secure communications, defense servicing, border and perimeter security systems, national and satellite communications, electric power control, drilling, and process control telemetering and control, not have an in-house capability? One Middle East country already has this. Can the others just sit idly-by?

Conclusions

In the Middle East, as in the rest of the world, information systems are playing an increasingly important role. Electro-optics is not only for pick-up and display, but for transmission, storage and processing for high V-I-U applications. (Even in medical-chemical sensors such as Serum Galactosyltranferase for early cancer detection electro-optics and electronics can provide faster devices for mass screening in medical information systems.) Electro-optics is not only for new technology devices themselves, but for the penetration of information about how best to utilize the devices and systems for all users, both professional and lay persons. Electro-optics permits better training and education in where and how to use systems. That is, instructions on software and effective utilization of any information system. As such it is often the intimate companion of electronics.

The rationale for cost effective fiber communications and electro-optic information systems in the Middle East may differ from that of developing countries elsewhere and from industrial or industrializing countries elsewhere (as discussed in my first two papers of this series at Chicago and Washington in September 1978), but its V-I-U and cost effectiveness is critical to rationalizing problems in the Middle East.

In order to have effective total information systems, electro-optics will play a major role in the hardware and software. It fundamentally supplies a rationalization of components for complexity analysis type systems; and it provides or facilitates the educational hardware and software for training and social unification.

NEW TECHNICAL DEVELOPMENTS IN FIBER OPTIC WIRED CITIES

Dr. Seiichi Takeuchi
Manager, Visual Information System R & D Group
Sumitomo Electric Industries Ltd.
Yokohama, Japan

Abstract

Optical fiber communication systems are discussed from the viewpoint of practical application in the field of analog video transmission systems. Transmission characteristics of optical fiber CATV systems are described from practical experiences at Higashi Ikoma two-way optical fiber CATV field trials. Some techniques to achieve high performance of such transmission characteristics are described. Some new devices such as optical directional couplers and optical switches are examined. Video transmission experiments of multiplexing two optical signals were performed using two LED's with different wavelengths.

Introduction

In the development of highly efficient communication systems especially for television, telephone, computer and facsimile type systems, decreased cost with increased information bandwidth per channel is the most desirable combination for new devices and technology. Some new devices and technology can now be obtained by fiber optic communication systems. Some of the advantages of optical fiber transmission systems are listed as follows:

1. The cost of fiber cable will be much lower and will be essentially independent of using meter materials costs.

2. The number of optical repeaters for low attenuation cables needed to transverse long link distances is less than that required for coaxial systems. For links of less than 10 km no repeaters may be needed in optical networks.

3. Optical fibers and cables are much safer to use in explosive environments than wire cables and can be used to effectively eliminate short-circuit fire and explosion hazards.

4. Optical transmission networks can be made to have total electrical isolation. Equipment now used to protect electrical cables against grounding and voltage problems can be eliminated when using fiber cables, thus reducing the total system installation costs.

5. Optical cables offer a substantial savings in size and weight over wire cables which is especially important in regions where the existing transmission line structure are already over-crowded.

6. The information bandwidth capacity of fiber cables is much larger than an equivalently sized wire cable. Depending upon the frequency of light transmission and the fiber attenuation, the transmission capacity increase per unit cross sectional area is about 10^2 to 10^4 times that of coaxial cable.

7. Optical transmission has very small or virtually no cross-talk or signal leakage, whereas an equivalent coaxial system may have some significant cross-talk and echo problems.

These advantages clearly make optical transmission and communications systems highly desirable for numerous applications including telephone networks, computers, cable television, submarine cables, avionics, process control systems, medical information systems, industrial automation, satellite ground stations and for military and commercial ocean vessel applications.

Cable television (CATV) has made rapid technological advances in the past few years and is currently used worldwide for many types of consumer and military applications. CATV was first developed to solve the problem of poor reception of television signals in mountainous areas and in some large cities where large buildings often interfere with television reception and transmission. The next phase of cable television growth was to install systems in regions where only a few channels were available to make it possible to see programs aired by stations in nearby cities, thus expanding the available channel programming to the users. The more advanced CATV stations then offered one or more special broadcast channels containing information closely related to specific communities which would normally not be a part of large region programming. The next level of development of CATV uses the addition of several specialized independent broadcasting devices, keyboards, facsimile or other information and communications functions. This advanced level of programming provides re-transmission of several television channels, dedicated service channels for local community information and other transmission services such as interactive communications devices. This type of multi-functional CATV is called coaxial cable information system (CCIS) which uses branched type coaxial cable for transmission.

Hi-OVIS Implementation

The Hi-OVIS project consists of two phases. The objective of Phase I which was completed in November of 1976 was to study the feasibility of a large scale optical fiber video trans-mission system for wide band video signals consisting of analog video, FM audio and digital data signals. The Phase I system consisted of three main subsystems; the center control equipment with a minicomputer, the optical fiber cable transmission system with a 6 x 16 video switch, 500 m of a 12 fiber cable and 500 m of an 18 fiber cable, and the home terminal equip-ment with a television monitor and keyboard. The Phase II system consists of 160 home sub-scribers and 8 local extension terminal units located in various areas such as the local bank. The major center equipment consists of a data control computer, television retransmission equipment, studio equipment, video cassette recorder, cassette changer, character generator and display devices and the transmitter and receiver equipment.

The programming service provided by Hi-OVIS for neighborhood localities is to convey news, events, shopping information, train schedules and other information specifically geared to small areas and separate from typical national coverage. The programs are based upon infor-mation selected and edited by the local area inhabitants and system users. By using home installed cameras and microphones coupled with home keyboards and mobile broadcast centers individual interactive participation for a wide variety of programming can be accomplished. The television request service enables the user to select any program by home keyboard oper-ation. The shared FDM (frequency division multiplexed) architecture as commonly used for coaxial systems does not have enough capacity to handle the advanced Hi-OVIS individual service request traffic.

In the Phase II implementation of Hi-OVIS the total fiber length is approximately 360 km. This includes approximately one 6 km distribution cable of 36 fibers, 400 m of a 24 fiber cable, 500 m of an 18 fiber cable, 500 m of a 12 fiber cable, 5.5 km of a 6 fiber cable, 1.5 km of a 4 fiber cable and 31 km of 2 fiber cable. The longest transmission distance of the composite video signal consisting of the analog video and FM audio signals is approximately 4 km from the UHF receiving station at the top of Ikoma mountain to the control center at the Higashi-Ikoma station. No repeaters were used throughout the entire transmission network. The Hi-OVIS system was designed with 30 channels available for transmission, allocated as shown in Table 1.

Table 1. Hi-OVIS Channel Allocation

retransmission of TV broadcasts: 6 VHF + 3 UHF	maintenance: 1
video cassette recorders: 3 (dedicated)	telop system: 4
local studio center broadcasting: 1	ACC system: 1
still picture projection system: 1	character gen.: 1
video cassette recorders: 3 (for 6 programs)	character dis.: 6

The maintenance channel can be used for automatic protection services such as fire alarms, emergency medical service, police alarms and related services. Signals can be triggered auto-matically to end-terminals which can display the name and address and phone number of the problem location for immediate dispatching of services. It is also possible to connect gas and electric meter reading devices to this channel. Some legal problems present themselves in this area, concerning the responsibility of the station for guaranteed delivery of the emerg-ency services, which must be handled by local and national contingencies. An interactive net-work providing these services would also enable the user to achieve reductions in required insurance costs for home property and medical insurance.

Fiber Optic Transmission

A. System Network + Design

The overall design of fiber optic communications systems is determined by four basic user oriented criteria; the distance between terminals and user location, the desired data rate (bandwidth), the desired or required signal-to-noise ratio (SNR or BER), and the type of source information available (digital or analog). Once these basic requirements have been established the system design must consider numerous external variables such as physical and chemical environment, reliability, cost, size, weight, power requirements, and upgradability. Many of the variables that must be considered in optical communications design are interre-lated making the achievement of optimal designs rather difficult. The economic aspects of design are complicated by the rapidly changing price of the individual system components in response to the increase in marketing demand and potentials.

The basic schematic diagram of the Hi-OVIS interactive cable TV network is shown in Figure 1. The first Hi-OVIS network will have 160 home subscribers and 8 local subscribers. The system depicted in Figure 1 could handle up to 540 (180 x 3) subscribers if all three sub-center units were operational. The construction of subcenters should be avoided if possible due to the high costs for the building enclosure and the land. The subscriber drop is comprised of a two fiber cable, one for each of the upstream and downstream transmission lines, leading from an optical junction box where it is connected to the distribution cable.

B. Modulation Scheme + Encoding

Light sources can be modulated by externally modifying the emitted light after it leaves the source or by directly modulating the source usually by current variations. External modulators include electro-optic, acousto-optic and magneto-optic types. These components are currently under development. The most practical method for now is using direct source modulation. Light source encoding alters the information transmission by various methods to give more efficient transmission per channel. In digital systems minimizing the source bit rate can increase the energy available per transmitted symbol (signal) thus lowering the channel BER for a given level of available optical power. In analog transmission source encoding is not used and the signal is modulated by various techniques including amplitude modulation (AM), frequency modulation (FM), intensity modulation (IM), polarization modulation (PL) and phase modulation (PM). Intensity modulation (IM) was used for the cabled television network which is the most convenient method for analog transmission for existing fiber optics components and technology.

Three types of multiplexing techniques are available for optical system; frequency division multiplexing (FDM), time division multiplexing (TDM) and space division multiplexing (SDM). For long length transmission systems where analog signals must be transmitted through one or more repeater-amplifiers, the cumulative effects of signal distortion would be too large for FDM. TDM is commonly used to multiplex different individual digital sources or digitized analog sources into one channel, and can be adapted for optical systems. Wave division multiplexing (WDM) can be used in future optical systems once higher wavelength light sources and detectors have been developed. In the Hi-OVIS system multiplexing was not used and analog repeaters were not required due to the short link distances. Part of the theme of the Hi-OVIS project is to provide local broadcasting services and information. By careful selection of subcenter and station center locations to minimize cabling distances, it should not be necessary to use large numbers of repeaters for any Hi-OVIS type of installation. If long link distances are required it will probably be necessary to convert the IM format into a digital signal format which is more suitable for long distance, multi-repeater transmission. It is of course preferred to avoid this situation if possible by clever data business and cabling to eliminate the expenses of analog to digital type converters and high data rate digital repeaters with the accompanied increase in the system power consumption.

C. Fibers And Cables

The important characteristics to be considered in optical fiber transmission lines are listed as follows.

1. physical parameters such as size, weight, and ease of installation including splicing and coupling,

2. attenuation and its variance with wavelength, modal distribution and cable temperature and bend radius,

3. distortion and its variance with fiber length, bandwidth, modal distribution, signal amplitudes and waveforms, and fiber temperature,

4. environmental conditions including temperature and humidity variations and resistance to stress and chemical corrosion,

5. numerical aperture and the refractive indices of the fiber core and the fiber cladding materials.

In the typical attenuation versus wavelength profile for low loss fibers, two loss regions exist, one in the 830 to 850 nm region and one in the 1050 to 1070 nm region. Using existing LED's transmitting at 830 nm for the Hi-OVIS project, reasonably low attenuations were achieved. The total mechanical strength of the fiber is mainly determined by the surface condition of the glass. Small cracks and defects can propagate in the presence of applied mechanical stress or by attach from moisture. The purpose of fiber coating is to protect the surface from chemical attacks and mechanical damage. Fibers of the best mechanical integrity are made by coating the fiber during the drawing process with a thin plastic film of controlled diameter, colinear to the fiber to avoid fiber microbending losses. When properly applied the plastic coating causes no additional attenuation and actually increases the mechanical strength and stability compared to the bare fiber.

D. Connectors + Splices

In a typical fiber optics communications network there are many individual components which must be efficiently linked together such as light sources, detectors, repeaters, end devices, signal splitters and other components and devices. Each type of link usually presents unique engineering problems to achieve low loss connections. A splice is a form of a coupler that permanently joins two fibers or two fiber bundles. A connector links one fiber to another or to some device such as a transmitter or receiver. Most connectors are the demountable type which introduce slightly larger losses than permanent splices but enable devices and components to be independently removed from the fiber system with relative ease and efficiency. Splices and connectors require optically flat fiber ends which must remain clean during installation and maintenance work. Since fibers are thin and fragile with samll coupling areas it is difficult to achieve low cost, high strength, low loss and reproducible connectors and splices. Micro-alignment techniques, devices and materials are advanced enough to achieve consistent splices with losses less than 0.4 dB. In the design of fiber optic communications systems, the fiber cable and splicing and connector fabrication methods and designs must all be considered together to achieve low attenuations in the total network.

In the Hi-OVIS installation there are about 30 cable joints and 1500 fiber splices. Major efforts were devoted to the preparation of the process of joining cables and splicing fibers for field application. The detachable coupler used for the receivers and transmitters have a measured connector loss of less than 1 dB. This coupler (demountable connector) is shown in Figure 2. The permanent fiber splices were attained using a fiber welding technique. The measured fiber splice loss was less than 0.5 dB with typical losses of about 0.3 to 0.4 dB.

E. Optical Transmitter Characteristics (OTX)

Optical fiber communications systems require light sources with high electro-optical efficiency, long life time in use values, reasonably low cost, sufficient power output, capability to attain high frequency type modulations and physical compatibility with optical fibers. The two sources presently used are LED's and laser diodes (DL's). The most commonly used semiconductor material is GaAlAs which depending on the doping material composition can have a transmission wavelength ranging from 750 nm to about 1100 nm with an expected life time of $>10^5$ hours. The laser diodes can couple more optical power than LED's especially into low numerical aperture fibers, but they are more expensive and are more sensitive to temperature variations than LED's. The temperature dependence of the output can be controlled with both sources using specially designed feedback circuits. The modulation of both LED's and lasers is commonly achieved by using direct modulation of the input current. Intensity modulation (IM) is the most compatible modulation method with existing light sources. A comparison of LED and laser diodes for fiber optic communications networks is shown in Table 2. The typical LED characteristics for the interactive fiber system are as listed in Table 3.

Table 2.   A Comparison Of LED's And Laser Diodes

Light Source	Output Power	Bandwidth Range	Spectral Width	Rise Time	Emitter Area Range	Expected Lifetime
LED	0.5-7mW	10-200Mbps	300-450Å	3-10ns	.005-1.10mm^2	5x10^5 hrs
LD	1-30mW	10-800Mbps	20- 25Å	.1-3ns	.002-.008mm^2	10^5 hrs

Table 3.   LED Characteristics

active emitter diameter = 50 μm	axial radial intensity = 1 mW/sr
coupled optical power = -5 dBmW	radiance = 25 W/sr-cm^2 at I_f = 150 mA
peak emission wavelength = 830 nm	spectral half-width intensity = 420Å
forward voltage V_f = 1.7V @ 100 mA	reverse breakdown voltage = 1.0V
rise time = 4 ns @ I_f = 5 mA	3 dB bandwidth = 150 Hz @ 20 mA$_{p-p}$
thermal impedance = 20°C/W	

In the future Hi-OVIS type installations improved LED's may be used to achieve higher coupled power ratings and longer life time in use values. Eventually, laser diodes will be used for these cable television networks to achieve higher coupled power levels, smaller spectral half widths and faster rise times. Development work is proceeding on mass production of such laser diodes to achieve long life-time rated diodes, hopefully to be equivalent to the present LED's.

Figure 3 shows the optical transmitter module (123 x 25 x 286 mm) used in the Hi-OVIS project. The total operational temperature range of the transmitter units is about -10°C to +50°C. Within this range the output power variation is less than about 10%. The LED nonlinearity is compensated by emphasis-de-emphasis techniques and by diode compensating circuitry. The expected LED lifetime is about 10^5 hours. Figure 4 shows the transmitter LED with the optical connector attached. The optical modulation format (scheme) is intensity modulation (IM) and the upstream data format is start-stop synchronization. The total optical transmitter frequency response curve is approximately linear in the range of 20 Hz to 10 MHZ. The specified frequency characteristics are: 20 Hz to 4.5 MHz, ±1.0 dB and from 4.5 MHz to 7 MHz, ±2.0 dB. The differential gain was specified at <5%, with typical measured values <1.5%. The differential phase was specified at <5° with typical values <1.5°.

In the Hi-OVIS system the transmitter (OTX) and receivers (ORX-D) were kept at room temperature. The audio signal (A) input-output level was specified at -20 dBV. The frequency allocation diagram for the baseband transmission signals for the optical fiber video system is as shown in Figure 5.

F. Optical Receiver Characteristics

The optical detectors used in optical fiber transmission must meet several important physical and electro-optical requirements including having sufficient bandwidth and speed of response, nominal noise characteristics, peak sensitivity matched to the light source wavelength, stability to thermal changes and long life time in use values. Photodiodes are usually described by four basic quantities; response time, quantum efficiency, total noise equivalent power and responsivity. For avalanche photodiodes (APD's) an additional quality, multiplication or gain, must also be considered. The couplings of photodetectors to fiber ends is more easily accomplished than coupling light sources since alignment losses are not as significant. The use of antireflection coatings or epoxies can reduce insertion (coupling) losses to 0.2 dB to 1.0 dB depending upon the physical sizes of the detector area, the fiber core diameter and the type of coupler used.

PIN photodiode performance and linearity is effected by temperature variations which can be compensated by using an appropriate feedback loop signal circuitry. Even the best designed temperature compensated modules have linearities less than that of the individual components which can present problems for large bandwidth, high gain systems. APD's have high internal gains, high responsivity, fast response times with small active areas. As the gain factor increases the need for precise controls on voltage and temperature become more critical if a constant gain is to be achieved. A comparison of PIN and APD detector characteristics is shown in Table 4.

Table 4. Comparison Of PIN And APD Detectors

Detector Type	Responsivity At Peak	Maximum Data Rate	Sensitivity At 1 MHz	Rise Time	Active Area	Expected Lifetime
PIN	.4-.7 A/W	up to 1 GHz	-58 dBmW	1-5 ns	.3-3 mm^2	5x10^5 hr
APD	10-70 A/W	200 MHz	-70 dBmW	2-5 ns	.6-5 mm^2	10^5 hr

The optical receiver module used in the Hi-OVIS project is shown in Figure 6. The central component of the receiver is the PIN photodiode which is shown in Figure 7, with an attached optical connector. The type of preamplifier is a combination of a FET and bipolar amplifier. The data rate signal circuitry operates at 200 bps with a specified data output (D) level (interface) of ±8V. The video input-output (V) impedance is 75 ohms (unbalanced) with a video signal input-output level of 1.0 V_{p-p}. The actual measured load resistance for the Hi-OVIS receiver was R_L = 150 kohms. The receiver was designed for a power level input of -30 dBm with a specified SNR of greater than 46 dB. The design characteristics of the receiver in terms of the signal to noise ratio (SNR) are as listed in Table 5. By checking the module number with a corresponding address listing it will be possible in practical applications to rapidly identify problem locations to expedite immediate servicing and repairs.

The PIN specifications of the optical receiver are listed in Table 5. This PIN selection and receiver performance has proven to be a most satisfactory choice for the cabled television requirements.

Table 5. Hi-OVIS PIN Specifications

active chip size: 0.8 mm diameter (minimum)

maximum absolute ratings: V_R = 20 V, I_R = 1.0 mA, P_d = 10 mW
              T(storage) = -65° to 150°C, T(junc) = 150°C

test conditions: I_D(max) = 100 nA at V_R = 8.0 V
              C_d(max) = 3.5 pf at V_R = 8.0 V, f = 1.0 MHz
              Q.E( ) = 70-80% at V_R = 8.0 V,   = .80-.83 μm
              rise time = 10 ns at V_R = 8.0 V, R_L = 50 ohms

Summary

Optical fiber communication systems were discussed from the viewpoint of practical applications. Transmission characteristics of optical fiber CATV systems were described on practical experiences in the field trial of Hi-OVIS (Highly Interactive Optical Visual Information System) project in Japan, DG (Differential Gain), DP (Differential Phase) and S/N were measured typically less than 1%, less than 1° degree and more than 50 dB, respectively. Some techniques to achieve high performance of such transmission characteristics were described. Engineering trade-off between device characteristics and the required transmission characteristics of the optical fiber network were discussed. New optical devices such as optical directional couplers, optical switches were discussed. Using these new optical devices, a simple optical video transmission system was designed. The designed system was implemented with these new optical devices. The functions of this optical fiber video transmission system include multiplexing two video signals, optical splitter and optical switching. Experiments were performed to evaluate the transmission characteristics of WDM (Wave Division Multiplexing) using two LED's. The wave lengths of these two LED's were 0.795 μm and 0.875 μm.

Figure 1

Basic Hi-OVIS Schematic Diagram

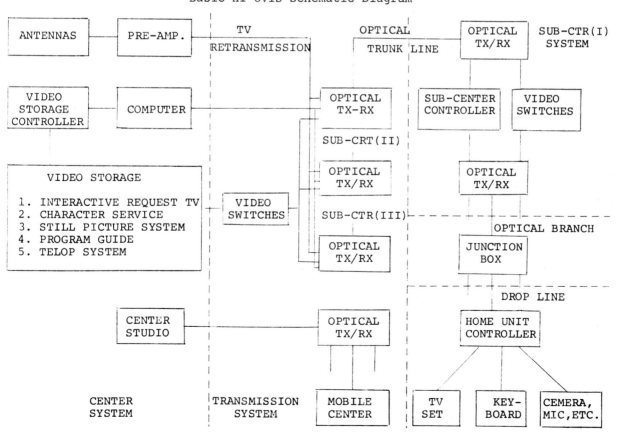

Figure 2

Detachable Optical Coupler For OTX + ORX

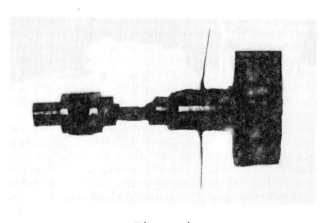

Figure 4

Transmitter LED With Optical Connector

Figure 3

Optical Transmitter Module

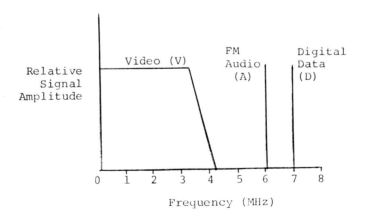

Figure 5

Frequency Allocation Diagram For Baseband Transmission

Figure 6

Optical Receiver Module

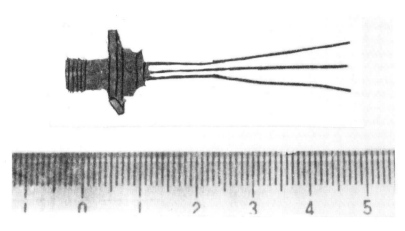

Figure 7

PIN Photodiode With Optical Connector

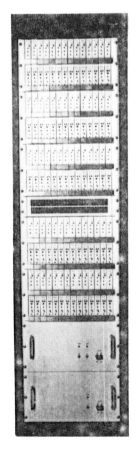

Figure 8

Optical Transmitter +
Receiver Rack

References

(1) T. Nakahara and H. Yanai: Fiber Optics in Japan,
Fiber and Integrated Optics, Volume 1, Number 1.
(2) T. Izawa et al.: Continuous Fabrication of High
Silica Fiber Preform: IOOC'77 Technical Digest
pp. 375-378.
(3) M. Kawahata: Hi-OVIS Development Project: IOOC
'77 Technical Digest pp. 467-471.
(4) S. Sakurai: Application of Optical Information
Systems: IOOC'77 Technical Digest pp. 463-466.
(5) T. Tokunaga, et al.: Rod-in-tube Method for CVD
Fiber Manufacture: Japan Electronic and Communication
Institute Convention, March 1978.
(6) T. Ono, et al.: Reliability Evaluation of Plastic
Clad Fiber: Japan Electronic and Communication
Institute Convention, March 1978.
(7) K. Asaya, and T. Kimura: Analog Modulation of
LED: J. of Applied Physics Japan, May 1978.
(8) I. Hayashi, Semiconductor Light Sources: IOOC'77
Technical Digest pp. 81-82.
(9) Y. Sakakibara and Y. Suematsu et al.: InGaAsp
High Speed Laser Direct Modulation: Japan Electronic
and Communication Institute Convention, March 1978.
(10) K. Tada and G. Yamaguchi, et al.: Optically
Activated Thin Film Optical Switch: Conference on
Lasers and Electro-optical Systems, Feb. 1978 San
Diego, Calif.
(11) M. Nakamura: Monolithic Integration of Distrib-
uted-feedback Semiconductor Lasers: IOOC'77 Technical
Digest pp. 227-230.
(12) T. Nakahara, H. Kumamaru, S. Takeuchi: An
Optical Fiber Video System: IEEE Transactions on
Communications, July 1978.

FUTURE PROSPECTS FOR OPTICAL HYBRID IMAGE PROCESSING

B. J. Thompson

College of Engineering and Applied Science, University of Rochester
Rochester, New York 14627

Abstract

Optical image processing has, with a few notable and important exceptions, been over-taken by digital image processing. However, current technology in incoherent-to-coherent converters, optical detectors, and microprocessors is producing renewed interest in optical methods. Future systems will probably be combinations of optical and electronic subsystems. The current trends in this field are discussed and examples are given of present systems so that future prospects can be evaluated.

Introduction

Optical signal processing has undergone a complete revitalization in recent years, and the 'new look' to these methods is extremely interesting and provides significant potential for future development and application. This transformation of the field has come about from several factors which include

(1) advances in detector technology have now made available to the electro-optical scientist and engineer single fast detectors, one- and two-dimensional arrays of detectors, and segmented detectors,

(2) the output of these detectors can be analysed in near real-time with the use of on-line minicomputors and microprocessors,

(3) a series of devices called light valves or spatial light modulators, have been developed that can be used in near real-time to convert an incoherent input to a coherent output,

(4) light valves have also been developed that can act as transducers from acoustic or electrical signals directly to a coherent optical field,

(5) some of these light valves can be used for real-time generation of spatial filters for coherent optical processing.

Systems that employ these new electro-optical techniques combined with traditional optical systems have been called hybrid optical systems.[1]

In this review, two basic categories of system will be described based on conventional optical Fourier transform methods. The first category involves detection in the Fourier plane followed by electronic analysis of the detected signals. The Fourier transform is produced optically from a variety of inputs that include the following:

(a) pre-recorded information on film;

(b) direct transformation of a special class of objects including wires, needles, aerosols, particulates, and periodic structures;

(c) an incoherently produced image that is converted to a coherent field whose amplitude is proportional to the incident image intensity. This conversion is produced by one of the several light valve devices;

(d) an electrical or acoustic input that is transduced to a coherent optical signal.

The transforms of the resultant coherent fields are produced optically in the traditional way. The transforms are then detected and the resultant electrical signals processed electronically in analogue or digital form. The second category of system does not detect in the Fourier plane, but the Fourier transform is modified (spatially filtered) and the detection takes place in an image plane. This is, of course, the classical coherent optical processing system originally introduced by Maréchal and Croce.[2] The difference is that the new techniques listed above are applied to this system. Thus all types of inputs discussed above (a-d) can be used as inputs to the optical processing system. In addition, the various light valves can be used for complex filter generation. Finally, the detector arrays make an ideal image plane detector, particularly if correlation peaks are to be detected.

Fourier Plane Detection

The use of optical methods for producing a Fourier transform are, of course, well known and the technique was perhaps first employed for measurement purposes by Thomas Young in his famous eriometer.[3] The basic system most often used is shown in Figure 1a. The advantages of this particular method are that the transform is spatially invariant with lateral position of the input; and if the illumination is collimated, then the scale of the transform is independent of the position of the input along the direction of the optical axis (i.e., independent of the value of z). Thus the system can handle both volume and plane inputs. If scaling is required, then for plane inputs, the input can be located on

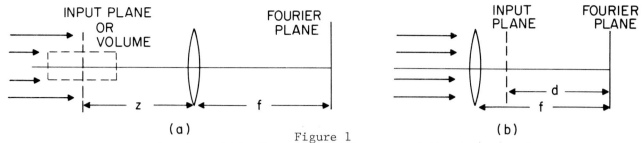

Figure 1
Conventional method of producing the optical Fourier transform

the other side of the lens; then the scale of the transform is proportional to the distance d (Figure 1b). We are, of course, interested here in the use of this basic system with the various new techniques mentioned earlier.

A block diagram of the system for Fourier plane detection and analysis is shown in Figure 2. The various inputs will be discussed separately in the sections to follow. The Fourier transform is produced optically as already described. The intensity associated with the transform may be detected by a variety of detector configurations. The first of these possibilities is shown in Figure 3, which uses a single silicon detector. The

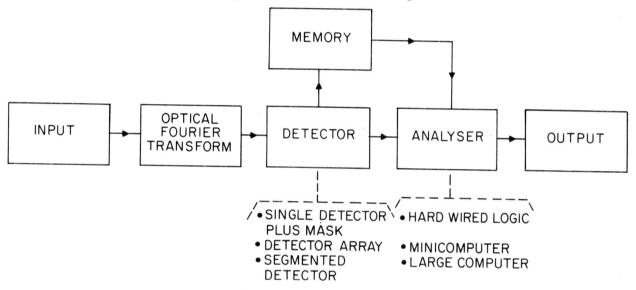

Figure 2
Block diagram of optical hybrid system for Fourier plane detection and analysis

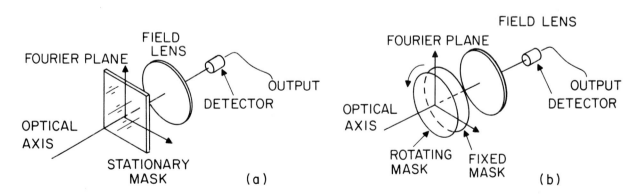

Figure 3
Fourier plane detection using a mask, field lens, and single silicon detector
(a) stationary mask   (b) rotating mask and fixed aperture

Fourier plane is sequentially sampled by a series of fixed stationary masks or a rotating mask in front of a fixed aperture. The flux passing through the mask is collected by a field lens and focused onto the detector. The output of the detector is then analyzed with an on-line minicomputer, microprocessor, or can be passed directly to a large computer on a time share basis. The results of this analysis are then displayed or produced as a hard copy printout. Several excellent examples of the use of this technique for the direct measurement of a class of objects including aerosols and wires will be given later.

The second method of detection that is available is shown in Figure 4. The principle

Figure 4

Fourier plane detection using an array of detectors; (a) two-dimensional rectangular array (b) typical construction of individual detectors (Courtesy, Reticon, Inc., Sunnyvale, California).

here is to use a one- or two-dimensional array of individual silicon photodetectors each with their own individual output.[4] The output of these detectors is usually digitized and multiplexed for analysis in a minicomputer or microprocessor. Various configurations are available including linear arrays of 1028 individual detectors and two-dimensional arrays of 50x50 elements.

The third method uses a specially designed segmented silicon detector, illustrated in Figure 5.[5] This segmented detector was developed specifically for Fourier plane analysis,

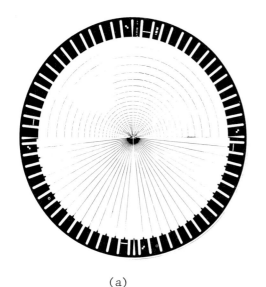

Figure 5

(a) segmented detector consisting of thirty two (32) wedge and thirty two (32) annular sectors (Courtesy, Recognition Systems, Inc., Van Nuys, California);
(b) the center of the detector magnified to show the four detectors used for alignment.

which explains its particular geometry. There are thirty two (32) individual wedge shaped detectors and thirty two (32) annular detectors. Clearly the wedges detect directional information in the transform and the annular detectors sample the spatial frequency content of the input averaged over all directions. This particular segmented detector has found considerable use in a variety of systems. The individual elements of the detector do not go into the center but stop short of that; the central region contains four individual detectors that can be used for alignment. As in the previous example, the detector outputs are separately amplified and then multiplexed, digitized, and then processed in a local small computer or microprocessor.

These techniques have allowed considerable versatility in the measurement processes using Fourier plane detection and subsequent electronic analysis. Naturally, at times it may be necessary to take the output of the detectors into a memory store before analysis as is indicated in the block diagram of Figure 2.

It remains now to discuss the various possible inputs to the system, shown in block diagram form in Figure 6.

INPUTS.

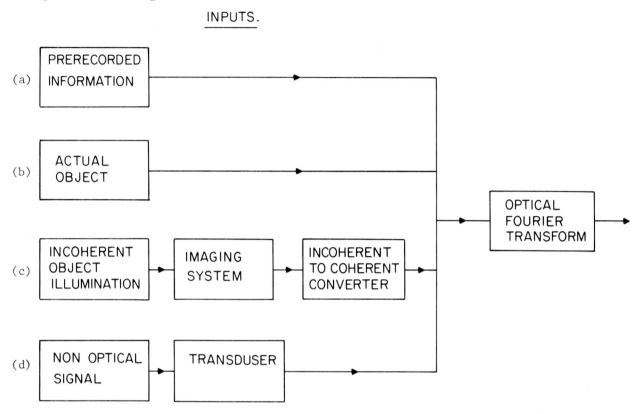

Figure 6
Inputs for Fourier plane detection systems
(a) pre-recorded information on film   (b) real objects direct transformation   (c) incoherent input with imaging and incoherent-to-coherent conversion   (d) non-optical input with appropriate transducer.

<u>Fourier Plane Detection - Recorded Inputs</u>

There is, of course, considerable interest in Fourier plane analysis of information (usually images) that have been prerecorded on film or other recording material. Thus the film becomes the input to the optical system; and since it is a two-dimensional record, it can be used in any plane in the optical system shown in Figure 1. Perhaps the largest single application in this class of system is for the analysis of aerial photographs for both military and civilian purposes. Lendaris and Stanley[6] were the first to devise such a system that used a series of Fourier plane masks and a single detector as indicated in Figure 3. In their work, the stationary masks were used sequentially and sampled the Fourier plane with concentric annular rings, a circular array divided into wedges and a series of parallel slits. The output of a single detector was fed directly to a large computer for analysis.

The segmented detector discussed earlier and shown in Figure 5 was devised specifically for applications of Fourier plane detection and forms the main component in a piece of instrumentation called a recording optical power spectrum analyser. The term 'optical power spectrum' is used since it is the intensity in the Fourier plane that is detected and analysed. This device has found a number of useful applications in both analysing aerial photographs and in image evaluation.[7] These methods could also have applicability in analysing other recorded information in medicine, biology, and photolithography.

<u>Fourier Plane Detection - Real Inputs</u>

Diffraction pattern analysis of small objects has always been an attractive proposition because the smaller the object the larger its diffraction pattern. However, such systems

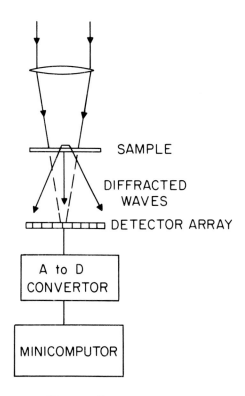

Figure 7

Schematic diagram of the system used by Kasdan and George for the measurement of small line-widths and gaps in photolithographic masks.

have been rather limited until recent times because of the difficulty of analysis of the patterns without using a human observer. The earliest application of this method was Young's eriometer for measuring the mean diameter of wool fibers and its subsequent application to the measurement of the mean diameter of blood cells.[8] The readers will no doubt recall performing, in their student days, a diffraction based measurement of the mean diameter of a sample of lycopodium powder particles.[9] The current advances in detector technology and minicomputers has allowed this relatively old subject to be completely re-vitalized and a variety of new systems have been developed to solve real measurement problems. Many of these systems are finding a ready acceptance in the marketplace. These applications can be most readily discussed by considering the nature of the object to be measured.

Single objects

In this application, the objects are a single droplet or particle, a hole, a wire or needle, etc. The object is illuminated with a coherent beam of light in one or the other of the configurations discussed in Figure 1, and the optical Fourier transform produced. For many of these objects, only simple logic steps are necessary to determine the size of the object from its detected Fourier transform. George and Kasdan[10] have made significant use of this method for a variety of applications. Figure 7 shows a schematic diagram of the system used by George and Kasdan for measuring small dimensions in photolithographic masks. It will be noted that the mask is placed behind the lens in the converging beam and the transform is detected by a linear array of silicon photodiode detectors. The output of these individual detectors is digitized and analysed in a minicomputer; an algorithm has been developed to give accurate information on the size of very small gaps and linewidth on the order of a micron or less.

In an extremely practical development, this same group of workers has developed a system for the on-line testing of hypodermic needles.[11] The segmented detector array shown in Figure 8 is used for this application and samples the Fourier transform of the needle point transilluminated by a narrow beam from a He-Ne laser. Thus the needle profile is being evaluated. Bad needles have flats or even hooks on them which are readily detected as directional information in the transform. Several of these systems are now being used by Becton Dickinson, the world's largest manufacturer of needles and syringes, who claim that this inspection system is now saving $300,000 a year in production costs.[12]

Many other applications will be forthcoming for this type of system, particularly since it can provide real-time feedback to the process under inspection so that, for example, the bad needles in the process discussed above can be rejected.

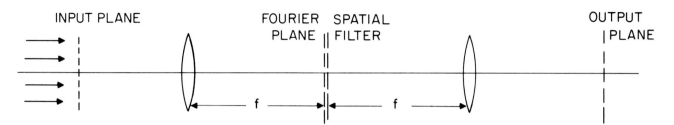

Figure 8
Conventional method of optical spatial filtering
This is but one of various configurations.

Multiple Objects in a Plane or Volume

An important application of Fourier plane detection is in the field of particle size analysis. The particles may be liquid or solid, in air or in a gas or liquid. In the configuration shown in Figure 1a, the particle can be moving within a volume. The required transform is stationary and the cross-terms resulting from the interference between the transforms produced by each individual particle are relatively small, but they are further averaged out because of the movement of the particles. The composite Fourier transform of the particles is appropriately sampled and detected, and algorithms used to determine, for example, the particle size distribution function. Again, it is the ready availability and relatively low cost of detectors and microprocessors that has changed this particular measurement problem in the last few years.

Cornillault[13] has designed a system that is now available as a commercial instrument. This instrument uses a 1mW He-Ne laser operating in a TEM_{oo} mode at 6328Å. An expanded beam from this laser illuminates a sample volume with a 2sq.cm cross-section. After the transform is produced optically, it is sampled with a rotating mask and single detector. A time sequence of pulses is thus provided for on-line analysis to give a number histogram of particle size.

Wertheimer and Wilcock[14] have also used a rotating mask and single silicon detector to produce a variety of outputs. Again a He-Ne laser is used for illuminating the sample. Outputs are produced that are proportional to the second, third, and fourth power of the radius, area mean radius, and area standard deviation. A rotating mask to give histogram information has also been designed. These methods have been fully instrumented and are available under the trademark $MICROTRAC^{TM}$.[14] In these systems, the output of the detector is amplified and digitized and then fed into a microcomputer. The microcomputer contains a programmable read-only memory, random access memory, and central processing unit; the output is either a LED display or a hard copy printer.

The segmented detector array shown in Figure 5 has also been used for instrumentation for particle size analysis by a group of workers at the University of Sheffield.[15] The annular ring detectors are used and the output of adjacent parts averaged, to give fifteen signals. These signals are fed into a PDP8 for analysis and a program fits the measured data to a Rosin-Rammler distribution function. This instrument is also available commercially.[16]

Fourier Plane Detection - Real Inputs Incoherently Illuminated

It is not possible to illuminate diffusely scattering objects with coherent light and input that scattered light into the coherent system. The uniqueness of position in the Fourier plane is lost. This has long been a serious problem with optical processing systems. However, this problem can now be overcome by using one of a series of devices that will allow the object of interest to be incoherently illuminated. The basic concept is shown in block diagram form in Figure 6c. The object of interest is illuminated incoherently and imaged in the usual way. This image is allowed to fall on an electro-optic device such that the image is recorded in the material of the device as, for example, a phase difference or as a voltage difference across an electro-optic material. This image can then be read out directly with a coherent beam of light so that the amplitude of the coherent beam is proportional to the incident intensity of the incoherent image. The coherent beam is then the input to the next element of the system which produces the Fourier transform; the detection and analysis process is as discussed previously.

It is not possible in this particular paper to discuss in detail the various devices for incoherent to coherent light conversion, but there are excellent papers that discuss these various light valves or spatial light modulators,[1,17] and the reader is referred to this literature for further details. There are several light valves that are most appropriate for this particular application. The first of these is the phototitus device which is an electro-optic device used with deuterated potassium dihydrogen phosphate (KD_2PO_4) as the active material. The crystal is approximately 150 microns thick and has a cross-sectional area of about 12sq.cm. One side of the crystal is coated with a dielectric mirror, selenium photoconductive layer, and a transparent electrode. A second electrode is located on the other face of the crystal. A voltage is applied across the crystal and an image formed on the photoconductor; carriers are produced which migrate to the crystal and create a voltage across the crystal proportional to the incident intensity. Since KD_2PO_4 is an electro-optic crystal, this image information can be read out coherently with a separate beam of light in a wavelength range in which the photoconductor is not sensitive.

The so-called Pockel's readout optical modulator also depends upon the Pockel's effect in an electro-optic crystal. This time the crystal is bismuth-silicon oxide about 200μ thick with a cross-section area of about 6sq.cm. The image is again stored as a potential difference across the crystal which is read out coherently.

Elastomeric devices have also been designed and fabricated. In these devices, a conducting layer is deposited onto a glass substrate and is covered with a photoconducting layer; a thin layer of elastomer is placed onto the photoconductor. A positive charge is placed on the surface and then the image projected onto the surface. Hole-electron pairs are produced; the holes migrate to the surface leaving bound electrons. Thus a mechanical force is produced which deforms the elastomer proportional to the incident intensity. The resulting surface relief image is read out coherently.

A fourth device is the hybrid liquid crystal device which consists of a 2μ thick layer of twisted nematic configuration. Like the other systems, it is a multilayer device including a photoconductive layer. When the incoherent light falls on the photoconductor, a voltage is produced across the liquid crystal. Readout is with a coherent beam of polarized light.

The majority of the applications of these devices in optical processing have been in the second category of system in which the detection and analysis are carried out in an image plane.

Fourier Plane Detection - Non-Optical Inputs

In a number of instances, the information to be analysed is not in an optical form but is, perhaps, acoustical or electrical. Transducers are now available so that the acoustic signal can be directly passed into a Bragg cell and read out optically. Scanning is, of course, needed to write a two-dimensional input, but one-dimensional analysis is often all that is required. For electrical signals, several devices are available (for details, see Reference 17). The first of these is another use of deuterated potassium dihydrogen phosphate (KD_2PO_4). The readout is similar to that described for phototitus; however, now the input is an electrical signal driving an electron gun. A second device is the General Electric light valve that is an oil film device. Charge is deposited on a thin oil film with a scanning electron beam; this charge causes a distortion of the oil film and hence a phase image which is read out coherently. Again the remainder of the system for detection and analysis is as previously discussed.

Image Plane Detection - After Spatial Filtering

This second class of system is based on the conventional optical processing system shown in one form in Figure 8. The basic system concept is that introduced by Marechal and Croce in which the Fraunhofer diffraction pattern (Fourier transform) of the input is produced, operated upon in amplitude and phase, and then this resulting field retransformed to an image plane for detection.

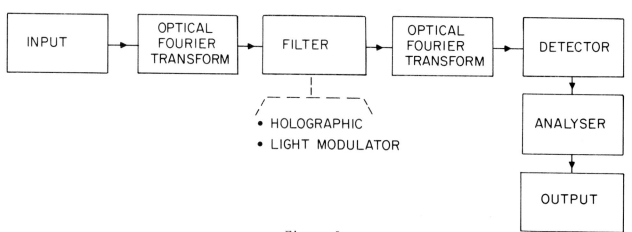

Figure 9
Block diagram of optical hybrid system for spatial filtering and image plane detection.

Given the various new technologies for the input and detection process, a block diagram of the system might consist of that shown in Figure 9. For the hybrid systems we are currently interested in, the input function could be any of those previously discussed and summarized in Figure 6. The filter may be a simple band-pass or band-stop filter, an amplitude filter for differentiation, a complex filter for aberration balancing, a complex filter for correlation. Naturally, these filters can be made by a number of processes, the most versatile of which is the holographic method. For repetitious processing with the same filter, the filter could be a physical mask or a hologram on photographic film. Some useful versatility can be added to the system if the filter element could be erasable and then a new filter written. Some of the light valves have this capability and have been demonstrated in this mode.

If correlation is to be performed, then a detector array in the image plane makes an ideal detection and readout system. Further analysis can then be carried out if required.

A number of systems of this type have been tested, and perhaps the most interesting in its potential uses an incoherent-to-coherent light converter. Gara and his co-workers[18] have produced a series of very interesting papers. In this work, a mechanical or electrical component is illuminated incoherently and imaged onto the liquid crystal light valve described earlier. This image is read out coherently with light from a He-Ne laser beam and this field then transformed; a previously recorded correlation filter prepared holographically is used to test the validity and integrity of the component under test. The final output is a correlation peak. This group has also done high pass filtering to produce an edge enhancement image.

The most recent work by Gara is concerned with real-time tracking of moving objects using optical correlation. The hybrid system is illustrated in Figure 10. The object is a connecting rod that is moving on a conveyer belt; the rod is illuminated incoherently and is imaged onto the image transducer with a demagnification of 60 by lens L_1. The transducer is a liquid crystal device and is read out with light from a He-Ne laser. The Fourier transform is produced by lens L_2 onto the holographic correlation filter. The correlation peak is produced in image space by lens L_3. The correlation signal is detected with a linear array of photodiodes. Part of this study was concerned with the tracking linearity. Figure 11 shows a plot of position determined by diode number against the true position in millimeters. Tests were carried out at speeds up to 250 millimeters per second. This work clearly has important implications in automation, quality control, and robotics. This example is intended to be illustrative of the kinds of processes that can now be accomplished.

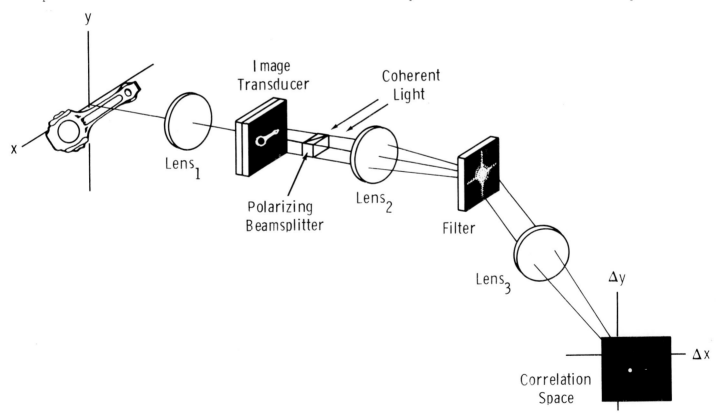

Figure 10
Hybrid system for real-time correlation analysis of a moving part on a conveyor belt (after Gara[18]).

TRACKING LINEARITY

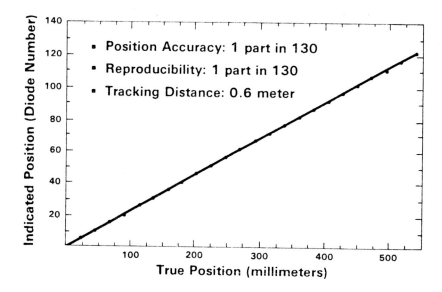

Figure 11
Results of a linearity study. The measured position of the object was determined by the diode number in the final image as a function of actual position (after Gara[18]).

Conclusions

It appears that these new hybrid optical and electronic systems have a very important future, since they obviously can be very competetive with other methods. Thus the well-known advantage of optical processing can now be taken advantage of.

Acknowledgements

I would like to thank my colleagues at Leeds and Northrup Corporation Research Labs and at General Motors Technical Center for freely discussing their latest results. The author would also like to acknowledge support of the U.S. Army Contract DAAK70-77-C-0190.

References

1.   D. Casasent, Opt. Eng. 13, 228, 1976.
     B. J. Thompson, Proc. IEEE 65, 62, 1977.
2.   A. Maréchal and P. Croce, Compt. Rend. 237, 607, 1953.
3.   T. Young, An Introduction to Medical Literature, Underwood and Blacks, London, 548, 1818.
4.   Reticon, Sunnyvale, California.
5.   Recognition Systems, Inc., Van Nuys, California.
6.   G. G. Lendaris and L. L. Stanley, Proc. IEEE 58, 198, 1970.
7.   N. Jensen, Photogram. Eng. 39, 1321, 1973.
     J. T. Thomasson, T. J. Middleton, and N. Jensen, Coherent Optics in Mapping, Proc. Soc. Photo-Opt. Instr. Eng. 45, 265, 1974.
     S. A. Armstrong and B. J. Thompson, Opt. Eng. 17, 273, 1978.
8.   A. Pijper, J. Lab. Clin. Med. 32, 857, 1947.
9.   B. L. Worsnop and H. T. Flint, Advanced Practical Physics for Students, 9th Ed., Nethuen and Company, Ltd., London, 390, 1950.
10. N. George and H. L. Kasdan, Proc. E.-O. Systems Design Conference, 494, 1975.
     H. L. Kasdan and N. George, Proc. Int. Opt. Comp. Conf., Capri, IEEE 120, 1977.
11. Recognition Systems, Brochure, Van Nuys, California.
12. Becton Dickinson, Report to Shareholders, January 1978.
13. J. Cornillault, Appl. Opt. 11, 265, 1972.
     Granulometric Type 226, Compagnie Industrielle des Lasers, France.
14. A. L. Wertheimer and W. L. Wilcock, Appl. Opt. 15, 1616, 1976.
     Microtrac[tm], Leeds and Northrup, North Wales, Pennsylvania.
15. J. Swithenbank, J. M. Beer, D. S. Taylor, D. Abbot, and G. C. McCreath, Proc. 14th Aerospace Sciences Meeting, AIAA paper number 70-09, 1976.

16. Particle Size and Droplet-Size Distribution Analyser, Type ST1800, Malvern Instruments, Malvern, England.

17. D. Casasent, Progress in Optics, Ed., E. Wolf, North-Holland, 1978.

18. A. D. Gara, Appl. Opt. 16, 149, 1976; Appl. Opt. 18, in press, 1979.
    R. W. Lewis, Appl. Opt. 18, 161, 1978.

A LASER ALPHA-NUMERIC CHARACTER GENERATOR FOR LINE PRINTER AND COM APPLICATIONS*

A. Podmaniczky

Computer and Automation Institute, Hungarian Academy of Sciences
H-1502 Budapest 112, P. O. Box 63

Abstract

Optical distortions araising from predeflection focusing are analyzed from the point of view of laser character generators. It is shown that distortions can be kept below an acceptable limit. The advantage of multi-beam recording is analyzed and the principles of operation of a new multi-beam acousto-optic modulator are described. Using this modulator a simple laser character generator has been developed for a line printer which can work at 12,000 lpm speed.

Introduction

Mechanical-impact line printers (LP) are the standard devices for recording computer output data on paper. But today, there is a real need to replace this technology to obtain higher speed, better print quality and quiter and more reliable operation. Laser based output devices offer better performance, but there are economical reasons, too, why laser LPs have good future: the user can use cheaper common office paper. Namely, the most economical way to record character lines generated by a laser character generator is to use a xerographic drum and well-known electro-photographic processes.

Till now, several laser LP models have been introduced into market. Among them there are the IBM 3800 [1], the Siemens ND 2 [2] and the Canon and Xerox models. The obtained highest speed is 21,000 lpm.

In the field of laser based computer-output microfilm (COM) systems a few companies such as Datalight Inc., Eastman Kodak, etc. are very succesful in developing new systems. The speed of these COM devices exceeds 20,000 lpm. At this point it is worth emphasizing that it is not the laser character generator that limits the speed of a laser recorder device, but the handling and processing paper or film moving at high speed.

Most of these devices have extremely good technical data, but at the cost of complexity and price. Our design goal was to develop a simpler system and to reach moderate price, but to maintain the advantage of high speed laser recording.

In this paper I should like to give a short description of the LCG-1 laser character generator which has a max. speed of 12,000 lpm.

Design Considerations

To avoid the use of complicated correction optics it is necessary to analyze that how large distortions arise when predeflection focusing and flat recording surface are applied.

Predeflection Focusing

Let us regard a scanning Gaussian laser beam (Fig.1) originating from an aperture D, which immediately follows the focusing lens. The distortions in the recorded spot come from the fact that the waist of the scanning beam moves on a circle instead of a line. Using simple geometry and relations for a Gaussian beam we get the following relations for the distortions.

Defocusing as a function of the scan arm R and scan angle α is given by:

$$\Delta R = R \left(\frac{1}{\cos \alpha} - 1 \right) \tag{1}$$

If we suppose long focal length for the focusing optics than the relative elliptical spot deformation due to the nonnormal incidence on the plane of recording is determined from the expression

$$\frac{\Delta d}{d_o} = \frac{1}{\cos \alpha} - 1 \tag{2}$$

where d_o is the beam diameter at $\alpha = 0$. The relative change in scanning speed along line

---

* The research work leading to this paper was performed under a contract with the National Council for Technical Development.

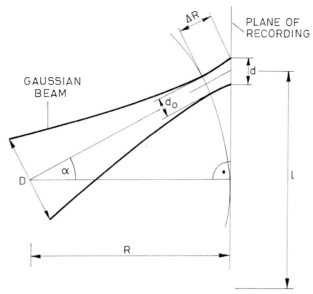

Fig. 1. Geometry for a
scanning Gaussian beam.

direction as a function of α is given by:

$$\frac{\Delta v}{v_o} = \frac{1}{\cos^2 \alpha} - 1 \qquad (3)$$

where v_o is the scanning speed at $\alpha = 0$.

Relations (2) and (3) are graphically illustrated in Fig. 2. If we want to print out 136 characters in a line in 7x5 point dot matrix form than the necessary resolution with a sufficiently large safety interval to get high modulation transfer should be N = 1500. Also, the length of lines ℓ is fixed for 280 mm. Than the half scan angle α_{max} is determined as

$$\alpha_{max} = \text{arc tan} \frac{\ell}{2R} \qquad (4)$$

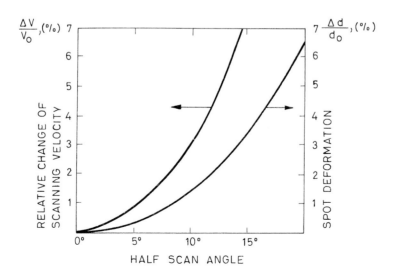

Fig. 2. Relative change of scanning velocity and spot deformation
as a function of half scan angle.

For these fixed values Fig. 3 shows the defocusing and half scan angle as a function of scan arm. Furthermore, if we do not want to overstep 5% increase in spot diameter due to defocusing, than for a Gaussian beam

$$\Delta R < \frac{0,08 \Pi}{\lambda} \frac{\ell^2}{N^2} \tag{5}$$

that is $\Delta R < 13,9$ mm for $\lambda = 633$ nm. It can be seen from the figures that if $R > 750$ mm all the resulting distortions will be well below 8-10%. Such distortions are difficult to see by eye and may be allowable for a line printer but, of course, are not allowable for a precision plotter.

If the value of R is determined, than the diameter of the laser beam at the scanner is given by:

$$D \geqslant \frac{4 \lambda R \; N}{\Pi \ell} \tag{6}$$

which gives $D > 3,2$ mm for $R = 750$ mm.

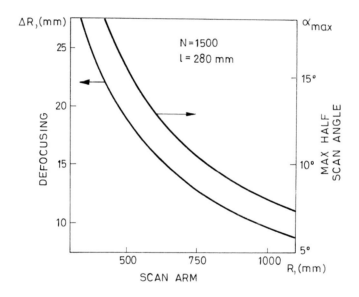

Fig. 3. Defocusing and half scan angle
as a function of scan arm

Multi-beam Contra Single Beam Recording

As every character line is composed from 7 raster lines the scanning frequency should be 7 times higher than the line frequency. This corresponds to 1400 Hz scanning frequency in case of 12,000 lpm recording speed. But this value is too large to realize it with a galvanometric mirror scanner having $\sqrt{2} D = 4,6$ mm mirror diameter and operating in saw-tooth scan mode. Galvanometric scanner are much cheaper than their rotating mirror counterparts. When multi-beam recording is used the necessary scanning frequency is 200 Hz, which is provided by commercially available scanners.

Furthermore, multi-beam recording is more preferable because the synchronization between the raster lines is automatically realized and thus there is no need for any complicated position sensing methods.

Acousto-optic Multi-beam Modulation

There is a well-known method [3] to get multi-beam modulation. This method is based on using many signals of different frequencies to drive the ultrasonic transducer of an acousto-optic light modulator. If, for example, seven signals are applied simultaneously to the transducer of a Bragg-cell than the incident laser beam is diffracted into seven beams, which can be individually turned on and off by a corresponding gate signal. But the adding network that adds the output signals of the 7 oscillators is complicated and causes a considerable signal loss. Also the power amplifier must have large bandwidth to cover all the frequencies used in the system and it must be highly linear to avoid intermodulation.

To overcome these difficulties a new multi-beam modulator has been developed [4] where there is no need for any adding network and wideband electronics. Fig. 4 shows this seven-beam acousto-optic modulator. Here the seven optical phase grids generated by seven transducer segments are spatially separated. The length of each segment in the direction of laser beam propagation is only 1 mm. Despite of this short acousto-optic interaction length the light diffraction taking place in the TeO_2 crystal is the so called Bragg diffraction, because the parameter Q is given by

$$Q = \frac{K^2 L}{k \cos \theta_o} \tag{7}$$

reaches 10, which is the Bragg limit. Here L denotes the interaction length, θ_o is the Bragg angle, K and k are the wavenumbers for the ultrasonic wave and the light wave resp.

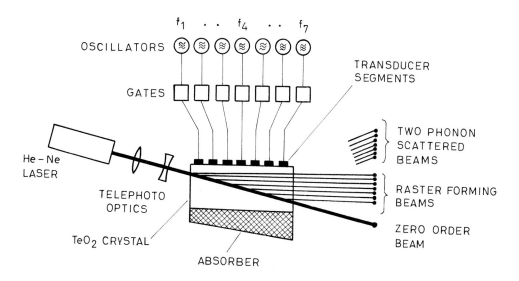

Fig. 4. Seven-beam acousto-optic modulator

Therefore the raster forming beams have most of the energy of the input laser beam. The crystal dimensions are 5x4x14 mm. The transducer is an x-cut $LiNbO_3$ shear wave transducer bonded on to the (110) plane of the TeO_2 crystal. The laser beam makes the Bragg-angle with the (001) axis. The transducer bandwidth between the -1,5 dB points is 27 MHz [5] and centered at 40 MHz. The seven frequencies are equally distributed in this interval. The number of raster beams separated to their e^{-2} intensity points is simply given by:

$$M = \frac{\Pi \tau \Delta f}{4} \tag{8}$$

where Δf is transducer bandwidth, τ is the ultrasonic transit time through the beam waist imagined into the modulator. Because τ is also the rise time of the optical pulse, it must be at least 0,7 µs for 12,000 lpm recording speed. Consequently, the max. number of raster beams is 14 in this case, but it is 28 for a medium speed (6000 lpm) printer.

One important parameter is the cross-talk between the modulation channels. As every phase grid has its own effect on the laser beam it is clear that every diffraction order coming from one of the phase grids will be rediffracted on a subsequent phase grid and so a composite diffraction pattern can be seen on a screen. This pattern contains not only the main raster forming beams appearing at $f_1 \ldots f_7$ frequencies but several other beams appearing at mixed frequencies. It is like an optical mixer. Furthermore, there is also a direct coupling between the channels through depletion of the zero order. Figure 5 shows the coupling between two channels which correspond to two diffraction cells. The effect of rediffraction processes on the intensity of the raster beams can be clearly seen. There is only one diffracted beam which can cause light noise, because it is diffracted at an angle corresponding

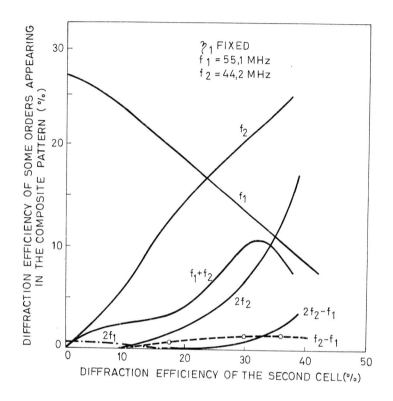

Fig. 5. Coupling between two channels

to $2f_2 - f_1$ frequency. This beam coincides with the raster beam of the third diffraction cell but its diffraction efficiency is well below 1%. Fig. 6 shows the diffraction efficiencies of the raster beams as a function of the numer of oscillators turned "on". The three curves in this figure show only the trends of change. It can be seen that it is not possible to get larger diffraction efficiencies in each of the seven channels than 7 or 8%. This is due to the existing intensive coupling which couples out considerable energy from the raster beams into unuseful diffraction orders.

The Optical System of the Character Generator

Fig. 7 shows the top view of the optical system. In front of the seven-beam modulator a telephoto optics is placed which gives long beam waist in the modulator. Making recordings with As-Se xerographic drums we found that 30 ... 60 µW power in each of the raster beams gave high constrast even at 12,000 lpm speed. So a 10 mW He-Ne laser was choosen as a light source. The main part of the spot forming optics is a camera lens which gives 200 µm spot diameter at the plane of recording. The galvanometer mirror works in saw-tooth scan mode with 3 ms fly-back time. The prism picks out the raster beams and the zero order diffracted beam scans across a photodiode. This diode gives synchronizing pulses for the page memory which controls the seven-beam modulator. The character dimensions are 2,0 x 1,5 mm. Using another camera lens after the prism microfilm recording has been obtained with 100 µm character height.

At last in Fig. 8 we show a small portion of the printout recorded with a specially designed xerographic machine at 12,000 lpm speed.

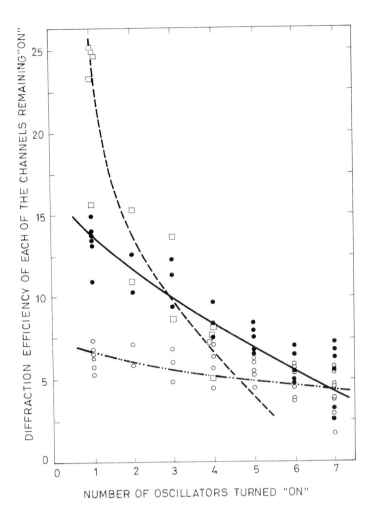

Fig. 6.
Diffraction
efficiency of
raster beams as
a function of
the numer of os-
cillators turned
"on"

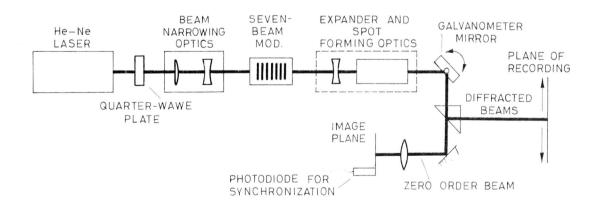

Fig. 7. Top view of the optical system of the LCG-1 laser character generator.

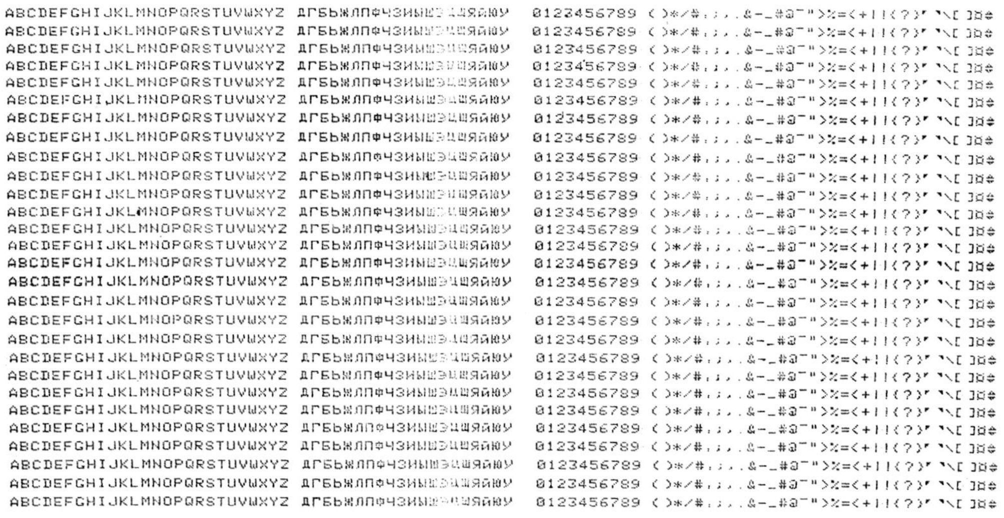

```
ABCDEFGHIJKLMNOPQRSTUVWXYZ  ДГБbЖЛПФЧЗИНЫЕ ЦЯЯАЮУ   0123456789 ()*/#,;..&-_#@-">%=<+!!(?)' `\[ ]¤¢
ABCDEFGHIJKLMNOPQRSTUVWXYZ  ДГБbЖЛПФЧЗИНЫЕ ЦЯЯАЮУ   0123456789 ()*/#,;..&-_#@-">%=<+!!(?)' `\[ ]¤¢
ABCDEFGHIJKLMNOPQRSTUVWXYZ  ДГБbЖЛПФЧЗИНЫЕЗЦЯЯАЮУ   0123456789 ()*/#,;..&-_#@-">%=<+!!(?)' `\[ ]¤¢
ABCDEFGHIJKLMNOPQRSTUVWXYZ  ДГБbЖЛПФЧЗИНЫЕЗЦЯЯАЮУ   0123456789 ()*/#,;..&-_#@-">%=<+!!(?)' `\[ ]¤¢
ABCDEFGHIJKLMNOPQRSTUVWXYZ  ДГБbЖЛПФЧЗИНЫЕЗЦЯЯАЮУ   0123456789 ()*/#,;..&-_#@-">%=<+!!(?)' `\[ ]¤¢
ABCDEFGHIJKLMNOPQRSTUVWXYZ  ДГБbЖЛПФЧЗИНЫЕЗЦЯЯАЮУ   0123456789 ()*/#,;..&-_#@-">%=<+!!(?)' `\[ ]¤¢
ABCDEFGHIJKLMNOPQRSTUVWXYZ  ДГБbЖЛПФЧЗИНЫЕЗЦЯЯАЮУ   0123456789 ()*/#,;..&-_#@-">%=<+!!(?)' `\[ ]¤¢
ABCDEFGHIJKLMNOPQRSTUVWXYZ  ДГБbЖЛПФЧЗИНЫЕЗЦЯЯАЮУ   0123456789 ()*/#,;..&-_#@-">%=<+!!(?)' `\[ ]¤¢
ABCDEFGHIJKLMNOPQRSTUVWXYZ  ДГБbЖЛПФЧЗИНЫЕЗЦЯЯАЮУ   0123456789 ()*/#,;..&-_#@-">%=<+!!(?)' `\[ ]¤¢
ABCDEFGHIJKLMNOPQRSTUVWXYZ  ДГБbЖЛПФЧЗИНЫЕЗЦЯЯАЮУ   0123456789 ()*/#,;..&-_#@-">%=<+!!(?)' `\[ ]¤¢
ABCDEFGHIJKLMNOPQRSTUVWXYZ  ДГБbЖЛПФЧЗИНЫЕЗЦЯЯАЮУ   0123456789 ()*/#,;..&-_#@-">%=<+!!(?)' `\[ ]¤¢
ABCDEFGHIJKLMNOPQRSTUVWXYZ  ДГБbЖЛПФЧЗИНЫЕЗЦЯЯАЮУ   0123456789 ()*/#,;..&-_#@-">%=<+!!(?)' `\[ ]¤¢
ABCDEFGHIJKLMNOPQRSTUVWXYZ  ДГБbЖЛПФЧЗИНЫЕЗЦЯЯАЮУ   0123456789 ()*/#,;..&-_#@-">%=<+!!(?)' `\[ ]¤¢
ABCDEFGHIJKLMNOPQRSTUVWXYZ  ДГБbЖЛПФЧЗИНЫЕЗЦЯЯАЮУ   0123456789 ()*/#,;..&-_#@-">%=<+!!(?)' `\[ ]¤¢
ABCDEFGHIJKLMNOPQRSTUVWXYZ  ДГБbЖЛПФЧЗИНЫЕЗЦЯЯАЮУ   0123456789 ()*/#,;..&-_#@-">%=<+!!(?)' `\[ ]¤¢
ABCDEFGHIJKLMNOPQRSTUVWXYZ  ДГБbЖЛПФЧЗИНЫЕЗЦЯЯАЮУ   0123456789 ()*/#,;..&-_#@-">%=<+!!(?)' `\[ ]¤¢
ABCDEFGHIJKLMNOPQRSTUVWXYZ  ДГБbЖЛПФЧЗИНЫЕЗЦЯЯАЮУ   0123456789 ()*/#,;..&-_#@-">%=<+!!(?)' `\[ ]¤¢
ABCDEFGHIJKLMNOPQRSTUVWXYZ  ДГБbЖЛПФЧЗИНЫЕЗЦЯЯАЮУ   0123456789 ()*/#,;..&-_#@-">%=<+!!(?)' `\[ ]¤¢
ABCDEFGHIJKLMNOPQRSTUVWXYZ  ДГБbЖЛПФЧЗИНЫЕЗЦЯЯАЮУ   0123456789 ()*/#,;..&-_#@-">%=<+!!(?)' `\[ ]¤¢
ABCDEFGHIJKLMNOPQRSTUVWXYZ  ДГБbЖЛПФЧЗИНЫЕЗЦЯЯАЮУ   0123456789 ()*/#,;..&-_#@-">%=<+!!(?)' `\[ ]¤¢
ABCDEFGHIJKLMNOPQRSTUVWXYZ  ДГБbЖЛПФЧЗИНЫЕЗЦЯЯАЮУ   0123456789 ()*/#,;..&-_#@-">%=<+!!(?)' `\[ ]¤¢
ABCDEFGHIJKLMNOPQRSTUVWXYZ  ДГБbЖЛПФЧЗИНЫЕЗЦЯЯАЮУ   0123456789 ()*/#,;..&-_#@-">%=<+!!(?)' `\[ ]¤¢
ABCDEFGHIJKLMNOPQRSTUVWXYZ  ДГБbЖЛПФЧЗИНЫЕЗЦЯЯАЮУ   0123456789 ()*/#,;..&-_#@-">%=<+!!(?)' `\[ ]¤¢
ABCDEFGHIJKLMNOPQRSTUVWXYZ  ДГБbЖЛПФЧЗИНЫЕ-ЦЯЯАЮУ   0123456789 ()*/#,;..&-_#@-">%=<+!!(?)' `\[ ]¤¢
```

Fig. 8. Small portion of the printout.

Conclusions

It has been shown that multi-beam acousto-optic light modulation and galvanometric mirror deflectors give good possibility to build laser character generators to be used in recording computer output data. Without using any scan-correction optics and complicated raster beam synchronization a simple system can be built which meets the requirements of high speed recording computer output data.

References

1. Vahtra, U. and Wolter, R.F., Electrophotographic Process in a High Speed Printer, IBM J. Res. Develop., Vol. 22, pp. 34-49. 1978.
2. Siemens Technical Bulletin on ND 2 Nonimpact High-Speed Printer, 1976.
3. Hecht, D.L., Multifrequency Acoustooptic Diffraction, IEEE Trans. Son. Ultrason., Vol. SU-24, pp. 7-18. 1977.
4. Podmaniczky, A., A New Multi-channel Acousto-optic Light Modulator in TeO_2 Crystals Proceedings of the Electro-Optics/Laser International'76 UK, pp. 94-98.
5. Podmaniczky, A., Some Properties of TeO_2 Light Deflectors with Small Interaction Length, Opt. Comm, Vol. 16, pp. 161-165. 1976.

AN OPTICAL TRACKING SYSTEM FOR LASER COMMUNICATION IN SPACE

L. Liebing, F. Kunstler

DFVLR-Institut fur Technische Physik

7000 Stuttgart 80, Federal Republic of Germany

Abstract

Space laser communication involves angular tracking accuracy of the order of 10 μrad, which can be accomplished by means of an autotracking system. The main components of a system are described and test results are presented.

Introduction

The application of lasers for space communication offers the advantage of high antenna gain and large information bit rates. Comparing, for instance, a laser in the infrared (Nd:YAG at 1.06 μm) with conventional S-band-microwave, the angular beam divergence is reduced by four orders of magnitude (at equal antenna dimensions) and by the same order of magnitude the potential bit rate could be increased. Therefore, laser technology for space communication has been considered for more than ten years. Hardware development was particularly made in the United States, where the essential components for the CO_2[1] and the Nd:YAG[2] systems are built and tested. In W-Europe some studies[3,4] were made and only little hardware[5] was built, and no commonly accepted concept or program does exist today.

The most interesting application for space laser is in the field of earth surveying with low altitude satellites[6]. If the optical capability of modern satellite telescopes would be fully utilized a continuous data stream of more than 10^{10} bit/s had to be transmitted to ground stations. This data rate would exceed μ-wave capability.- Another potential application for spaceborn lasers refers to the interconnection of geostationary communication satellites. For an optimum satellite system the interconnecting carrier frequency should be at least an order of magnitude higher than the uplink- or downlink frequency[7].

In order to make full use of the small laser beam divergence, the laser beam has to be tracked with high accuracy, a typical angular deviation must be less than 10 μrad. This tracking accuracy is more than three orders of magnitude higher than used in μ-wave technology and therefore laser tracking has to be particularly considered.

Description and Tests of an Autotracking System

Tracking at high accuracy is usually accomplished in an autotracking mode, which means that the transmitting station utilizes the received laser beam (of the partner-station) as a beacon which allowes the transmitted beam to be placed exactly into the direction of the received beam. A straight-forward optical arrangement for laser autotracking is shown in Fig. 1.

The transmitted[1] and the received[2] beam are both reflected by the gimbaled mirror[3]. The angular position of this mirror with respect to the direction of the received beam is detected (in a differential way) by means of the optical roof[4] and the two photodetectors [5]. The electrical signals from the detectors control the angular position of the gimbaled mirror in a closed loop way by means of a control electronics and special linear motor drives (Fig. 2). The optical roof is adjusted such that the transmitted beam is parallel to the received beam if the detectors measure equal intensities.

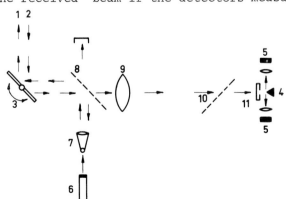

Fig. 1. Optical arrangement for laser autotracking (one axis). 1, 2 transmitted and received beams; 3 gimbaled mirror; 4 optical roof; 5 detector; 6 laser; 7 transmitting optics; 8 beam splitter; 9 receiving optics; 10 beam splitter for second axis; 11 filter and aperture.

An autotracking-system corresponding to Fig. 1 was built at the DFVLR. The characteristics of the main optical components are listed in Table 1. The gimbaled mirror with its linear motor drives is shown in Fig. 2 and a detector-modul in Fig. 3. With these components the system was expected to have a tracking accuracy of better than 10 μrad.

Table 1. The main optical components of the autotracking system

laser (6)	He Ne, 10 mW output
transmitting optics (7):	3 cm diameter output beam
gimbaled mirror (3):	17 cm diameter, accuracy of 1λ , linear motor drives, moment of inertia: 2.10^5 g cm^2
receiving optics (9):	10 cm diameter, 1 m focal length
optical roof (4):	10^{-4} - 10^{-3} cm radius of curvature

Fig. 2. Gimbaled mirror with linear motor drives. Each drive consists of a coil which moves inside a magnetic field. The field is produced by permanent magnets located at the ends of curved iron strips.

Fig. 3. Adjustable modul with optical roof, detectors etc.

Tests of the system were performed with a retro-reflector acting as the partner station. The tracking distance was varied from 100 m to 1 km. It turned out, however, that at a distance of more than 100 m the test was severely disturbed by atmospheric turbulence. A typical measurement at a tracking distance of 100 m is shown in Fig. 4. To demonstrate the tracking accuracy the retro-reflector was stepwise moved at steps of 30 μrad. The upper trace is the signal from the detectors with the gimbaled mirror blocked at a fixed position, such that the angular sensitivity of the detecting optics is demonstrated. The noise like shape of the signal is due to atmospheric turbulence. This turbulence makes the laser beam changing its direction and structure in a statistical fashion. The lower trace of Fig. 4 shows the same test but with the gimbaled mirror in tracking action. The mirror now follows the stepwise motion of the retro-reflector and even reacts at most of the angular fluctuations of the laser beam. The comparison of upper and lower trace indicates a tracking accuracy of better than 10 μrad.

Another performance characteristic of the tracking system is its resonance frequency which determines the tracking stiffness. This resonance frequency increases with increasing amplifier gain of the control electronics but there is an upper limit at which the system becomes instable. It usually is difficult to find the source of the main instability and when this instability is eliminated, there always appears another one. Because of the relatively large moment of inertia of the gimbaled mirror (Table 1) the resonance frequency of the described tracking system was rather low, about 5 Hz.

If an additional gimbaled mirror of smaller size is used, the resonance frequency can be drastically increased. This mirror, which is frequently referred to as "Image Motion Compensator (IMC)" is closely located at the optical roof (Fig. 1). A photograph of a partly disassembled IMC is shown in Fig. 5.

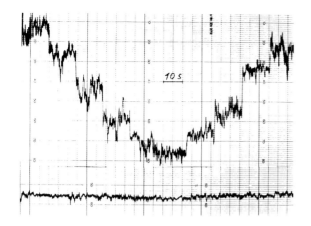

Fig. 4. Detector Signal at stepwise motion of a retro-reflector at 100 m distance. Angular motion of one step: 30 μrad. Upper trace: gimbaled mirror is locked. Lower trace: gimbaled mirror is tracking.

Fig.5. Image-Motion-Compensator (partly disassembled). The centre magnet with radial magnetic field is gimbaled, and is controlled by four magnet coils. A small mirror will be fastened to the magnet.

A test of the tracking-system with an IMC is shown in Fig. 6. The measurement which corresponds to Fig. 4, demonstrates a tracking accuracy of better than 3 μrad. The IMC follows the atmospheric turbulence completely. The resonance frequency was about 1 kHz.

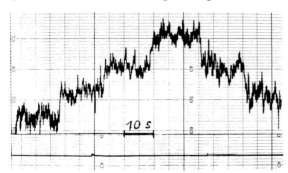

Fig. 6. Test of the tracking system with an IMC. Conditions correspond to Fig. 4.

The Electro-Optical Receiver

There are three problem areas associated with the receiver: drift stability, noise and detector sensitivity.- Zero point drift can be practically eliminated if the received laser beam is modulated. The modulation frequency has to be set to a relatively high value to avoid instabilities in the tracking system due to phase shifting of the mirror positioning signal. For the described system the modulation frequency was 300 kHz and the receiver bandwidth was limited to 30 kHz, which is far higher than the upper stable resonance frequency of the tracking system.

The receiver performance is different for various types of detectors: A straight-forward receiver electronics for a pin-diode (D) is shown in Fig. 7 with amplification at T and a filter at F. The performance limitation at room temperature is due to thermal noise at R and internal transistor noise (T). For an optimum wavelength λ_o , the equivalent noise power at the optical input due to total system noise Ps and due to quantum noise Po differ by more than 3 orders of magnitude, indicating the gap between actual performance and the basic quantum noise limit. If the pin-diode is replaced by an avalanche-diode or a photo-multiplier the performance is no longer limited by electronic components and the quantum noise limit is further approached.- Besides noise-performance, the quantum efficiency η of the receiver ist important. At their optimum wavelength, η_o is rather high for the considered detectors. Typical performance data for the receiver of the tracking system are listed in Table 2.

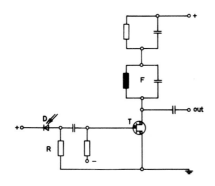

Fig. 7. Receiver for direct detection of modulated laser radiation at 300 kHz. R = 200 kΩ, input capacitance of T:4 pF, bandwidth of filter F : 30 kHz.

Table 2. Typical receiver performance of the tracking system for various detectors at their optimum wave-length λ_0 and at room temperature. The modulated laser beam (300 kHz) is directly detected with 30 kHz bandwidth. η_0 : quantum efficiency; P_S/P_Q : ratio of equivalent input-noise to quantum noise.

	pin-diode with electronics	avalanche-diode	photo-multiplier
λ_0	0,9 μm	0,9 μm	0,4 μm
η_0	0,9	0,9	0,25
P_S/P_Q	4000	300	10

Conclusions

The angular accuracy of the described autotracking-system meets the demands of space laser communication. If an "image motion compensator" is incorporated in the optical system, a relatively high resonance frequency of 1 kHz can be reached at full system stability. The performance of the electro-optical receiver is good with photo-multiplier at optimum wave-length and decreases with avalanche- and pin-diodes.

Acknowledgements

The authors like to thank M. Kling for designing optical components and R. Dufner for technical assistance.

References

1. J. H. McElroy et al., CO_2 Laser Communication Systems for Near-Earth Space Applications, Proce. of the IEEE, Vol. 65, No. 2, February 1977.
2. M. Ross et al., Space optical Communications with the ND:YAG Laser, Proc. of the IEEE, Vol. 66, No. 3, March 1978.
3. Battelle-Institut, Anwendungsmöglichkeiten der Laser in der Raumfahrt, RV I-TP-5/72.
4. Battelle-Institut, Phase A Studie über ein Spacelab-Experiment zur Breitband-Nachrichtenübertragung und Lageregelung mit Laser, Juli 1977, BF-R-63.224-1.
5. Laser Focus, April 1974, p. 40
6. Kraemer, A. R., Free-Space Optical Communications, Signal, October 1977, p. 26.
7. Liebing, L., Künstler, F., Laser-Kommunikation in der Raumfahrt, Laser-Strahl-Tracking, 1. Teil, Interner DFVLR-Bericht 452/76-2, S. 10.

LASER IN UNDERWATER OPTICAL COMMUNICATIONS

Goran Stojkovski

Faculty for Electrical Engineering
Zagreb, Yugoslavia

Abstract

In this paper some fundamental parameters of sea water as a transmitting media for optical communications have been analysed. There are mean values of ε (total attenuation coefficient), calculated distances for transmission of broadband information using argon laser. There are also measuring results for indirect measurements of ε for two near, but different points, in Adriatic sea, and experimental results of TV signal transmission for chosen model for distances to the 30 m.

Introduction

One of the most important thing of the present time is exploatation of the sea wealths. Very significant part of this welth can be found on the continental shelf - to the depths of some tenths of meters. Research work in this depths requires reliable communications between the ship and the underwater equipment.

For this purpose two transmitting medias can be used - ultrasonic waves and electromagnetic waves. Ultrasonic wave attenuation in sea water is sufficiently low (Fig.1), but there is a problem of broadband information transmission. Generally it is possible to transmit TV information using upper part of ultrasonic wave spectra, and eliminating of redundancy. However, in this case attenuation is very similar to that of electromagnetic waves in visible spectra. Using of electromagnetic waves of extremely low frequences (ELF) it is possible to transmit information to the some hundreds of meter in the sea water, but it has to be used the prolonged time (some hours) for transmission of one TV picture, so that this part of spectra can be used only for special communications (communications with submarines /4/).

Sea water structure and its optical parameters

The fundamental optical parameters which are important in our case are: dispersion parameter σ, attenuation coefficient k and their sum $\varepsilon = \sigma + k$ - total attenuation coefficient. In this work we shall not deal with a foton life time probability, dispersion indicatrice and refraction coefficient n, although they can be very important in some case.

Sea water structure /3/

Generally sea water is destilated water with added (melted) sault, gas, organic material, mud and in some cases added air bubbles.

Destilated water . In the sea, destilated water is in form of free molecules H_2O or molecular connections $(H_2O)_x$ in which there is always transformation process from one form to the other - $x(H_2O)_2 \quad (H_2O)_x$.

Dispersion coefficient σ^x of destilated water can be found by the equation /3/:

$$\sigma = \frac{8\pi^3(n^2-1)^2 \eta k\,T}{\lambda^4} \quad \frac{2+\Delta}{6-7\Delta} \tag{1}$$

or for visible spectra

$$\sigma = \frac{1{,}42 \cdot 10^4}{\lambda^4} \tag{2}$$

(λ = wavelenght in free space, $\eta = 45{,}7 \cdot 10^{-12}$ barr^{-1}, k = Baltzmanns constant, T = apsolute temperature , Δ = polarization effect).

Attenuation coefficient k can be found by transmission of paralel light through water and measuring of transmission coefficient T:

$$T = e^{-\varepsilon h} = e^{-(\sigma+k)h} \tag{3}$$

where h is the transmission distance.

Attaining k by this way gives wide dispersion of results, so that differences of 20% can be tolerated.

For the wave lenghts from 250 to 700 nm coefficient k is more significant than σ_1 and in green-blue part of visible spectra it has a minimum value in order of 0,01 dB m^{-1}.

Melted gas. The air above the sea consists of 78,08% N, 20,95% O, 0,03% CO_2 and 0,93% Ar. These gas are represented in water in various degrees: N - 8-17 cm^3/l, O - 5-10cm^3/l, CO_2 to 50 cm^3/l, Ar - 0,23-0,41 cm^3/l.

Melted gas change the optical characteristics of sea water indirectly by the influence on the other parts which are in sea water.

Salt. Dry rest of sea water consists of 78,32% NaCl, 9,44% MgCl, 1,69% KCl, 6,4% MgSO$_4$, 3,94% CaSO$_4$ plus rest 0,21%.

The mean saltness of sea water is 35% and in the most cases has the values from 33% to 37%. There are very big differences especially in basins like Black Sea (18%) and Baltic (2-7%), and hot seas (Red Sea and Persian Bay - 42%).

Melted salt generally changes the optical characteristis in very little values. They have small influence on the refraction index n and dispersion and, in visible spectra, even smaller on the attenuation coefficient. Extremely clear sea water has total attenuation coefficient about 0,016 m^{-1}. In that case light attenuates for order of value for distance of 144 m. It means that for this case the day light can propagates in sea water to the depths of 2 km.

Melted organic materials. In the sea water there is about 0,001% to 0,005% of organic materials.

They practicaly do not change dispersion, and attenuation falls exponentially with increasing λ. It causes the water to become yellow.

Mud . Sea water mud consists of particles of organic and anorganics origin. Anorganic materials are consequences of rivers, winds, volcanos and cosmic influences. Organic parts are the rest of fitoplanktons , zooplanktons, detrits microbes, etc.

The main contibution to the optical characteristics of sea water is from dispersed parts. Its number changes in wide range by changing of seasons, daytimes, meteorogical situation and that leads to the big changes and unstability of characteristics. Mean concentration in sea water is in order 1 mg/l and changes in ocean waters from 0,03 mg/l to 2 mg/l. In coastal waters it can reach a value of 10 mg/l. Number of particles depends of measuring systems and author from 10^4 to 10^{10} particles/l. So big differences in measurement results for the same aquatoires are a consequence of determining of physical values that are cosidered.

Mean value of refraction index for mean and big ungles is n = 1,15 but for the most important minerals /6/ the value is n = 1,51 - 1,58, and n$_o$ = 1,165 - 1,19.

In order to attain usable results in dispersion analysis one can use the fraction parameter $\rho = 2\pi r/\lambda$ - particles are normalized on the wave lenghts values. It gives a possibility to make tables for very different dimensions of particles and wave lenghts /5/.

Transmission

The most important parameter with setting of underwater optical communication systems is the coefficient of the total attenuation $\varepsilon = \sigma + k$. As it was said before in visible part of electromagnetic spectra the most important is attenuation coefficient k. The participation of parameter σ is of order 0,5 - 2,5 % although there can be very big differences from this value.

Minimal value for k reachs 0,025 to 0,035 m^{-1} and, in special cases, gets value 0,012m^{-1} which can be compared to that of destilated water.

Attenuation of electromagnetic waves depending on λ in sea water is shown on Fig.2 and 3. On the Fig. 2 there is attenuation of ultrasonic waves too (b). One can sea that total attenuation reachs very low values for visible spectra. On the Fig.3 there are some typical ones for individual aquatories. It can be seen that total attenuation coefficient changes in great degree depending on wave lenght and there are minimal values for blue-green part of spectra. One can also sea that there are significant differences for more than order of magnitude for individual aquatories.

Optical signal propagates through the sea water by attenuation according to exponential law.

$$B = B_o \, e^{-\varepsilon h} \tag{4}$$

In measurements in this work parameter ε is substituted by attenuation coefficient k, which gives the mistake that can be neglected.

Laser

Transmitting characteristics given by Fig.3. shows that the best results can be obtained by using of laser working in blue-green region of visible spectra i.e. for wave lengts between 440 nm and 480 nm. Fig.4. shows partial spectra for some kind of laser which can be used for this purposes.

In our case it was the argon laser with spectral line on 4879,9 A^O i.e. for transition $4p^2D^O_{5/2}$ $4s^2P_{3/2}$ /2/, which can be very efficiently used for transmition.

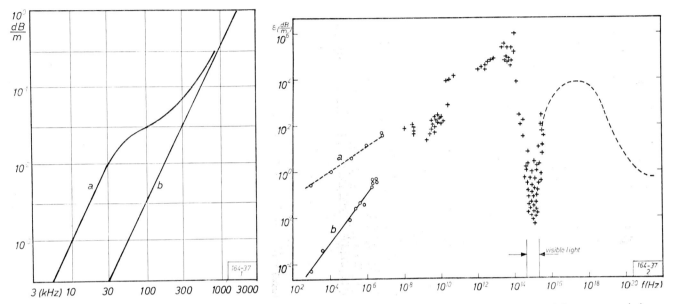

Fig.1 - Ultrasonic waves attenuation in water (b) and sea water (a)

Fig. 2 - Attenuation of electromagnetic waves (a) and ultrasonic waves (b) in sea water

Modulation

At the beginning of this work there was a question which kind of modulation has to be chosen for this purpose. For digital transmission total attenuation is something lower according with:

$$B = B_o e^{-\varepsilon h} \{ 1 + \frac{\omega}{4\pi} \rho(o) h \} \simeq B_o e^{-(\varepsilon - \Delta\varepsilon)h} \qquad (5)$$

where is $\Delta\varepsilon = \frac{\omega}{4\pi} \rho(o)$.

But, the digital transmission requires more complexequipment and this requirement cant be compensated by some longer transmission distances. So, the equipment was completed by argon laser end electrooptical modulator. Driving voltage for our modulator is more than 150 V_{pp} and this fact causes the problem to make good driving circuit for modulator. Since the total capacitance is more than 100 pF the circuit has upper frequency (- 3 dB point at 7 MHz. Detector was made with a PIN diode with low-noise preamplifier. For transmission we used the test signal of RTV Zagreb and signal from transportable TV camera. Received signal was observed on a modified TV monitor.

Calculated results

For 1W argon laser and receiver working in accordance with Neuman-Pearsons statistical criteria there are the calculated results /6/ for possible distances in broadband TV information transmission. These results are given in the table 1. and fig. 5.

As we can see from the table 1. and fig. 5. transmitting system consisting from 1W argon laser and receiver working as a photoelectron counter can ensure transmitting distances for broadband information from 22m to 26m. It is generaly enough for research of continental shelf which is in depht ranges from 0 to 30 m.

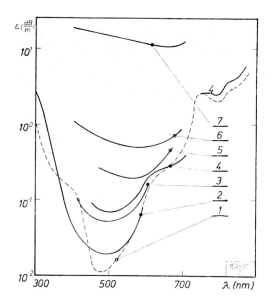

Fig.3. Some typical values of parameter ε for: 1- Destilated water, 2- Carribean Sea, 3- Atlantic, 4- Sargas Sea, 5- Black Sea, 6- Wainwright Bay, 7- rivers water /3/

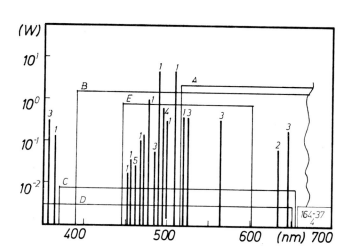

Fig. 4. Lasers lines usable for underwater communications (1- Argon laser, 2- He-Ne laser, 3- Crypton laser, 4- Xenon laser, A- Dye laser pumped by flash lamp, C- Dye laser pumped by N-laser, D- Nd::YAG laser - LN parametric)

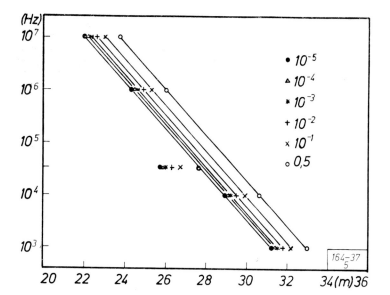

Fig. 5. Calculated distances for 1 W argon laser

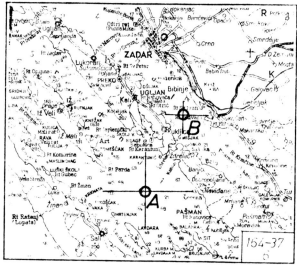

Fig. 6. Measuring points in Kornati archipelago (A) and Pašman Channel (B)

Table 1. Calculated distances for 1W argon laser

	P	0,5	10^{-1}	10^{-2}	10^{-3}	10^{-4}	10^{-5}
$\Delta f(Hz)$							
10^3		33	32,2	31,8	31,5	31,4	31,2
10^4		30,6	29,9	29,5	29,2	29,1	28,9
$4 \cdot 10^4$		27,6	26,7	26,3	26,0	25,9	25,7
10^6		26,0	25,3	24,9	24,6	24,5	24,3
10^7		23,7	23,0	22,6	22,3	22,2	22,0

Experimental results

To get the total attenuation parameter values we made some measurements of the sea water transparency for two points expecting that there shall be significantly different results, for acceptable geographical distances - points A and B (fig.6)
Their coordinates are:
A - $15^{\circ}13'$ E $43^{\circ}58'$ N   in Kornati archipelago
B - $15^{\circ}17'$ E $44^{\circ}4'$ N   in Pašman Channel

The transparency was measured by method "white disk" with folowing results (table 2)

Table 2. Transparency measurements (A)

N⁰	1	2	3	4	5	6	7	8	9	10	11	12	13	14	15	16	17	18
m	33,5	36	34	33	36,5	33,5	35,5	36,5	37	36	36,5	38,5	38	37	38,5	39	38	4C

19	20
39,5	39

$d_{mean} = 36,78 = 37m$

date: 6.8.1978; start: 10 AM ; duration: 200 min; resolution: 0,5m; weather:slightly cloudly
step: 10 min
For given meteorological parameters the most expected value for coefficient β is 5. For this case attenuation is:

$$\varepsilon_{mean} = \frac{\beta}{d} = 0,135 \text{ m}^{-1}$$

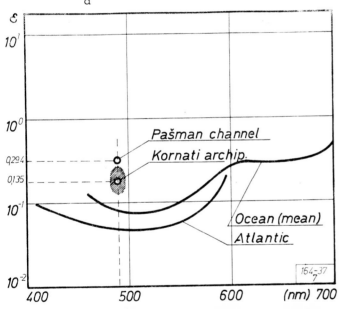

Point B
At this point we made only one measurement to compare result with results at point A.

$$\varepsilon = 0,294 \text{ m}^{-1}$$

These results are shown on the fig. 7 and compared by mean value of the ocean water and Atlantic . Man can see very good agreement despite on that the measurements was made by approximative method.
Results of TV signal transmission are presented on the fig. 8, for test signal from the TV receiver (a), and TV camera (b).

Fig. 7. Calculated values of the total attenuation parameter ε for Kornati archipelago (A) and Pašman Channel (B)

Fig. 8a. Test signal transmissin results for 14m and 22m

Fig.8b. Camera signal transmission results for 14m and 22m

Conclusion

TV and any other broadband signal can be transmitted to the distances of abouth 25 meters in standard sea water using laser working in the blue - green part of optical spectra. Results depend of many parameters like a basin, season, weather,daytime etc.

References

1. L.H.Dulberger: Will the laser succesed sonar for undersea electronic? Electronics, 1961, 34, N° 23, p. 24
2. C.G.B. Garret: Gas - Laser, R. Oldenbourg Verlag, München und Wien, 1969
3. A.P. Ivanov: Fizicheskie osnovi gidrooptiki, Nauka i tehnika, Minsk, 1975
4. G. Rowe: Extremly low frequency (ELF) communication too submarines, Trans.IEEE, vol-COM N° 4, 1974
5. K.S. Shifrin, N. Salganik: Tablici po svetoraseyaniyu, Gidrometeoizdat, leningrad, 1973
6. G. Stojkovski: Laseri u podmorskim komunikacijama,Proceeding of symposium ETAN in marine Zadar, 1976
7. G. Stojkovski: Optičke karakteristike morske vode i njihov utjecaj na podmorske komunika cije, Symposium ETAN in marine, 1977
8. G.I. Vlasov & alt.: Beskabelnaya peredacha signalov televidenija pod vodoi, Tehnika kino i televideniya, 2, 64, 1971

FOURTH EUROPEAN ELECTRO-OPTICS CONFERENCE

Volume 164

SESSION 3

ELECTRO-OPTICS IN CHEMISTRY

Session Chairman
Dr. Dirk J. Kroon
Philips Research Laboratories
Eindhoven, The Netherlands

NEW PERSPECTIVES IN INTERFERENTIAL SPECTROMETRY
APPLICATION TO ATMOSPHERIC POLLUTANTS DETECTION

A. Marechal and G. Fortunato

Institut d'Optique, Orsay and ENSET Cachan, France

Modern spectrometry offers rich possibilities by convenient use either of interferometric devices (Fourier transform spectrometry) or of selective modulation (grid spectrometry). Progresses in resolution and luminosity have been spectacular in the last decade, and new domains of high resolution and low luminosity have been explored. In those cases of extreme performances, the instruments are generally highly sophisticated, delicate, and expensive. Nevertheless, Fourier transform spectroscopy is now developing for pratical applications. We have examined the possibilities of interferometric devices to routine spectral analysis in chemistry, biology, pollution, detection etc ... and are now aware of the interesting characteristics of those mountings by the fact that they are luminous, flexible and very simple. They need no computer and are very suitable for low resolutions. We shall describe first the basic principle, and later focus on the various possibilities resulting from the direct access to the interferogram and the application of the mathematical properties of the Fourier transform : Fourier derivation, Fourier correlation with a reference spectrum, Fourier correlation of derivatives etc ... Moreover when the spectrum has a quasi periodic structure a very simple interferometric device can be used and has proved to be very efficient for the detection of some atmospheric pollutants (SO^2, NO^2).

A) BASIC PRINCIPLE OF THE SIMS

The "Spectromètre Interférentiel à Modulation Sélective" (SIMS) combines the use of interferences and selective modulation. Its principle has first been pointed out by R. Prat (1), and G. Fortunato (2) has shown that it is the only way to obtain a high "étendue" and consequently a hith optical signal.

The leading idea of the device is
- the production of very luminous interference fringes in a plane.
- The analysis of those fringes by a moving grid and the production of photoelectric signal obtained by synchronous detection.

1 - The Interferometric Device
In order to obtain luminous fringes with an extended source, it is necessary to manage in order that the position of the fringes should not depend upon the position of the emitting point on the source. Fortunato has shown that the only solution is the Prat mounting comprising
- a doubling device producing two laterally shifted images of the source ; on Fig. 1 this is performed by a couple of Wollaston prisms W_1 W_2 ,
- a converging lens.
The fringes located in the focal plane of the lens, and produced by various coherent coupled images of the points of the source, do not move when the point source moves ; as a consequence, the flux can be important by the fact that there is no fundamental limitation on the solid angle of the beams.

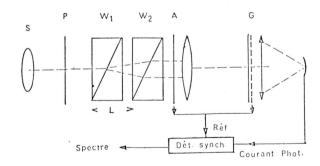

Fig. 1

2 - Selective Modulation

If now we put a periodic grid G in the fringes and move it or if we rotate polarizer P or analyser A, the outgoing flux is modulated only for the wavelength for which the fringe separation is matched to the grid spacing : it is then possible to modulate selectively one wavelength (or small spectral region) and use synchronous detection in order to produce the spectral signal. The modulated flux will be proportional to the source luminance for the wavelength for which fringe separation $i = \lambda/\alpha$ (where α is the angle between the two interfering rays), is matched to the grid period. Changing the wavelength can easily be performed by acting on the angle which depends upon the lateral shift produced by the interferometer : we use now a very simple polarizing interferometer invented by G. Nomarski and based on the properties of two Wollaston prisms W_1 W_2 located between polarizer and analyser. The adjustment of the distance between the two prisms allows the variation of the wavelength.

3 - Luminosity

The high gain in "étendue", with respect to classical devices, leads to appreciable advantages : as an example, the fluorescence spectrum of anthracene has been obtained on a bench mounting down to concentrations of 10 ppB. Nevertheless, we have to take into account the increase of noise due to the unmodulated flux, and a detailed discussion done by Fortunato has led to various encouraging conclusions.

4 - Resolution

The response of the apparatus to monochromatic light (laser) has shown be in agreement with theoretical predictions : we perform the analog Fourier analysis of a sample of N fringes, and the resolution is equal to N and does not depend directly on the luminosity : if we increase the number of fringes, we increase the resolution at the same time, and in fact, luminosity and resolution do no depend upon the same parameters and are practically independent.

B) OPTICAL DATA PROCESSING ON THE SIMS

It turns out that the SIMS is a luminous spectrometer, but another feature can also be significant : we have in the focal plane of the lens the interferogram of the source, i.e. the Fourier transform of the spectrum. It is then possible to operate on this interferogram in order to obtain various signals representing linear operations on the spectrum. In other words, we have at our disposal the Fourier transform and we can take advantage of this situation.

1 - The spectrum

If $S(\sigma)$ is the spectrum (as a function of the wave number σ) the interferogram is $I(\Delta)$, where Δ is the optical path difference and they are related by

$$i(\Delta) = I(\Delta) - I_0 = \int S(\sigma) \cos 2\pi\sigma\Delta d\sigma$$

where $i(\Delta)$ is the variable part of the interferogram and I_0 the average of I. If we only need $S(\sigma)$, we operate an analog Fourier analysis of $i(\Delta)$ by using a movable periodic grid.
The response function is normally a sinc function, but procedures of apodisation can be applied by using a convenient smoothing screen on the interferogram.

2 - Derivation

If, instead of wishing to obtain the spectrum $S(\sigma)$, we prefer to obtain directly one of its derivatives, we can use the general properties of the Fourier transform ; as an example, the F.T. of $S'(\sigma)$ is proportionnal to $\Delta I(\Delta)$, which means that, in order to obtain a signal representing S', we should multiply the interferogram by Δ, what is very simple : we put in the interferogram a "two triangles" mask, and replace the ordinary periodic grid by a grid made up of two zones : for $\Delta > 0$ and $\Delta < 0$, the black and white bars are interchanged in order to multiply the signal by a negative factor for $\Delta < 0$. Fig. 2 represents the result obtained on the transmission spectrum of NO_2, where the curves $S(\sigma)$ and $S'(\sigma)$ are represented.

Fig. 2

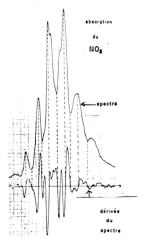

We should notice that happily the Fourier derivation does not operate the derivation of the noise which is also the case of synchronous detection methods for obtaining the derivative.

3 - Correlation

In order to recognize a reference spectrum S (σ), it is useful to correlate the observed spectrum S (σ) with the reference S_R. This means that we should multiply the interferograms i (Δ) and I_R (Δ), which is possible if we put in the interferogram plane a screen representing I_R (σ), i.e. the interferogram of the reference spectrum. If we move the interferogram, the signal obtained is a linear combination of the signals produced by the various wavelengths and we act at the same time on the various elements composing the spectrum.

4 - Correlation of Derivatives

Spectral signals S or S_R are always positive. This means that even if S and S_R are totally different, they can be both non-zero in some spectral domain and the correlation will be positive, which is misleading. On the other hand, derivatives are positive or negative and it seems to be safer to correlate the derivatives rather thant the spectra ; in order to perform this correlation, we have to multiply Δ i (Δ) by Δ $i_R(\Delta)$ which means that we have to use a Δ^2 filter on the interferogram i_R. It is noticeable that this operation tends to eliminate the central part of the interferogram, which brings no useful information, and use the parts of the interferogram that correspond to an appreciable value of Δ .

5 - The Case of Quasi Periodic Spectra

In the case of spectra having a quasi periodic structure (for example molecular spectra having a vibrational structure) the interferogram has an appreciable contrast for a given optical path difference. When associated with broad band spectral filtering, this O.P.D. can be considered as a pratical criterion for detection of a given pollutant so that it is useless to treat the whole interferogram : only one fragment of the interferogram in the vicinity of the characteristic O.P.D. is useful. This led us to the following ideas for the detection of SO^2 or NO^2 .

Fig. 3 represents the spectrum of SO^2 and Fig. 4 the interferogram where the enhancement of contrast of fringes in the vicinity of $\Delta \sim 100\lambda$ is in evidence. A possible set up is represented on Fig. 5 where the characteristic Optical path difference is introduced by a birefringent plate L but the device can still be simplified as on Fig. 6 : plate L produces in the focal plane of lens O an interference pattern (of hyperbolic type) in the center of which a circular hole can be put in order to isolate a region where the OPD is stationary (and almost a constant) ; this type of device has been experimented and has proved to be very efficient in the case of the detection of SO^2 in a chimney stack : the sensitivity of this mounting is better than 10 ppm and the response is quick. Moreover the linearity has been checked as well as the agreement with conventional chromatographic methods.

Fig. 3

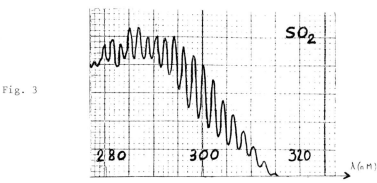

Fig. 4

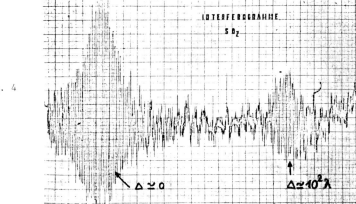

Fig. 5

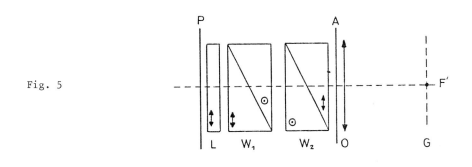

Fig. 6

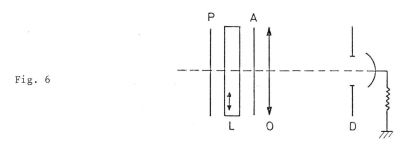

CONCLUSION :

New developments in the field of interferential spectroscopy can be expected ; the possibility of analog treatment of the luminous interferogram obtained by a simple device opens new perspectives of realisation of specific optical detectors ; in the case of quasi periodic spectra (SO^2, NO^2 ...) the apparatus can be remarquably simple and shall be available in the near future for the detection of atmospheric pollutants.

BIBLIOGRAPHY

1. Prat, R., Japan J. Appl. Phys. Suppl. 1,448 (1965) ; Optica Acta 13,2,73 (1965).

2. Fortunato G. et Maréchal A., C.R. Acad. Sci (France) 274 B (1972) 931 and 276 B (1973) 527.

3. Fortunato G., Thesis Orsay June 76, to be published in the Nouvelle Revue d'Optique.

4. Maréchal A.   Applic. of holography and optical data processing Pergamon Press (1977).

5. Maréchal A.   ICO 11, Madrid, Septembre 1978.

CORRELATION SPECTROSCOPY

K. Muller

Infrared Physics Group, Solid State Physics Laboratory, ETH
8093 Zurich, Switzerland

Abstract

Today, the interpretation of spectroscopic data with the aid of time-dependent autocorrelation functions (acfs) has found widespread application to molecules in solids, liquids and dense gases. In this paper the two dominant relaxation processes influencing the acfs of rotational and vibrational relaxation are reviewed. The influence of Coriolis coupling on the rotational acfs of degenerate vibrations is emphasized. Experimental acfs are discussed for the rotational relaxation of HCl in tetrachlorides, OH- and OD- ions in single crystals of alkalihalides, gaseous chloroform and of methanes in liquid argon as well as for intramolecular vibrational relaxation in the metal carbonyls $Cr(CO)_6$, $Mo(CO)_6$ and $W(CO)_6$.

1. Introduction

Infrared and Raman band contours of molecules in a dense medium contain a great amount of information on molecular dynamics. Kubo's (1) linear response theory has demonstrated that this information can be extracted in favourable cases by simply performing a Fourier transform of the band contours. The result of this procedure is a time-dependent autocorrelation function (acf) of the corresponding transition dipole moment for infrared spectra and of the polarizability tensor for Raman spectra (2). These acfs give a detailed statistical description of the various relaxation processes that occur in a liquid or a dense gas. The process which called most attention is the rotational relaxation of the individual molecule in its surroundings.

For small molecules in diluted gases the infrared and Raman bands show a discrete rotational structure. Hence, rotational energy levels can be discussed directly. Correlation spectroscopy, which means interpretation of spectroscopic data with the aid of auto-correlation functions, is not required in this case. Yet, for large molecules in gases the discrete rotational lines cannot be resolved, and for molecules in dense media the rotational structure is completely quenched. In both cases the band profiles represent continuous lines. An interpretation in terms of rotational energy levels is not possible. Therefore, correlation spectroscopy is the appropriate spectroscopic method for the interpretation of infrared and Raman spectra of molecules in a dense medium. In contrary, most of the earlier spectroscopic work on molecules in liquids has dealt with the measurement of line widths and relaxation times derived therefrom, yielding a single parameter to describe the complicated molecular behaviour.

In the following we restrict ourselves to infrared spectra. Raman spectra have been discussed by other authors in several papers (17, 18). In chapters 3 and 4 rotational relaxation is assumed to be the dominant relaxation process. In chapter 5 vibrational relaxation is considered.

2. Theoretical models of molecular dynamics in dense media

In order to obtain all information on molecular rotational motion in a liquid or a dense gas contained in measured auto-correlation functions, it is necessary to develop a suitable theoretical model for molecular dynamics in the medium considered. A large number of papers has been published on this subject (3-10). These rotational models extend from the free rotation, studied by St.Pierre and Steele (9) to the pure rotational diffusion introduced by Debye (5). However, the most widely used rotational model was created by Gordon (6). He applied it to diatomic molecules in liquids. Because of its widespread application we repeat its assumptions:
(i) The molecules undergo binary collisions in a very short time compared to their mean rotation time. (ii) In the time between two collisions the molecules rotate freely. (iii) During a collision the spatial orientation of a molecule is not changed, but the total angular momentum is randomized. (iv) The collision probability is represented by a Poisson distribution.

This model was extended to symmetric-top and spherical-top molecules by St.Pierre and Steele (10), and by McClung (8). Further, it was applied to linear polyatomic molecules by Dreyfus et al. (12) and by Lévi, McClung et al. (13).

Comparing the acfs calculated by Gordon's model and its extensions with experimental results, we obtain reasonable agreement by proper choice of the parameters involved. But there is one exception. The experimental acfs of degenerate vibrations in symmetric-top and in spherical-top molecules cannot be approximated by the acfs calculated by the authors mentioned above. In degenerate vibrations Coriolis coupling of first order may occur (14). In chapter 4 we demonstrate the drastic changes which can be induced by Coriolis coupling in acfs of infrared spectra. The influence of Coriolis coupling on the acfs of Raman spectra has been treated theoretically by Drifford et al. (17, 18).

Keller and Kneubühl (4) introduced group theoretical considerations in correlation spectroscopy with the aid of a correlation matrix $\langle R_{ij}(\omega(t))\rangle$. $R_{ij}(\omega(t))$ represents the orthogonal matrix of the molecular orientation $\omega(t)$ described by suitable parameters. This allowed them to construct a relaxation ellipsoid, which describes the time-development of the rotational diffusion tensor. This rotational diffusion tensor gives a more intuitive picture of the actual three-dimensional average rotation than single acfs.

3. Experimental acfs for non degenerate vibrations of molecules in liquids and solids.

Experimental acfs, published in the past 12 years, have been determined for a large amount of solute-solvent systems. They range from liquids of highly polar molecules, studied by Rothschild (21), to molecules in the gas phase under high pressure, investigated by Mme Vincent et al. (19, 20).

Among the earliest presented experimental acfs are those by Keller and Kneubühl (4) for HCl dissolved in liquid tetrachlorides and for OH^- and OD^- in single crystals of alkalihalides.

The solution of HCl in liquid tetrachlorides represents the opposite of the models, which describe pure rotational diffusion according to Debye (5). The probe molecules are much lighter than the neighbouring host molecules. Because of the relatively large cavities in the liquid, the time intervals between collisions exceed the duration of the collisions in accordance with the assumption in Gordon's diffusion model (6). The resultant acfs for this solute-solvent system give evidence for the existence of considerable torques, which hinder the rotation of the HCl molecule. With a simple classical model the time of flight of the HCl molecule in the cavity between two collisions is found to be about $2 \cdot 10^{-13}$ s (4).

OD^- - and OH^- - ions in single crystals of alkalihalides represent a quite different system. The fields acting on the ions are far stronger than those influencing molecules in liquids. At low temperatures all acfs oscillate and there by indicate a librational motion of the OD^- - and OH^- - ion. With increasing temperature the oscillations as well as the complete correlation functions are progressively damped. As an example, for OD^- in KJ the oscillations collapse near 530 K indicating a rotational barrier corresponding to this temperature.

4. Influence of the Coriolis coupling on the acfs of symmetric-top and spherical-top molecules

The effect of the Coriolis coupling on degenerate vibrations of symmetric-top and spherical-top molecules was first discussed by Teller and Tisza (14). The vibrational motion produces an angular momentum in the molecule with degenerate vibrations. This vibrational angular momentum couples to the rotational angular momentum. Teller (14) introduced the Coriolis coupling constant ζ as the corresponding coupling factor.

At first we confine our considerations to symmetric-top molecules. According to Teller (14) Coriolis coupling in isolated symmetric-top molecules is equivalent to a rotation of the transition dipole of the degenerate vibration in the molecule-fixed frame with the circular frequency

$$\Omega = - \zeta \cdot K_z / I_z \qquad (1)$$

K_z and I_z designate the z-component of the rotational angular momentum $\vec{K}$ and the moment of inertia around the z-axis, respectively. Thus, the angle of rotation of the transition dipole in the molecular plane of degeneracy can be written as

$$\phi(t) = \phi(o) - \zeta \cdot K_z / I_z \cdot t \qquad (2)$$

for the isolated molecule and as $\quad \phi(t) = \phi(o) - \zeta \cdot \frac{1}{I_z} \cdot \sum_j K_z^j \cdot \Delta t_j \qquad (3)$

for a molecule in a dense medium, which can be described by Gordon's model. K_z^j and Δt_j denote the z-component of the rotational angular momentum $\vec{K}$ during the jth period of free rotation and the duration of this period, respectively. Müller et al. (15, 22) introduced eq. (2) and eq. (3) to the calculations of the acfs of degenerate vibrations in symmetric-top molecules by St. Pierre and Steele (10). The result is illustrated by fig. 1 and fig. 2. In fig. 1 the theoretical acf for $\zeta = 0.96$ is plotted corresponding to the ν_4 - transition of chloroform with the ratio of the moments of inertia $b = \frac{I_x}{I_z} - 1 = - 0.481$ together with the experimental acf of the ν_4 - transition of $CHCl_3$ dissolved in gaseous N_2 at a temperature of 300 K and at a pressure of 1 atm. They are compared with the theoretical acf of chloroform for $\zeta = 0$ of the hypothetical transition corresponding to the rotation of the molecular frame only. The agreement between the experimental acf and the theoretical acf is excellent. In fig. 2 the experimental acf of the perpendicular ν_4 - transition of CD_3H dissolved in liquid argon at a temperature of 84 K is plotted together with the best fitted theoretical acf. The experimental curve was measured by Marsault et al. (23). The corresponding theoretical curve for the reorientational process of the molecular frame only is added. There is a pronounced improvement of the agreement between experiment and theory for times larger than $3 \cdot 10^{-13}$ s.

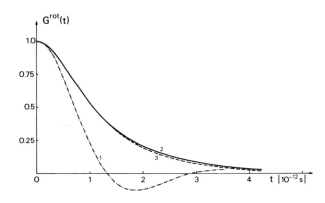

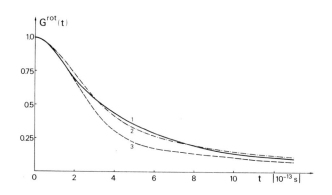

Fig. 1. Theoretical rotational correlation functions of chloroform for $\not{3}$ = 0.0(1) and $\not{3}$ = 0.96 (2) compared with the experimental rotational correlation function of the ν_4-transition of gaseous $CHCl_3$ (3) at a temperature of 300 K.

Fig. 2. (1) Experimental rotational correlation function of the vibration ν_4 of CD_3H, dissolved in liquid argon at a temperature of 84 K, determined by Marsault et al.
(2) Corresponding theoretical rotational correlation function with the parameters β = 1.43, $\not{3}$ = 0.14 and b = -0.19.
(3) Theoretical rotational correlation function neglecting Coriolis coupling with the parameters β = 1.43, $\not{3}$ = 0.0 and b = -0.19.

Coriolis coupling in triply degenerate vibrations of spherical-top molecules can be treated in a similar manner as for symmetric-tops. In spherical-tops the component of the transition dipole $\vec{\mu}$ perpendicular to the total angular momentum $\vec{J}$ performs a rotation, which depends on the magnitude of the Coriolis coupling constant ζ. In (22) we demonstrated, that this rotation can be described by the following equation of motion

$$\frac{d\vec{\mu}}{dt} = (1 - \zeta) \cdot (\vec{J} \times \vec{\mu}) \tag{4}$$

in a laboratory fixed coordinate system. For $\not{3}$ = o this equation describes the ordinary rotation of the molecular frame. Thus, Coriolis coupling simply can be introduced by replacing $\vec{J}$ by $\vec{J} \cdot (1 - \zeta)_2$ or T by T $\cdot (1 - \zeta)^2$ in the derivation of the acfs before averaging over the Maxwell distribution of $\vec{J} \cdot$ T designates the temperature in Kelvin.

With the above substitutions the acf $G^{rot}(t, \beta, \zeta \neq o)$ for triply degenerate vibrations in spherical-tops can be expressed by the acf $G^{rot}(t, \beta, \zeta = o)$ from McClung (8).

$$G^{rot}(t, \beta, \zeta) = G^{rot}(t \cdot (1-\zeta), \frac{\beta}{1-\zeta}, \zeta = o) \tag{5}$$

β designates the normalized collision frequency. This equation includes the case of isolated molecules by β=o.

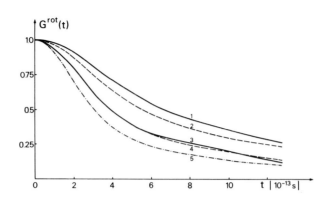

Fig. 3. Experimental rotational correlation functions of the vibrations ν_4 (1) and ν_3 (3) of CD_4 in liquid argon at a temperature of 91 K, compared with corresponding theoretical rotational correlation functions with $\beta = 1.43$ and the ζ-values equal 0.34 (2) for the ν_4-transition and 0.16 (4) for the ν_3-transition. The experimental curves are by Cabana et al. (5) Corresponding theoretical rotational correlation function with $\beta = 1.43$ and $\zeta = 0.0$.

For illustration of eq. (5) the acfs of the ν_3- and the ν_4- vibration of CD_4 dissolved in liquid argon at a temperature of 91 K are plotted in fig. 3. These acfs have been determined by Cabana et al. (24). They are compared with the corresponding theoretical acfs calculated with eq. (5). The lowest curve is the curve for the rotation of the molecular frame only, that is for $\zeta = 0$. The normalized collision frequency is chosen to be 1.43 to give a best fit between theoretical and experimental curve for the ν_3-vibration.

The agreement between experiment and theory for the ν_4-vibration with the same β as for the ν_3-vibration is not conclusive. Possibly the ζ-values, which are taken here from the gas phase, change in the liquid, especially for values near + 1. Nevertheless eq. (5) provides a clear improvement of the interpretation of the above experimental acfs in comparison with the curve for $\zeta = 0$.

5. Vibrational relaxation

In molecules larger than the methanes discussed in chapter 4 rotational relaxation need not to be the dominant relaxation process. Vibrational relaxation can occur in an order of magnitude comparable to rotational relaxation. Morawitz and Eisenthal (11) have studied this problem first. They assume a weak coupling between the vibration and the translational and rotational degrees of freedom of the neighbouring molecules. Thereby, they can represent the acf, resulting from the Fourier transform of an infrared or a Raman band as the product of a rotational relaxation function, described in chapters 2 - 4, and of a vibrational relaxation function. The assumption of Morawitz and Eisenthal is expected to be valid for molecules in liquids with no or at least small permanent dipole moment. The factorization of the acf is widely used for the interpretation of experimental data.

There are two main types of vibrational relaxation. On one side, vibrational relaxation can be caused by intermolecular interaction, where soft collisions of the vibrating molecule with its neighbours produce a dephasing of the vibration without loss of the vibrational energy. On the other hand, vibrational relaxation can occur by intramolecular interaction by anharmonic coupling of a vibration to the low-frequency modes in the molecule. Vibrational relaxation has caught great attention in the recent past. Different theoretical models have been proposed by Bratos (25), Rothschild (27), Fischer and Lauberau (26) and others.

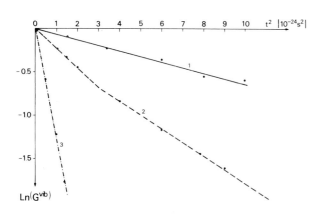

Instead of discussing these theories, we show an experimental example. We measured the triply degenerate infrared bands ν_6, ν_7 and ν_8 of the three metal carbonyls $Cr(CO)_6$, $Mo(CO)_6$ and $W(CO)_6$ dissolved in latm N_2 at room temperature. In the relevant time range of some 10^{-12} s the molecules endure a negligible number of collisions at the supposed pressure of latm. Therefore the factorization of the acf into a rotational part G^{rot} and a vibrational part G^{vib} can certainly be performed. The rotational relaxation function G^{rot} is calculated by eq. (5) in chapter 4 with the collision frequency $\beta = o$. (8). Thus, the experiment allows to determine G^{vib}. Fig. 4 represents the intramolecular vibrational relaxation functions of the vibrations ν_6, ν_7 and ν_8 of $Cr(CO)_6$ (28). Astonishingly, the vibration ν_6 with the highest frequency shows the slowest and the vibration ν_8 with the lowest frequency the fastest decay. All three relaxation curves have a common feature, they can be well approximated by one or two Gaussian functions.

Fig. 4. Experimentally determined vibrational relaxation functions of the vibrations $\nu_6(1)$, $\nu_7(2)$ and $\nu_8(3)$ of $Cr(CO)_6$ dissolved in gaseous N_2 at a pressure of 1 atm.

6. Conclusions

The concept of correlation spectroscopy, originally formulated for rotational relaxations of diatomic molecules and later extended to nondegenerate vibrations in polyatomic molecules has been completed by the introduction of Coriolis coupling. Thereby, all effects of first order affecting the rotational relaxation are included. Reasonable agreement is obtained between theoretical and experimental acfs if rotational relaxation is the dominant relaxation process or as far as the vibration and the translational-rotational degrees of freedom of adjacent molecules are weakly coupled. In the cases where these assumptions are not valid, other dissipative and nondissipative relaxation phenomena, as vibrational relaxation etc. have be taken into consideration.

In gaseous molecules correlation spectroscopy presents a unique possibility to determine exactly the shape of the intramolecular vibrational relaxation functions by separation of the theoretically calculated rotational relaxation part from the total acf. This simple method might have possible applications in laser chemistry, where vibrational relaxation functions are of great interest.

References

1. Kubo, R., J. Phys. Soc. Japan, 12, 570. 1957.
2. Gordon, R., J. Chem. Phys., 43, 1307. 1965.
3. Shimizu, H., J. Chem. Phys., 43, 2453. 1965.
4. Keller, B. and F. Kneubühl, Helv. Phys. Acta, 45, 1127. 1972.
5. Debye, P., Polar Molecules, Dover Publ., New York. 1954.
6. Gordon, R., J. Chem. Phys., 44, 1830. 1966.
7. Bratos, S., J. Rios and Y. Guissani, J. Chem. Phys., 52, 439. 1970.
8. McClung, R.E.D., J. Chem. Phys., 51, 3842. 1969.
9. St.Pierre, A.G., and W.A. Steele, Phys. Rev., 184, 172. 1969.
10. St.Pierre, A.G. and W.A. Steele, J. Chem. Phys., 57, 4638. 1972.
11. Morawitz, H. and K.B. Eisenthal, J. Chem. Phys., 55, 887. 1971.
12. Vincent-Geisse J. and C. Dreyfus, Molecular motions in liquids, ed. J. Lascombe (Reidel Publishing Company, Dordrecht. 1974).
13. Lévi, G., J.P. Marsault, F. Marsault and R.E.D. McClung, J. Chem. Phys., 63, 3543. 1975.
14. Teller E. and L. Tisza, Z. Physik 73, 791. 1932.
15. Müller K. and F. Kneubühl, Chem Phys. Lett., 23, 492. 1973.
16. Müller K., P. Ethique and F. Kneubühl, Chem. Phys. Lett 23, 489. 1973.
17. Gilbert, M. and M. Drifford, J. Chem. Phys. 65, 923. 1976.
18. Gilbert, M., P. Nectoux and M. Drifford, J. Chem. Phys. 68, 679. 1978.
19. Dreyfus C., E. Dayan and J. Vincent-Geisse, Molecular Physics 30, 1453. 1975.
20. Dreyfus, C., L. Berreby, E. Dayan and J. Vincent-Geisse J. Chem. Phys. 68, 2630. 1978.
21. Rothschild, W. G., J. Chem. Phys. 51, 5187. 1969.

22. Müller, K. and F. Kneubühl, Chemical Physics 8, 468. 1975

23. Marsault, J. P., F. Marsault-Herail and G. Lévi, private communication.

24. Cabana, A.,R. Bardoux and A. Chamberland, Can. J. Chem. 47. 2915 1969

25. Bratos, S., Guissani and J.C. Leicknam, Molecular Motions in Liquids, edited by J. Lascombe (Reidel Publishing Company 1974).

26. Fischer S.F. and A. Lauberau, Chem. Phys. Lett. 35, 6. 1975.

27. Rothschild, W.G., J. Chem. Phys. 65, 455. 1976.

24. Müller, K., P. Ethique and F. Kneubühl, in Molecular motions in liquids, ed. J. Lascombe (Reidel Publishing Company, Dordrecht, 1974).

ELECTRICALLY INDUCED FLUORESCENCE CHANGES FOR CHARACTERISING MACROMOLECULES

P. J. Ridler and B. R. Jennings

Electro-Optics Group, Physics Department, Brunel University
Uxbridge, Middlesex, U. K.

Abstract

Fluorescence involves the absorption and reemission of light, in which incident linearly polarised radiation is generally depolarised depending upon the directions in space of the corresponding molecular transition moments. An applied electric field imposes order on macromolecules in dilute solution and the polarised components of the fluorescence change. An apparatus is described in which the changes in the four polarised components are measured for solutions of dye-tagged nucleic acids, polymers, pigment particles and liquid crystals. Using pulsed electric fields, information on the nature of dye binding and the size of the macromolecules is obtained.

Introduction

Electro-optical devices are commonplace in laser technology. Generally, they utilise the changes in one or other optical property of highly anisotropic crystals when these crystals are subjected to electric fields. It is the anisotropy of the crystal structure which provides the foundation of the phenomena. Yet if one considers the structure of many macromolecules such as polymers, proteins, viruses and the like, the molecular array is generally anisotropic. Crystallisation of such materials to optical quality is impractical. If these molecules can be dissolved or dispersed in a fluid medium and the solute molecules aligned using an externally applied force field one can, at least in principle, obtain a 'dilute crystal' which embodies to some degree the inherent properties of the constituent molecules. This concept can be used to switch and modulate electro-optic systems much in the way that Kerr cells use the electrical orientation of the molecules in a pure liquid.

It is the converse of this line of reasoning which is finding increasing use in certain branches of bio-physics, chemical physics, polymer science and colloid science, namely, that the recording and measurement of electro-optical phenomena of dilute solutions of macromolecules leads to useful information on the structure of these interesting yet complicated molecules. To date, electric birefringence[1], scattered intensity[2], scattered intensity spectroscopy[3], dichroism[1], optical rotation[4] and fluorescence[5,6], have all been harnessed for the characterisation of macromolecules. Furthermore, with the advent of electric generators capable of giving pulsed outputs, one can study the phenomena as rate processes, monitor the rotation and deformation of the molecules in the resistive medium of the solvent and calculate molecular sizes. Using pulses, the methods become extremely fast. Following the application of the electric pulse to the sample under test, a transient electro-optic response is obtained (see figs. 2 to 5) which may be classified by three regions. The first signifies the establishment of the transient as the molecules respond to the applied field. The second is the plateau region which corresponds to the statistical equilibrium orientation of the molecules in the applied field. This is analysed for the amplitude of the electro-optic phenomenon. The third region is that in which the effect decays back to its pre-field value once the applied pulse has terminated. The rate of this decay process is a manifestation of the disorientation of the molecules and is very dependent upon the viscosity of the medium and the size of the molecules. It has been used often as a rapid means of particle size analysis[1].

Electric field fluorescence effects are novel. Recently, an apparatus was assembled in this laboratory which far exceeds the sensitivity of anything designed heretofore. In this paper we indicate how transient changes can be induced in each of the four polarised components of fluorescence using electric fields. Furthermore, illustrative data reveal the value of the method for the study of dilute solutions of dye-tagged nucleic acids, synthetic polymers, polysaccharides, liquid crystals and non-tagged pigment crystallites.

The Apparatus

The apparatus[6] is schematically represented in fig. 1. An argon-ion laser (Spectra Physics 170) provides a high power, low noise laser beam with a few selectable wavelengths in the ultra-violet and blue spectral ranges. After suitable attenuation by a bank of neutral density filters, the beam passes through a polarisation adjusting system consisting of a pair of Fresnel rhombs in the back-to-back configuration and a Glan laser calcite polariser. Suitable azimuth setting of this combination enables one to transpose the inherent vertically plane polarised light into a horizontally polarised state as required. The light then passes through the sample cell and is lost in a suitable absorber. The cell is of rectangular or cylindrical design. The cylindrical cell is shown inset in fig. 1. In its base is set an electrode. A second horizontal electrode lowers into the cell and forms the upper boundary of the test solution which is typically only some 1 ml, in volume. Entrance and exit windows inset in the cell prevent spurious multiple reflections.

On either side of the cell, and generally set at right angles to the incident beam, are two limbs which hold the detection optics. Each of these consists of a diaphragm to define the acceptance angle of the

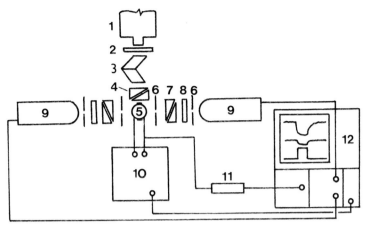 

Fig. 1. Diagram of Apparatus: Components designated by, 1 - Argon laser, 2 - neutral filter,
3 - Fresnel rhombs, 4 - polariser, 5 - cell (also inset photograph), 6 - apertures,
7 - analysing prism, 8 - cut off filter, 9 - photomultiplier, 10 - pulse generator,
11 - probe, 12 - storage oscilloscope.

system, followed by polarisers, set to receive quadrature states of linearly polarised light. One receives
light polarised in the vertical and the other in the horizontal plane. Optical cut-off filters follow these
polarisers and allow light to reach the photocathodes of a 14 stage, end-on photomultiplier (EMI type 9816
KB).

After the solution has been placed in the cell and the laser and cut-off filters adjusted as required for
wavelength, the procedure is as follows. With the incident light on the cell adjusted to be vertically (V)
plane polarised, a pulsed electric potential is applied between the cell electrodes. The changes ΔV_V and
ΔV_H in the vertical and horizontal components of the fluorescence are recorded on each of the photomultipliers. The procedure is then repeated with the incident light horizontally polarised so as to record the
changes ΔH_V and ΔH_H in the components H_V and H_H. In each case, the responses of the photomultipliers are
recorded on a storage oscilloscope and either photographed or fed directly into a transient recorder for
analysis. Typical examples of the transient changes in the four polarised components of the fluorescence
are shown in the following figs. (2-5).

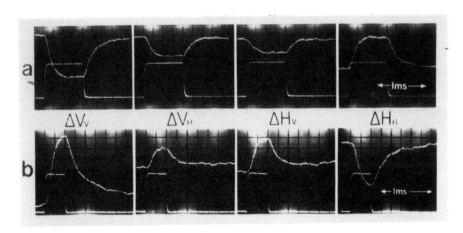

Fig. 2. Electrically induced changes in the fluorescent components for native DNA plus dye.

    (a) Proflavine dye: $\dfrac{\Delta V_V}{V_V} = -0.33;$ $\dfrac{\Delta V_H}{V_H} = -0.15;$ $\dfrac{\Delta H_V}{H_V} = -0.15;$ $\dfrac{\Delta H_H}{H_H} = +0.23$

    (b) Hoechst 33258 dye (a benzimidazole derivative):

    $\dfrac{\Delta V_V}{V_V} = +0.15;$ $\dfrac{\Delta V_H}{V_H} = +0.05;$ $\dfrac{\Delta H_V}{H_V} = +0.07;$ $\dfrac{\Delta H_H}{H_H} = -0.10$

For both sets, E = 2.7 kV cm^{-1}, with aqueous DNA of 4.5 x 10^{-5}g.ml^{-1} concentration.

Illustrative Results

To illustrate the method, measurements have been made on a range of materials as itemised below.

Biopolymers

Many dye molecules bind readily to biopolymers, often due to the highly polyelectrolytic nature of the latter. This phenomenon is utilised in many biomedical studies. For example, various dyes form antitumour, antiviral, antimalarial and antitripanosomal agents in the chromosomes of human cells[7], presumably through their binding to nucleic acids. An important biochemical quest is the elucidation of the geometry of binding of the dye molecules to the nucleic acid helices. The electric field fluorescent method is able to assist in this realm.

In the first set of transients in fig. 2, one sees the changes in the polarised components of the fluorescence for a dilute solution (4.5×10^{-5} g.ml^{-1}) of high relative molecular mass (5×10^{6}) DNA from calf thymus. Proflavine is a dye which is known to cause mutations in the cells of bacteria and animals[8]. It has been added up to a concentration corresponding to 1 molecule for each 130 base pairs of the DNA.

With 458 nm light used to excite the fluorescence and selected filters in the recording limbs to accept light of 495 nm and greater wavelength, the transients were obtained. It is known that in water, DNA molecules align in an electric field parallel to the electric field direction[1]; in our apparatus this is the vertical direction. When the field is applied, both fluorescence components for vertically polarised incident light decrease. The absorption transition moment (μ_a) is thus predominantly associated with the horizontal plane. However, the decrease in the horizontally polarised output is less than the decrease in the vertical component indicating that the emission transition moment (μ_e) is also predominantly in the horizontal plane. Now the acridine dyes are planar molecules with both μ_a and μ_e in the face of the molecule[9]. Hence, one can deduce that the proflavine molecules bind along the DNA helix with their planes perpendicular to the helix axis.

Hoechst 33258 is a new fluorochrome which is excited by ultra-violet radiation and is used for identification of adenine and thymine-rich regions in chromosomes[10]. It binds strongly to DNA. Experiments on solutions of DNA with this dye added to a concentration of 1 dye molecule to 1,200 base pairs of DNA give quite different fluorescence component changes than the DNA/acridine orange system. They can be seen in the second set of fig. 1, for which incident radiation of 351 and 364 nm excited the fluorescence, which was detected for wavelengths exceeding 400 nm. In this case, the changes are consistent with a predominant binding of the dye molecules parallel to the DNA helix. A comparison of the photograph series of fig. 1 indicates how, with the trained eye, one can quite quickly obtain some idea of the way these and other medically important dyes bind to nucleic acids. In this laboratory, a whole host of dyes and their interactions with DNA, RNA and the synthetic polynucleotides have been studied.

Synthetic Polymers

The method also has potential for the study of synthetic polymers, some of which are naturally fluorescent. The vinyl polymers are examples. Many other polymers react with dye molecules to produce fluorescent species. It is of interest (a) to know the geometry of dye binding and (b) to use the properties of the appended fluorescent group to indicate the behaviour of the polymer molecules in solution. With the apparatus described above, quite striking changes can be evidenced in the polarised components of the fluorescence upon application of electric fields.

Polystyrene sulphonic acid is a polymer of commercial interest as it is used extensively in water purification and ion-exchange systems. Fig. 3 shows the electrically induced fluorescence changes for a dilute aqueous solution of polymer (1.35×10^{5} relative molecular mass) in the presence of added rhodamine B. This dye is itself of current commercial importance in dye-laser technology. In these studies, fluorescence was excited with 488 nm radiation. Detection was for radiation of greater than 515 nm wavelength. From fig. 3, and knowing that the flexible polymer molecules extend and align with their major axes along the electric field direction[11], it is evident that the rhodamine molecules bind with their molecular planes predominantly parallel to the polymer axes.

An important factor here however, is that the fluorescence from the bound dye molecules enables one to follow the rate process whereby the polymer molecules revert to their field-free condition in the absence of any field. The molecular rotation and deformation is indicated by the relaxation time (in this case of some 70 μs) obtained from the decay of the fluorescence transients after the termination of the electric pulse. Monitoring of the amplitudes and the rates of the fluoescence changes should thus prove of value in following the conformation and behaviour of polymer molecules during laboratory and industrial processes.

Polysaccharides

These molecules are the objects of significant current scientific and technological study. In particular, there is current medical interest in the mucopolysaccharides (glycosaminoglycans) which are found in the

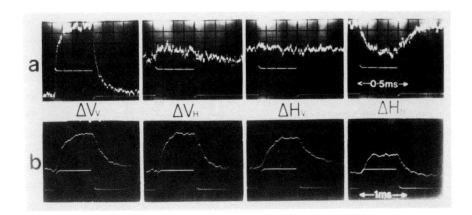

Fig. 3.   Transient fluorescent polarisation changes for polymer and polysaccharide solutions.

   (a)   Polystyrene sulphonic acid with rhodamine B in water. $c = 4.5 \times 10^{-5} \text{g.ml}^{-1}$,
         $E = 6 \text{ kV cm}^{-1}$, dye : monomer = 1 : 250.

   (b)   Proteoglycan (ex-pig laryngeal) with acridine orange in water. $c = 3 \times 10^{-4} \text{g.ml}^{-1}$,
         $E = 5.7 \text{ kV cm}^{-1}$.

cartilage and various fluids of the body. Often, their molecules adopt shapes which resemble bottle brushes. Extended molecular backbones carry flexible 'hairy' side chains[12]. A typical example is the proteoglycan molecule which is thought to be an important agent in the arthrotic processes[13]. Here again, the electric field fluorescence method can be used to study this complex molecule and the way it binds important dye molecules. Using 488 nm wavelength incident light, and detection limb filters to cut-off light below 515 nm wavelength, the transient changes in the polarised components of fluorescence shown in fig. 3 were obtained for a solution of some $3 \times 10^{-4} \text{g.ml}^{-1}$ concentration, tagged with acridine orange dye. A comparison with the previous examples shows this system to give different responses. The transient changes were all positive and this is probably an indication of a change in the structure of the molecule and its flexible side chains in the presence of the electric field. The clarity of the recorded transients does indicate that electric field fluorescence can be used to study such complex molecules as these.

Crystalline Particles

   Pigments are of extensive importance in the colouring, ink, paint and printing industries. The pigments are generally produced as precipitates in dispersion at the end of their production process. Certain materials have inherent fluorescent properties with the result that they are used as 'brightening agents'. Electric field fluorescence is again of value in studying the structure and properties of such suspensions.

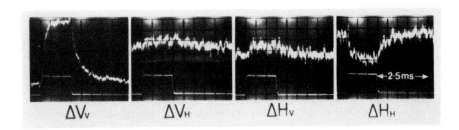

Fig. 4.   Transient changes in the fluorescence polarisation components for an aqueous yellow
          diazo-pigment sol. $c = 2 \times 10^{-5} \text{g.ml}^{-1}$, with slight amount of surfactant for
          stabilisation. $E = 5.3 \text{ kV cm}^{-1}$.

By way of an example, fig. 4 shows the manifestation of the transient changes for an aqueous suspension of a yellow diazo-pigment when excited by blue light (488 nm) and fluorescing above 515 nm. The changes in the polarisation and the intensity of the fluorescence indicate that, within the particles, both μ_a and μ_e are not randomly directed, and thus have predominant directions within the particle structure. This indicates that the particles are highly crystalline. Such measurements should provide a quantitative measure of the degree of crystallinity for pigment particles. In addition, the relaxation time from the decay of the transients leads to information on the size and size distribution of the particles in the sol.[14]

Liquid Crystals

These materials are of current interest as electro-optic switch and device media owing to their dynamic scattering properties. Historically, the nematogens have featured largely in this usage[15]. Dyes have recently been incorporated in these materials to produce greater contrast devices through the observation of dichroism, pleochromism and fluorescence[16,17].

Electric field fluorescence studies have been made on hexylcyanobiphenyl (K 18 from Messrs. BDH of Poole, Dorset) in the presence of added rhoduline red dye. Excitation was with 515 nm light and the fluorescence recorded for wavelengths exceeding 590 nm. Because of the opacity of these materials, a different cell was employed. It consisted of a thin rectangular cell of 1 mm thickness, holding a pair of electrodes with a 2 mm gap so that the field was applied in the horizontal plane, transverse to the optical beam. The field induced changes are shown in fig. 5. From these we note that the transient changes are significant, with a fast establishment time when compared with other liquid crystal dynamic phenomena. This may be worthy of commercial exploitation. The speed depends upon the field strength (E) and the thickness of the cell. The greater E or the thicker the cell, the more rapid the turn-on process. In fact, there are two processes, a fast and a slow one, whose relative magnitudes vary with these two parameters. Auxiliary scattering studies show that field induced depolarised scattering is a major contributor to the observations. This complicates detailed molecular interpretation.

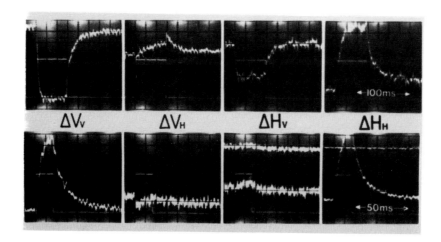

Fig. 5. Similar transients obtained with liquid crystals, observed at 15° to the forward direction.

(a) Nematogen - hexylcyanobiphenyl with rhoduline red. Dye concentration of 0.06%; temp. = 23°C, E = 5 kV cm^{-1}.

(b) Lyotropic - polyethylene oxide based system with tetracene. Dye concentration of 0.01%; temp. = 24°C, E = 4.5 kV cm^{-1}.

Hitherto, liquid crystal devices and displays have predominantly engaged thermotropic materials. There is a growing interest in the so-called lyotropic materials[18]. Fig. 5 also shows that the electric field fluorescence method is equally viable for the study of this class of liquid crystal. In fact, for the polyethylene oxide based system used here, with tetracene as the added dye, the speed and amplitude of the effects are comparable with those obtained with the cyanobiphenyl thermotropic material. Commercial exploitation of the electro-optical properties of the lyotropic materials would appear to hold some promise.

Acknowledgements

The Science Research Council is thanked for a grant which funded much of the apparatus and provided a fellowship for PJR.

References

1. Fredericq, E. and Houssier, C., 'Electric dichroism and electric birefringence', Clarendon Press, Oxford, 1973.

2. Jennings, B. R., in 'Light Scattering from Polymer Solutions' (ed. M. Huglin), 529, 1972.

3. Ware, B. R. and Flygare, W. H., J. Coll. Int. Sci., Vol. 39, 670, 1972.

4. Tinoco, I. and Hammerle, W. G., J. Phys. Chem., Vol. 68, 1619, 1956.

5. Weill, G. and Sturm, J., Biopolymers, Vol. 14, 2537, 1975.

6. Ridler, P. J. and Jennings, B. R., J. Phys. E. (Sci. Instr.), Vol. 10, 558, 1977.

7. Gale, E. F., Cundliffe, E., Reynolds, P. E., Richmond, M. H. and Waring, M.J., 'The Molecular Basis of Antibiotic Action', Wiley Interscience, 1972.

8. Peacocke, A. R., 'Acridines' (ed. R. M. Acheson), Ch. 14, p.723, Interscience Publishers, 1973.

9. Jakobi H. and Kuhn, H., Z. Electrochem., Vol. 66, 46, 1962.

10. Moutschen, J., Prog. Biophys. Mol. Biol., Vol. 31, 31, 1976.

11. Ridler, P. J. and Jennings, B. R., Polymer, Vol. 19, 627, 1978.

12. Rees, D. A., 'Polysaccharide Shapes', Chapman & Hall, 1977.

13. McDevitt, C. A. and Muir H., Ann. Rheum. Dis., Vol. 34, Suppl. 2, 137, 1975.

14. Foweraker, A. R., Morris, V. J. and Jennings, B. R., 'Particle Size Analysis', Proceedings of Conf., Salford, Sept. 1977, p.147, Heyden Publ.

15. de Gennes, P. G., 'The Physics of Liquid Crystals', Oxford Univ. Press, London, 1974.

16. Uchida, T., Shishido, C., Seki H. and Wada, M., Mol. Cryst. Liq. Cryst., Vol. 39, 39, 1977.

17. Sackmann, E. and Rehm, D., Chem. Phys. Lett., Vol. 4, 537, 1970.

18. Winsor, P. A., Chem. Rev., Vol. 68, 1, 1968.

EXPERIMENTAL INVESTIGATION OF SMALL SINGLE AND MULTIPLE
FREE JETS BY LASER RAMAN SPECTROSCOPY

Alfred Leipertz and Martin Fiebig

Institut fur Thermo- und Fluiddynamik, Ruhr-Universitat

D-4630 Bochum 1, FRG

Abstract

CO_2 and N_2 jets issuing into air from small rectangular channels of 2x25 mm cross section were investigated. Concentration measurements were obtained by cw-laser Raman spectroscopy; gradients larger than 1o vol.%/1oo μm could easily be resolved; spatial resolution was better than $1o^{-2}$ mm^3 and the measured fluctuations of the Raman signals show satisfactory agreement with Poisson statistics. Axial and radial profiles are reported for single and multiple jets and comparison with theory is given, where possible.

Introduction

Single and multiple free jets have many technical applications. Air conditioning, drying processes, chemical and gasdynamic mixing lasers are some technical fields where jets are used. A large body of literature therefore exists on jets; pertinent monographs were written by Pai[1] and Abramovich[2],[3]. While the literature on single jets in a quiescent surrounding is extensive, investigations on initially laminar jets and multiple jets are rather limited. Here the papers by du Plessis et al.[4], Champagne and Wygnanski[5], and Rapagnani and Lankford[6] are to be mentioned.

The purpose of our work is to investigate experimentally the influence of initial profiles on the development of jets and the interaction of multiple jets of different composition. Small jets are best suited for such investigations if an appropriate diagnostic tool is available. For velocity measurements hot wire and laser anemometers are applicable. For concentration measurements laser Raman spectroscopy is suitable because it is molecule specific, non-perturbing and can give high spatial resolution and accuracy. It is applicable for nearly all fluid constituents as each molecule has at least one allowed Raman band. Such concentration measurements have been reported for example by Hartley[7], Lederman et al.[8], Williams and Lewis[9], and Lederman[10] in recent years.

The emphasis of this paper is on concentration measurements in single and multiple jets. The fluid dynamic aspects of the investigation will be reported separately.

Raman Diagnostics

The Raman effect has its physical background in the inelastic scattering process between photons and molecules. Fig. 1 shows schematically the generation of allowed Raman lines of diatomic molecules, which are of major interest for fluid dynamic applications. The incident photons may induce vibrational and/or rotational transitions in molecules and the photon may furnish energy to the molecule (STOKES transitions) or may receive energy from it (ANTI-STOKES transitions). By STOKES transitions the scattered photons will have a lower energy than the incident photons. Their frequency is shifted to longer wavelengths; these are the STOKES Raman lines. The opposite holds for ANTI-STOKES transitions and the corresponding ANTI-STOKES Raman lines. For diatomic molecules only allowed transitions are those which result in changes in vibrational quantum number v by $\Delta v = o, \pm 1$ and in rotational quantum number J by $\Delta J = o, \pm 2$.

Because of its relative high intensity we use the Q-branch of the STOKES vibrational band ($\Delta v = +1$ and $\Delta J = o$) for concentration measurements. The scattered Raman intensity I_{Ri} due to an incident laser intensity I_o is given by

$$I_{Ri} = k \cdot I_o \cdot n_i \cdot (\frac{d\sigma}{d\Omega})_i \cdot \Omega \cdot 1 \tag{1}$$

where

n_i = number density of gas component i under observation $[cm^{-3}]$
$(d\sigma/d\Omega)_i$ = differential Raman scattering cross section of i-th component $[cm^2/sr]$
Ω = solid angle of registration optic [sr]
1 = length of sample in direction of laser beam [cm]
k = constant factor given by the experimental setup

Equation (1) shows the direct dependence of Raman intensity (and the count rate N_{si}, when photon counting is used) on the number density of the gas component of interest. At higher temperatures, the broadening of Q-branch shape may be taken into account by an additional temperature dependent setup factor f(T) in (1).

In practice no absolute concentration measurements are made. A given Raman setup is calibrated for a fixed temperature by using gas components of known number densities or a selected reference gas, mostly nitrogen, to establish the relation between I_{Ri} and n_i. For identical experimental conditions the measured signal count rate N_{si} for the gas component i of interest is a measure of the concentration. When calibration has been done with the same gas component the concentration c_i is simply the count rate N_{si} divided by the count rate of the pure component $N_{i,1oo}$

$$ c_i = \frac{N_{si}}{N_{i,1oo}} . \tag{2} $$

When calibration has been done with pure reference gas ($N_{r,1oo}$) the different cross sections of reference gas $(d\sigma/d\Omega)_r$ and gas component of interest $(d\sigma/d\Omega)_i$ must be considered. In addition differences in the wavelengths dependence of the experimental setup must be evaluated. The optical transmission of monochromators or interference filters differ at different wavelengths; this may become important for larger spectral distances between the Raman bands of the reference and investigated molecules, and is expressed by a factor F. The concentration is then given by

$$ c_i = F \cdot \frac{N_{si}}{N_{r,1oo}} \cdot \frac{(d\sigma/d\Omega)_r}{(d\sigma/d\Omega)_i} . \tag{3} $$

For several gas components the cross section ratio in (3) for N_2 as reference gas has been reported by many authors, see for instance Refs. 11-14 and the compilation given by Stursberg[15].

Experimental Setup

Figure 2 shows the experimental setup. A cw-argon-laser is focused by a f=12o mm biconvex lens into a jet issuing from a small rectangular channel. The smallest beam diameter for the best local resolution has been measured to be below 5o μm (the theoretical value[16] for λ_o = 5145 Å ist 26 μm). The laser has usually been operated at a power of about 4 W. After the beam has crossed the test area laser power is measured by a power meter as reference value for Raman signal normalization with respect to laser power. The spontaneous Raman scattered light of the fluid constituents is observed perpendicular to the laser axis. The sample is enlarged by an image ratio of 4.4:1 at the entrance slit of a double monochromator. The image ratio, based on the width or the height of the entrance slit, and the beam diameter determine the spatial resolution; it was always better than $1o^{-2}$ mm^3. The monochromator selects the STOKES-Q-branch of the gas component of interest with high spectral purity better than $1o^{-1o}$ at 2o cm^{-1} away from the adjusted wavenumber. In practice the slit width is choosen to the corresponding band pass of the Q-branch. The exit slit is imaged on the cathode of the photomultiplier tube. A schematic of the data acquisition is shown in Fig. 3. The PMT signals are amplified and led to either a photon counting system, which we use for profile measurements, or a multichannel analyzer, used mainly for initial adjustments and for statistical investigations. Behind the analog outputs we usually use a multichannel recorder for signal registration. The externally measured laser power is used as a reference value. In addition a standard hot wire anemometer has been used for velocity measurements. Quasi two dimensional parallel jets were generated by rectangular channels (Fig. 4 a). The inner channel had a cross section of 2x25 mm, the cross sections of the outer channels could be varied. The channel arrangement was mounted on a three dimensional traversing mechanism. The gas supply, shown in Fig. 4. b, could be regulated separately for the three channels.

Calibration

The count period of all measurements was 1 second; in general 2o measurements per calibration point were taken.

The linear dependence of count rate on laser power, see Eq. (1), is shown in Fig. 5 for N_2, O_2, and CO_2. To deduce the differential cross sections of O_2 and CO_2 relative to N_2, the count rates of O_2 and CO_2 had to be adjusted corresponding to each Q-branch band pass entrance slit width of the monochromator. The results were for O_2:1.2o±o.o5 and for CO_2: 1.38±o.o8 at an exciting wavelength of 488o Å. Comparison with previously reported values (11-14) of 1.22±o.o6 and 1.46±o.o4 shows good agreement. The linear relation between count rate and molar concentration can be seen in Fig. 6 for a premixed CO_2-N_2 jet and an adjusted laser power of 2 Watt. For constant laser power this linear dependence follows from (1) because $N_{si} \sim T_{Ri} \sim n_i \sim c_i$ (molar or volume concentration). The error bars give the standard deviation assuming Poisson statistics.

For error analysis the assumption of Poisson statistics[17] was tested in a pure CO_2 jet. 3o cycles have been measured (abscissa of Fig. 7). 2o of these cycles consisted of 1oo measurements, two of 5oo measurements and eight of 25 measurements. The count rates N_{si} have been varied from cycle to cycle; the limits are 1ooo and 52oo counts/sec. The ratio of the calculated Poisson standard deviation σ_p-square root of the average count rate of each cycle - to the measured standard deviation σ_m is shown as ordinate in Fig. 7. For 12 of the 3o cycles the measured standard deviation is less predicted by Poisson statistics, for the others it is larger but never more than 1o % of the Poisson value. Poisson statistics can therefore be assumed to be valid; this implies, that the relative standard deviation of a number density measurement (statistical standard deviation devided by the average count rate of the measurement) is for example ± 1 % for count rates of 1oooo counts/sec and ± 1o % for 1oo counts/sec.

A further test was done by measuring the concentration of N_2 and O_2 in air. Fig. 8 shows part of an analog output. The pure N_2 jet led to a N_2 concentration in air of 78.54 vol.% and to an O_2 concentration of 21.21 vol.%. The deviations from the literature values[18] (78.o9 vol.% for N_2 and 2o.95 vol.% for O_2) are about 1 %. The deviations are well within the expected statistical uncertainties.

Experimental Results

The velocity profiles in the exit plane (z=o.2 d) of the rectangular channels are shown in Fig. 9. For the parallel channel the experimental profile is compared with the Hagen-Poiseuille solution. The agreement is excellent as expected at a Reynolds number of 15o. The outer channels represent diffusors with a half angle of 15°, so that separated flow regions of low velocity will occur near the walls. In these regions the flow velocities were so low that they could not be detected by the hot wire anemometer; the expected profile shape is indicated by dotted lines. The corresponding concentration profiles are step functions.

Figure 1o shows the radial and the axial concentration profiles for CO_2 and N_2 jets issuing into quiescent air from a parallel channel. Each point corresponds to the average of 2o count rate measurements of one second duration. The very large concentration gradients, larger than 1o vol. %/1oo μm, could easily be resolved with high accuracy, better than 3 % for the highest gradients (Fig. 1o a). The spreading characteristics are similar for CO_2 and N_2 jets. The normalization of the concentration profiles with respect to the half-width distance, the distance at which the concentration has half the centerline value, Fig. 1o b, shows that for large distances from the orifice the profiles become similar. In Fig. 1o c the centerline concentration decay is graphed. The CO_2 jet decays initially less than the N_2 jet. The reason lies in the higher momentum and higher initial Reynolds number of the CO_2 jet. Further downstream the initially laminar jet will become turbulent. Large deviations for different count rate measurements occured in regions, where an intermittent turbulent structure was expected. This would account also for the measured large standard deviation (formal errors) in some regions of the jet.

Fig. 11 a,b shows multiple jet results. The influence of the outer divergent N_2 jets on the inner parallel CO_2 jet and vice versa can be seen. The spreading between the different jets is clearly diminished (Fig. 11 a). This follows directly from the continuity and the momentum considerations. The centerline concentration decay (Fig. 11 b) is enhanced for the inner and the outer jets at the small distances from the exit. This holds also for the outer N_2 jets in the farfield, where this trend is reversed for the inner CO_2 jet. This behaviour can be explained by comparing the single and the multiple jets with respect to their spreading characteristics and the magnitude of the concentration and the velocity gradients.

Conclusions

For concentration measurements in small jets laser Raman spectroscopy gives high local (5o μm) and spatial ($<1o^{-2}$ mm^3) resolution with an accuracy given by Poisson statistics. When experimental results show large deviations from Poisson statistics, they are caused by the flow structure, i.e. turbulence. This then can be used to gather information on the structure of the flowfield, for example intermittency.

In conjunction with hot wire or laser anemometry, laser Raman spectroscopy is ideally suited to investigate the behaviour of small single and multiple jets.

References

1. Pai, S.I.; Fluid Dynamics of Jets, D. von Nostrand Co., New York 1954
2. Abramovich, G.N.; Theory of turbulent jets, M.I.T. Press, Cambridge 1963
3. Abramovich, G.N. (Ed.); Turbulent Jets of Air, Plasma, and Real Gas, Consultants Bureau, New York 1969
4. du Plessis, M.P.; Wang, R.L.; Tsang, S.; Development of a submerged round laminar jet from an initially parabolic profile, Trans. of ASME 6 (1973) 148-154
5. Champagne, F.H.; Wygnanski, I.J.; An experimental investigation of coaxial turbulent jets, Int. J. Heat Mass Transfer 14 (1971) 1445-64

   6. Rapagnani, N.L.; Lankford, D.W.; Time dependent nozzle and base flow cavity modeling of C-W chemical laser flowfields, 2nd Int. Symp. Gas-Flow and Chemical Lasers, 11-15 Sept. 1978, VKI, Rhode-St. Genese, Belgium; Proceedings in press; Hemisphere Publishing Co.
   7. Hartley, D.L.; Experimental gas mixing studies utilizing laser Raman spectroscopy, AIAA-Paper No. 71-286 (1971)
   8. Lederman, S.; Bloom, M.H.; Bornstein, J.; Khosla, P.K.; Temperature and specie concentration measurements in a flow field, Int. J. Heat Mass Transfer 17 (1974) 1479-1486
   9. Williams, W.D.; Lewis, J.W.L.; Hypersonic flowfield measurements using laser Raman spectroscopy, AIAA- Journal 13 (1975) 1269-1270
   10. Lederman, S.; Some applications of laser diagnostics to fluid dynamics, AIAA-Paper No. 76-21 (1976)
   11. Fenner, W.R.; Hyatt, H.A.; Kellam, J.M.; Porto, S.P.S.; Raman cross section of some simple gases, Journ. Opt. Soc. Am. 63 (1973) 73-77
   12. Fouche, D.G.; Chang, R.K.; Relative Raman cross section for N_2, CO, CO_2, SO_2, and H_2S, Appl. Phys. Lett. 18 (1971) 579-580
   13. Hochenbleicher, J.G.; Klöckner, W.; Schrötter, H.W.; Messung von Raman-Streuquerschnitten an Gasen, Lectures at Laser Meeting, Essen (FRG) 1974, pp. 191-204
   14. Penney, C.M.; Goldman, L.M.; Lapp, M.; Raman scattering cross sections, Nature 235 (1972) 110-111
   15. Stursberg, K.; Intensität, Winkelabhängigkeit und Polarisationszustand der spontanen molekularen Streustrahlung, DLR-FB 77-20 (1977)
   16. Barrett, J.J.; Adams III, N.I.; Laser excited rotation-vibration Raman scattering in ultra-small gas samples, Journ. Opt. Soc. Am. 58 (1968) 311-319
   17. see for example: Goulard, R.; Laser Raman scattering applications, in: Laser Raman Gas Diagnostic, edited by M. Lapp and C.M. Penney, Plenum Press, New York 1974, p. 3-14
   18. Kohlrausch, F.; Praktische Physik, Band 3, Teubner Verlag, Stuttgart, 22. Auflage 1968, p. 39

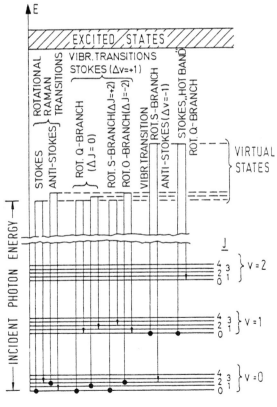

SCHEMATIC ENERGY LEVEL DIAGRAM AND
RAMAN TRANSITIONS FOR DIATOMIC MOLECULES

Figure 1

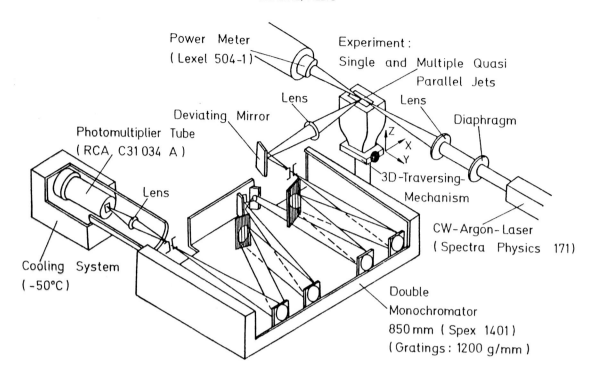

Schematic of Experimental Set-up for CW-Laser Raman Spectroscopy

Figure 2

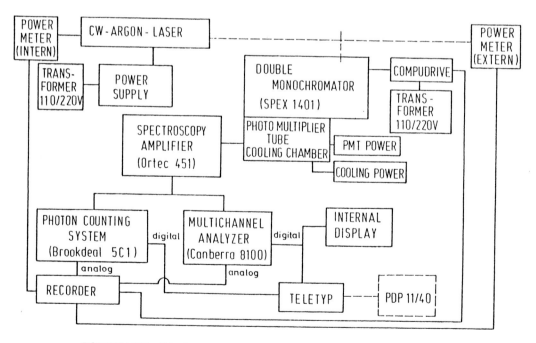

SCHEMATIC DIAGRAM FOR RAMAN DATA AQUISITION

Figure 3

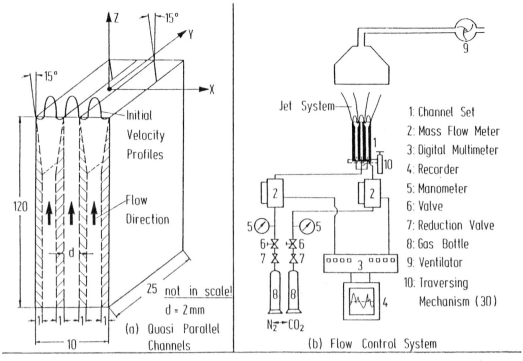

Schematic of Flow Control System for Jet Generator (Quasi - Twodimensional Channels)

Figure 4

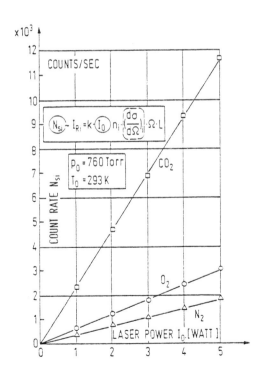

Dependence of Count Rate on Laser Power for fixed CO_2, N_2, O_2 Concentrations (each point represents the mean of 20 measurements of 1 sec ducation)

Figure 5

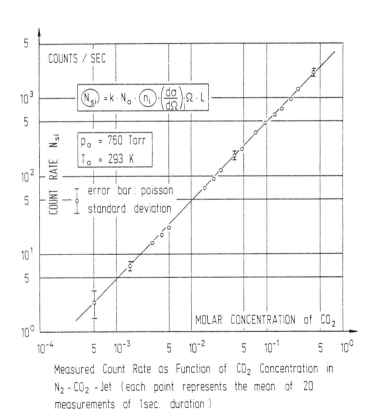

Measured Count Rate as Function of CO_2 Concentration in $N_2 - CO_2$ - Jet (each point represents the mean of 20 measurements of 1sec. duration)

Figure 6

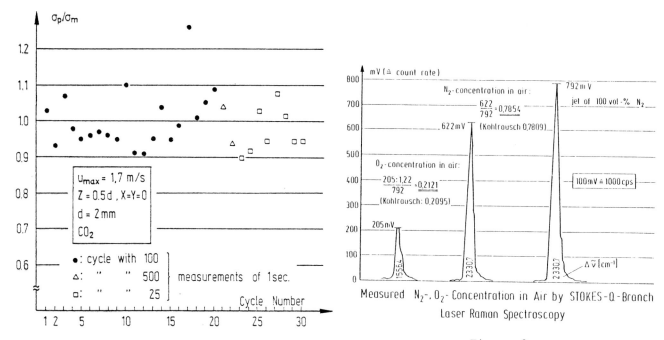

Ratio of Calculated Poisson and Measured Standard Deviation for Several Counting Cycles in CO_2 Jet

Figure 7

Measured N_2-, O_2- Concentration in Air by STOKES-Q-Branch Laser Raman Spectroscopy

Figure 8

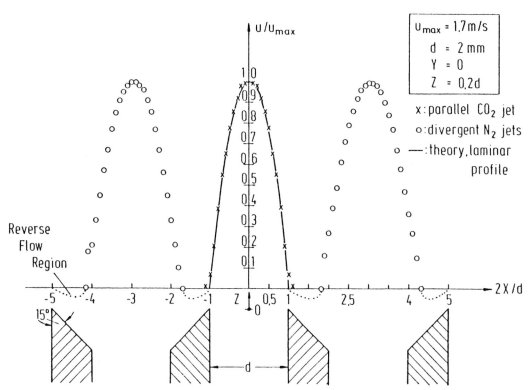

Initial Velocity Profiles for Free Jets(Measurements with Hot Wire Anemometer)and Comparison of Parallel Channel Profile with Hagen-Poiseuille Profile

Figure 9

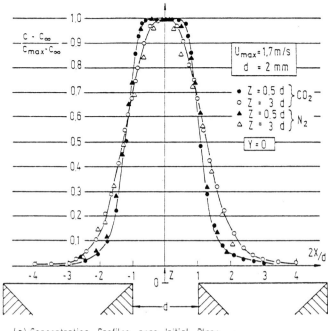

(a) Concentration Profiles near Initial Plane

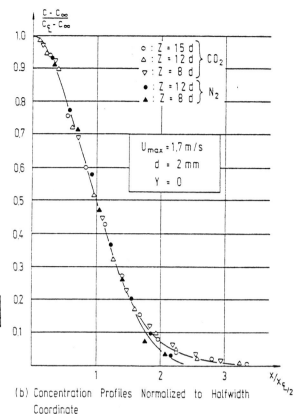

(b) Concentration Profiles Normalized to Halfwidth Coordinate

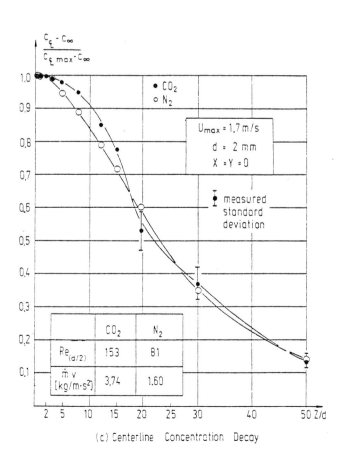

(c) Centerline Concentration Decay

Figure 10

Concentration Profiles and Centerline Decay of Single CO_2 and N_2 Jets Issuing from Parallel Channel into Quiescent Air

(a) Concentration Profiles near Initial Plane
(b) Concentration Profiles Normalized to Halfwidth Coordinate
(c) Centerline Concentration Decay

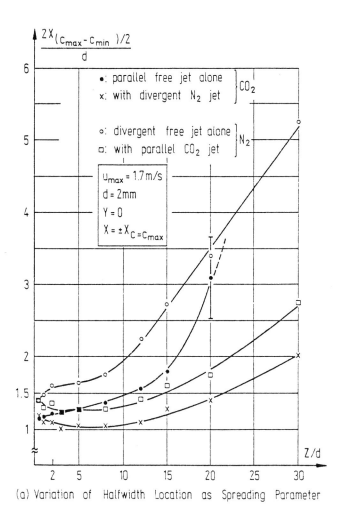

(a) Variation of Halfwidth Location as Spreading Parameter

Figure 11

Spreading and Centerline Decay of Unaffected and Affected Parallel CO_2 and Divergent N_2 Jet Issuing into Quiescent Air

(a) Variation of Halfwidth Location as Spreading Parameter
(b) Normalized Centerline Concentration Decay

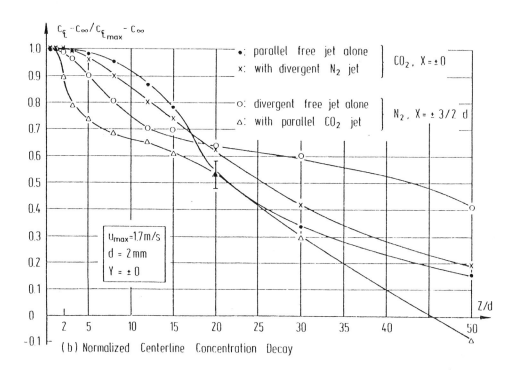

(b) Normalized Centerline Concentration Decay

MOBILE LIDAR SYSTEM
FOR THE MEASUREMENT OF GASEOUS POLLUTANTS IN ATMOSPHERIC AIR

C. Weitkamp, J. Harms, W. Lahmann, W. Michaelis
Institut fur Physik
GKSS-Forschungszentrum
D-2054 Geesthacht, Germany

Abstract

A mobile system for the remote measurement of polluting gases such as NO_2, SO_2 and HCl has been designed and tested. The detection is based on differential absorption and scattering of infrared radiation. The essential components of the system are a deuterium fluoride laser operated in multiline mode, a 0.6 meter diameter receiver optics, monochromator, and cooled multielement detector array for the simultaneous detection of light backscattered at different frequencies. Range-sensitivity adjustment is optimized using geometrical signal compression techniques. For maximum mobility and versatility the system is mounted in a standard 20 foot container. Design characteristics and special features of the system are reported, ancillary research and development work is described, and applications are proposed.

Introduction

Despite considerable effort in recent years, pollution of ambient air continues to be an important issue in environmental policy and engineering. The chemical and metallurgical industry, fossil-fuel power plants, home heating and automotive traffic all contribute to the emissions that deteriorate air quality in urban industrial centers. The most important air pollutants occur in the form of both particulate matter and toxic gases such as sulfur dioxide, nitrogen oxides, carbon monoxide, hydrogen chloride, etc.

For the assay of actual air quality and the localization of major emission sources, air parameters must be monitored periodically or continuously in a large number of places. This is generally accomplished by using a number of stationary installations that sometimes measure automatically the basic parameters, average over predetermined time intervals, and relay the results to a central unit. A different approach is the use of manually-operated mobile laboratories mounted in a van or trailer; these units are better suited for special tasks and generally less efficient for routine-type surveillance measurements. In any case, a compromise must be found between the number of parameters to be determined and the spatial density of measurement points and, on the other hand, the capital investment and operating funds available. The terms of this compromise have usually been unsatisfactory.

An alternative approach has therefore been investigated that relies on the remote, yet space-resolved determination of the concentration of gaseous air pollutants with laser techniques. Recent progress in the development of lasers and associated optical and electronic components has brought this possibility near at hand. In the present paper a brief description of the principle of operation of such systems is given. An actual setup constructed at GKSS is described; the solutions of special problems essential to this development are presented, and some fields of research are identified where progress could improve or simplify the present design. Various applications of the system are proposed.

Principle of Operation

The most sensitive remote measurement of gaseous pollutants has been achieved with so-called bistatic systems which consist of an emitter (laser) and, up to several kilometers apart, a receiver system that measures the attenuation of the laser beam as the primary information. For convenience, folded transmission lines are usually employed in which the emitter and receiver are at the same location and a retroreflector is installed at the remote end. Where reduced sensitivity can be tolerated, topographic targets are also used instead of the retroreflector. Measurements under different azimuth angles and different ranges are only possible in this measurement scheme if the emitter/receiver overlooks concave terrain or a large number of objects at different distances and in different directions appears at the horizon. Usually the result of measurements made with these systems is an *integrated concentration* given in such units as *ppm kilometers*; an *average concentration* can be calculated from the data with no information about the variation of concentration with distance, or depth.

This shortcoming is removed in measurements with lidar systems where a short pulse of laser radiation is emitted and the backscattered light is detected and analyzed. Generally, the wavelength, intensity, and time behaviour of the return signal is typical for the nature, concentration, and distance of the substances to be measured. According to the physical process that governs the transformation of emitted light into backscattered radiation, we distinguish Raman lidar, fluorescence lidar, and differential absorption and scattering lidar (sometimes abbreviated DAS lidar, or DIAL). Whereas Raman lidar suffers from small cross sections and, correspondingly, low sensitivity and range, fluorescence lidar is based on a process that usually involves relatively long delay times before reemission of the absorbed radiation and, consequently, provides inadequate spatial resolution. These restrictions are not present in DAS lidar where the return signal is mainly due to backscattering by dust and fog particles as described by Mie's theory. A signal measured, say, at wavelength λ_0 is compared to a signal at a slightly different wavelength λ_1 where radiation is strongly absorbed by the gas of interest (Fig. 1). If a cloud of this gas is located at some distance X, that part of the signal that corresponds to distances $x < X$ will be the same for λ_0 and λ_1, but for $x > X$ the signal at λ_1 will be weaker than at λ_0 due to absorption of λ_1 in the cloud (Fig. 2). The ratio of the two signals which is constant for $x < X$ will therefore show a marked decrease at distance X, after which it again remains constant out to the next cloud of polluting gas. It thus directly reflects the concentration of the gas as a function of distance. If the wavelength λ_1 is changed to λ_2 such that another gas (no. 2) strongly absorbs at λ_2, the resulting signal is a measure for the concentration of gas no. 2.

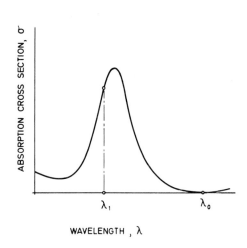

Fig. 1. Symbolic representation of two laser wavelengths showing no absorption (λ_0) and absorption (λ_1) by polluting gas

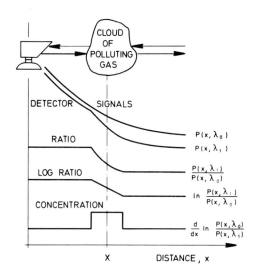

Fig. 2. Principle of extracting concentration information from signals $P(x,\lambda_0)$ and $P(x,\lambda_1)$ measured at two different wavelengths

A simplified mathematical formulation of the phenomenon is given by the so-called lidar equation [1]

$$P(x,\lambda) = \frac{c}{2} E(\lambda) A \eta(\lambda) \frac{\beta(x)}{x^2} \exp\{-2 \int_0^x [\alpha(\xi) + N(\xi)\sigma(\lambda)]d\xi\} \tag{1}$$

which relates the power P measured at wavelength λ from distance x with the density N of the pollutant gas molecules. The constants E, A and η are the energy of the transmitted laser pulse and area and detection efficiency of the receiver; note that E and η are usually wavelength-dependent. β is the Mie backscattering coefficient and α the extinction coefficient of the atmosphere, $\sigma(\lambda)$ is the extinction cross section of the polluting gas at wavelength λ, and c is the velocity of light. If P as measured at the two wavelengths λ_0, λ_1 is inserted in eq. (1), the density of the absorbing molecules is computed as

$$N(x) = \frac{1}{2[\sigma(\lambda_1) - \sigma(\lambda_0)]} \frac{d}{dx} \ell n \frac{P(x,\lambda_0)}{P(x,\lambda_1)} . \tag{2}$$

Design Specifications

The design of the system described below was governed by a number of requirements some of which turned out to be hard to fulfill simultaneously with present-day technology and reasonable financial effort. A compromise has therefore been adopted that seems to be acceptable for future users of the system for some time to come.

Under normal weather conditions, a *range* of about 5 km is desired allowing to cover an area up to 80 km². The number of polluting gases that can be measured ought to be as large as possible and should include hydrogen chloride, which is generated in large amounts in the incineration of polyvinyl chloride and other chlorinated hydrocarbons. The *sensitivity* is based on the requirement that unsafe levels of pollution can be detected and excessive emitters of certain gases localized. As to the *measuring time*, half-hourly results for the circle or sector under consideration are desirable. For the *spatial resolution* a value of 300 m appears sufficient for the usual density of industrial installations. * The complexity of the system and the necessity of day-and-night operation require that the device be amenable to a high degree of *automation*. The measured primary data should be converted to concentrations and the final values archivated retrievably in some mass storage device. It is essential, however, that the results be also presented in a quickly readable format, preferably *contour maps* containing lines of equal concentration superimposed on a normal map of the region of interest. As the interpretation of the results is greatly facilitated if the *meteorological data* are known, these data must also be measured and recorded simultaneously with the concentration measurement. Finally, the whole system should be *mobile* to allow its use in different places.

Measurement Device

The choice of the wavelength, i.e., the laser to be used, is governed by the number of gases that are to be detected and by the transmission of the atmosphere. Excellent transmission is given in the visible and near ultraviolet, and DAS lidar has indeed been proved to be feasible with visible light [2], although the method suffers from limited performance at daytime and also from eye safety problems. However, very few of the interesting gases show sufficiently strong absorption bands in this part of the spectrum. Most of the intense absorption bands of gaseous pollutants lie in the infrared; IR lidar thus offers a wider range of application and better flexibility [3, 4]. A deuterium fluoride laser is therefore used that emits about 30 lines between 3.5 and 4.1 μm. As can be seen from Fig. 3, the transparency of the atmosphere in this wavelength range is quite good. A serious alternative is also a carbon dioxide laser emitting between 9 and 11 μm; the recently available high-pressure version of the CO_2 laser [6] appears particularly attractive due to its continuous tuneability over certain regions of the CO_2 emission bands. However, the DF laser can be made to optionally emit one, a few, or all lines of the spectrum simultaneously, thus greatly facilitating the implementation of the differential absorption concept even under quickly varying atmospheric conditions. Also no wavelength stabilizing device is necessary with the DF laser because of its inherently small linewidth.

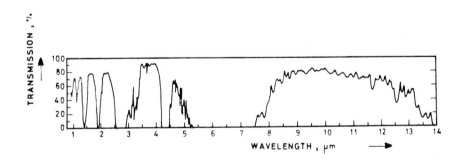

Fig. 3. Atmospheric transmission, per sea mile, at 15.5°C (60°F) for a visual range of 14 km and 70 percent relative humidity [5]

---

* Note that only the depth resolution is critical as the azimuthal resolution can be made as small as necessary for any practical purpose.

A schematic diagram of the setup is shown in Fig. 4. The infrared beam from the DF laser is transmitted into the atmosphere via a Cassegrain telescope that, by expanding the beam diameter, matches the divergence of the beam to the field of view of the receiving optics

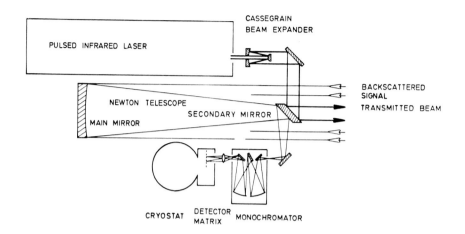

Fig. 4. Schematics of the transmitting and receiving optics

and also reduces the irradiance well below the maximum permissible values with respect to eye safety regulations. Two 45° mirrors shift the transmitted beam into the receiver axis to provide coaxial geometry. Some reasons for this design of the transmitter-receiver optics will be discussed below.

The receiver consists of a parabolic mirror, 45° flat secondary mirror, monochromator for the separation of the different wavelengths, and detector. For the simultaneous detection of scattered light with different wavelengths separated in angle by the monochromator (whose exit slit has been removed for this purpose), a multielement detector array is positioned in the exit focal plane, or (as indicated in Fig. 4) an image of the focal plane is projected onto the detector array. The physical data of the transmitter-receiver system are summarized in Table 1.

Table 1. Physical Data of Transmitter-Receiver System

Wavelength	3.5 - 4.1 μm	Beam diameter after expansion	150 mm
Pulse energy @ 1 Hz, all lines	900 mJ	Divergence after expansion	0.25 mrad
Pulse width, FWHM	≤ 500 ns	Diameter of receiving optics	600 mm
Pulse-width limited depth resolution	75 m	Focal length of receiving optics	3000 mm
Beam diameter before expansion	38 mm	Detector material	InSb
Divergence before expansion		Number of detector elements	20
(full angle, 50 % of beam energy)	1 mrad	Dimensions of detector elements	0.2 x 0.4 mm²

The signals from the detector elements are amplified, digitized and fed to an on-line computer for analysis. In order to allow processing of the amplified signals in a single digitizer, a special unit has been developed that delays each of the signals by a different, fixed amount of time and multiplexes all signals onto a single output line. This multiple-delay analog multiplexer is described in some detail below. In order to reduce the statistical scatter of the noisy signals, the results from several laser shots are added together. This can either be done by the computer or by a hard-wired signal averager. The schematics of the data flow is represented in Fig. 5. The figure also shows that the control of the whole measurement system including the mechanical motion of the setup, initialization of the laser, switching and timing of the data acquisition electronics, and accomodation of the devices for the measurement of the meteorological parameters is performed by the computer.

The system has been tested in the laboratory and a 420 m test range, and performed satisfactorily although a few components need modification as is partly described in the following paragraphs. Field measurements will be started soon; a container is being built for this purpose to house the complete measurement system and associated equipment.

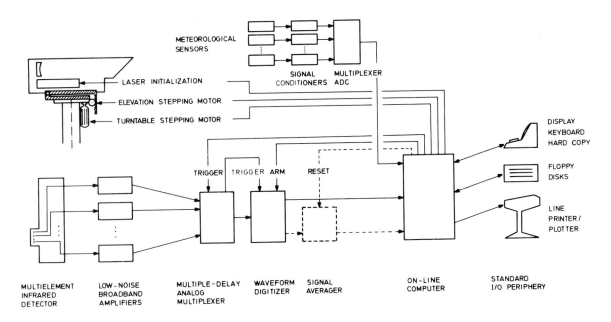

Fig. 5. Block diagram of the data acquisition and control

Special Features

Commercially available components have been used as far as possible. To meet the design specifications, however, a few components needed considerable modification, some were newly designed, and the problem of the signal dynamics was solved by a new approach.

Geometric Compression of the Return Signal Dynamics

The intensity of the return signal as given by the lidar equation (1) varies with distance x roughly as $x^{-2} \exp(-2\,\alpha x)$, or by a factor of about 100 000 for distances between 100 m and 10 km. To our knowledge, no waveform digitizer is available with a resolution adequate for digitizing this kind of signal with the precision and speed required. All methods presently used to solve the problem such as the use of detectors with time-variable sensitivity [7] or logarithmic or switched amplifiers [8] are either not applicable to the present case, or present serious shortcomings as to the performance. Instead, a method has been applied in the present system that makes use of the fact that light from an object at distance x produces an intensity distribution in the focal plane of the receiving system the area of which is, to first approximation and up to a certain distance, proportional to x^{-2}. Objects at different distances x with the same total radiant power per unit solid angle thus generate intensity patterns in the focal plane *that have the same peak irradiance,* but extend further out for smaller x. It is thus possible to compensate the x^{-2} dependence of the lidar equation by positioning a small detector in the focal plane of the receiver; the exact size is determined by the range up to which a flat response is desired.

For illustration the calculated response function of the present system is shown in Fig. 6. Evidently the use of a small detector results in a response with considerably reduced dynamic range as compared to the range dependence according to the lidar equation. A 1.6 mm - diameter detector produces a flat signal out to about 1 km, and an 0.8 mm detector to about 2 km. Still smaller detectors reduce the field of view of the receiving optics below the divergence of the laser beam, and thus result in incomplete collection of the back-scattered light from *any* distance. Note that in the present setup the effective detector area is the diaphragm at the monochromator entrance, and an optics with magnification < 1 is placed between the exit focal plane of the monochromator and the detector; it is thus possible to match the detector size as dictated by the geometric compression concept to the actual size of the individual detector elements.

Details of the calculation and numerical results are published in Ref. [9]; the formalism has also been extended to non-coaxial geometries and to systems with central beam obstruction as by a secondary mirror [10].

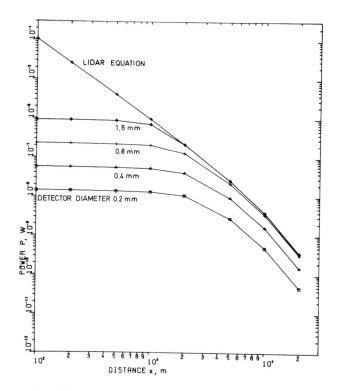

Fig. 6. Power incident on the detector. Upper curve: no geometric compression. Lower curves: geometric compression with detector diameters as indicated. Meteorological conditions chosen are as follows: midlatitude summer, clear atmosphere (visibility 23 km), $\alpha = 0.05$ km^{-1}, $\beta = 0.001$ km^{-1}sr^{-1}. Geometric data are the same as given in Table I. It is evident from the figure that the optimum detector diameter is 0.8 mm corresponding to a field of view of 0.27 mrad (to be compared to the beam divergence of 0.25 mrad).

Amplifier

The requirements imposed on the detection system include sensitivity, low noise, speed, and linearity of response. The first three of these are closely related by the fact that the ultimate sensitivity is limited by the background from radiation emitted by every room-temperature surface "seen" by the detector; this background fluctuates statistically, and the fluctuations average out for a longer period of observation (i.e., smaller bandwidth). An amplifier that does not add significant noise of its own under these conditions is said to allow background-limited infrared photodetection, or BLIP. For the wavelength region of interest a photovoltaic InSb detector is best suited as it combines high sensitivity, fast response and good linearity if operated in the current mode, i.e., the short-circuit current of the photodiode is measured. For this purpose a low-noise high-speed current-to-voltage converter had to be developed since no commercial product was available. This amplifier allows near-BLIP operation up to a bandwidth as large as 0.8 MHz.

Multiple-Delay Analog Multiplexer

Following each laser shot up to 16 individual return signals must be registered, digitized and processed. Although this could in principle be accomplished with 16 parallel amplifier-digitizer-averager chains, a more economic solution appeared preferable in which the digitization and averaging of the different signals is done sequentially with a single waveform digitizer and pulse averager. This is possible because of the low repetition frequency of the laser (1 to 10 Hz) and the short duration of the return signal (100 μs for 15 km). A multiple-delay analog multiplexer has been designed that allows the simultaneous acquisition, storage, and sequential transfer of the detector signals.

The device consists of 16 independent parallel analog delay lines [11] with 100 capacitive storage elements each. Associated to each storage element is a multiplex switch; these switches are activated by a shift register operated as a ring counter. Each delay line sequentially samples the analog input signal from one detector element and current-to-voltage converter until the total information from the laser shot is transferred. Upon completion of the read-in cycle, the information is stored for a predetermined, but different amount of time for each of the 16 delay lines. During readout, the contents of the 100 storage elements of each delay line are transferred sequentially to the common output line, thus generating a sequence of 16 signals that is digitized by the large-channel-number waveform digitizer and processed by the averager or computer as if it were a single pulse. The evaluation of the different components and subsequent analysis occurs under software control.

Monochromator

While the focus of the Newton telescope is circular in shape, most monochromators do not provide stigmatic, i.e., point-by-point imaging of the entrance slit into the exit focal plane; in fact, no stigmatic IR monochromator with appropriate dispersion and resolution was available.

Therefore, a low-cost commercial monochromator was modified to provide stigmatic imaging. The use of lenses, although possible in principle, requires careful design and presents difficulties during checkout and alignment due to chromatic aberration. This problem is eliminated if mirrors are used instead of lenses. Therefore an alternative approach has been adopted in which off-axis parabolic mirrors are utilized.

Dimensional Stability

The small detector area necessary for the geometric compression of the lidar signal dynamics requires both careful alignment and mechanical and thermal stability of the optical system in order to ensure complete overlap of the transmitted beam and receiver field of view at large distance. In field measurements, however, considerable temperature variations must be encountered, and during transportation shocks and vibrations are unavoidable. On the other hand, excessive weight of the optical part of the system would require even heavier lift and pivot gear, with the resulting problems of handling, installation, power supply etc.

The optical part has therefore been mounted in a rack characterized by small temperature coefficient, high stability and light weight. It consists of two frames on either side made of carbon-fiber reinforced plastic to provide the dimensional stability and low temperature coefficient, and four aluminium honeycomb-invar sandwich plates at the top, bottom and two ends for rigidity and shear stability. Invar mounts are used for critical components.

Container

A container has been designed for accommodation of the whole measurement device. Because of its standard outside dimensions of $6.1 \times 2.4 \times 2.6$ m^3 ($20' \times 8' \times 8.5'$) it can easily be transported by rail or road. The rack with the optical part of the system is hoisted out of the container roof through a hatch. For ancillary equipment such as battery, hydraulics, gas storage, laser scrubber etc. and also for noisy components like the vacuum pump for the laser an extra compartment has been provided. This compartment also contains the meteorological mast with the sensors that allow the measurement of air temperature and humidity, barometric pressure, precipitation, visual range, radiation balance, and wind direction and speed up to 10 m above ground. A schematic drawing of the container is given in Fig. 7.

Ancillary Research and Development

Cross Sections

With a very few exceptions, the cross sections needed in order to determine the polluting gas concentration from the measured signals according to eq. (2) are not known because the resolution of conventional measurements is far too low with respect to the small laser linewidth. Simultaneously with the development and test of the mobile lidar system, measurements of the cross sections for the most important pollutant gases are therefore under way in the laboratory. It is essential that this determination is performed under the same conditions (temperature, pressure, mixing ratios) as in the field measurement in order to avoid errors due to Doppler (temperature), resonance (concentration) and pressure (foreign gas) broadening of the spectral lines. Under these conditions, the absorption of the gas mixtures is usually very small so that its measurement is not straightforward.

Two methods are being used in this research, an optoacoustic measurement that allows the determination of very small absorption coefficients [12], but yields relative values only, and a direct measurement with a multipass absorption cell as shown schematically in Fig. 8.

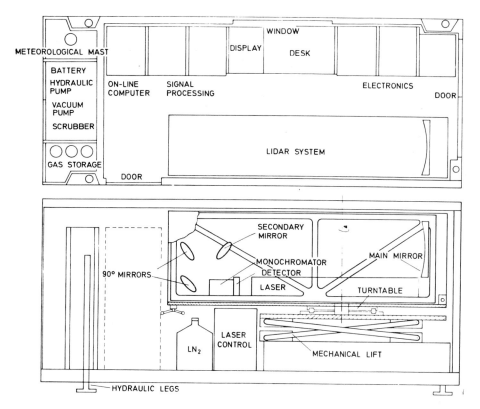

Fig. 7. Container

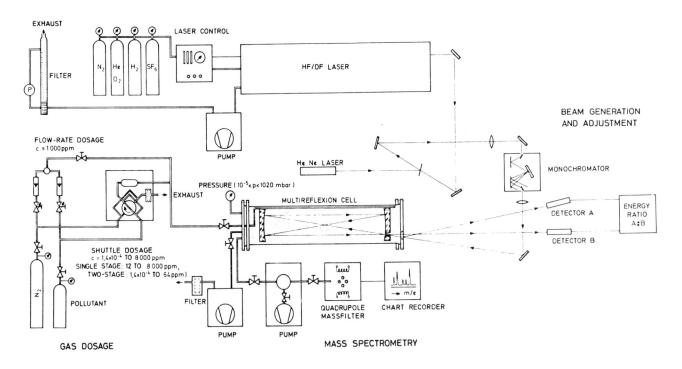

Fig. 8. Setup for the measurement of cross sections with the multipass absorption cell

The cell, with a length of about 1 m, can be adjusted for more than 100 passes; the reflectivity of the mirrors rather than the geometric alignment has up to now been the limiting element for the effective path length.

Single lines of the laser radiation are selected by means of a monochromator. The intensity of the laser pulse attenuated in the cell is compared, for normalization, to the intensity of the light pulse obtained by reflexion from the entrance window of the cell. Great care has been given to the production, preservation, and analysis of the gas ratios.

Detection Schemes

Sensitivities for the detection of the different pollutants depend upon many factors including weather conditions and will be determined experimentally. The present data indicate that ppm levels can be measured for gases with moderate absorption cross sections. In favorable cases much better sensitivity seems possible.

At present, a severe limitation of IR DAS lidar is due to the large photon background in the wavelength region of interest. As longer measurement times or smaller detection bandwidth (and, consequently, poorer depth resolution) are not desirable, improvements may be obtained by use of larger telescopes (to increase the collected fraction of the backscattered light), smaller detectors (to reduce the photon noise associated with the detector area) and more powerful lasers (to simply override the noise). Even a combination of these, however, is not likely to yield more than one order of magnitude if methods are used that appear technically and economically feasible in the near future. Alternate detection schemes have therefore been proposed that rely on the shift of the IR wavelength out of the region of maximum thermal background to longer wavelengths, e.g., by heterodyne spectrometry [13], or to shorter wavelengths by upconversion [14].

It is this latter approach that is being actively pursued in our laboratory. In upconversion, the infrared radiation to be detected (frequency ω_{ir}) is mixed in some nonlinear crystal with intense radiation from a local oscillator (the pump radiation, frequency ω_p) to form the signal ω_s. As both the energy and momentum balance

$$\omega_s = \omega_{ir} + \omega_p \quad \text{and} \tag{3}$$

$$\vec{k}_s = \vec{k}_{ir} + \vec{k}_p \tag{4}$$

must be closed, a way has to be found to match the indices of refraction of the crystal to fulfill the condition (4). This phase-matching can be achieved by using the temperature dependence or the angle dependence of the refractive index in an anisotropic medium; however, in both cases usually collinear phase-matching ($\vec{k}_s \parallel \vec{k}_{ir} \parallel \vec{k}_p$) has been used in which, for a given ω_p, the condition (4) is met for a single narrow frequency band ω_{ir}. If noncollinear k vectors are allowed, a much larger frequency band ω_{ir} appears to fulfill eq. (4). Calculations have shown that nearly the whole spectrum of DF laser frequencies may be upconverted with the same crystal temperature and orientation; this requirement is mandatory if such a scheme is to be useful for the present application. Experimental work in this direction is in progress.

Laser Gas Recycling

Only a small fraction of the laser feed gas mixture (H_2 or D_2, SF_6, $He-O_2$) is consumed in operation, whereas more than 90 % leaves the laser volume unreacted. Not only would a closed loop to which only the consumed fraction of the gases is added reduce the frequency of gas cylinder changes and thus facilitate a prolongation of the periods of unattended, automatic operation; the gas consumption also contributes significantly to the operating cost of the system so that substantial savings can be expected from a better utilization of the feed material. This requires removal of the reaction products, especially HF (or DF), which must not be discharged into the atmosphere and are already retained in the present system. Based on experience by other groups [15, 16] work is being started on a filter system that allows recycling of the unconsumed feed.

Applications

The lidar system described offers a number of advantages compared to locally measuring devices. The properties of remote pollution measurements include the lack of any sampling; thus no sampling errors, no change of samples during storage and transport, no errors due to treatment or conditioning of samples and no labelling errors can occur. The possibility of remote measurement enables the operator to fully cover an area of interest with a grid of mesh points almost as fine as desired, without prior permission of the tenants and without notice to the tenants. The immediate availability of the results offers the possibility of timely detection and countermeasures, and the measurement of several pollutants in different places with a single instrument not only produces a large flow of data at moderate cost, but also ensures consistency of the different results with each other. Numerous application can therefore be devised for which the system is ideally suited:

Air Quality Monitoring

By the possibility of large-area monitoring of urban air quality restrictions by inaccessible terrain are eliminated. Although the need for locally measuring devices, e.g., for measurements of traffic-generated pollution in downtown business areas, will subsist, a general reduction of the number of locally measuring stations accompanied by a reduction of investment and operating cost appears possible.

Correlation of Air Quality and Meteorological Conditions

As the measurement system is equipped with a weather mast for the determination of the meteorological data of interest, correlations of weather and pollution data are possible. Under consideration of the diurnal and annual variations of these correlations, principles for a better exhaust and waste gas management by certain industries may be established.

Local and Regional Early Warning System

The coincidence of unfavorable events such as industrial production campaigns with large amounts of gaseous waste or unusually dense automobile traffic with adverse weather conditions like calms, inversions, low temperatures with increased emission from home heating etc. can lead to intolerable concentrations of pollutants in the ambient air. Still larger concentrations have occurred in some areas due to accidents with railroad cars or other large-volume transport or storage containers with toxic liquefied or pressurized gases. A remote measurement system is certainly valuable for both early detection and the determination of the spatial distribution and variation of such areas of increased concentration.

Preparation of Air Quality Maps

When the performance of the system as to precision, accuracy, and freedom of the observed data from interferences has been sufficiently established, the time-space-dependent concentration values of the different pollutants may be archivated and averaged over longer periods of time. The resulting maps may be useful for multiple purposes including the local and regional planning of land utilization and industrial development, e.g., in the event of installation of new power plants or other industries.

Acknowledgement

The main mirror could be borrowed from the observatory of the Universität Hamburg; this is gratefully acknowledged.

References

1. R.T.H. Collis and P.B. Russell in E.D. Hinkley, ed., Laser Monitoring of the Atmosphere, Springer-Verlag Berlin - Heidelberg - New York, 1976, p. 71.
2. K.W. Rothe, U. Brinkmann and H. Walther, Appl. Phys. 3 (1974) 115.
3. K.W. Rothe and H. Walther in A. Mooradian, T. Jaeger and P. Stokseth, eds., Tunable Lasers and Applications, Springer-Verlag Berlin - Heidelberg - New York, 1976, p. 279.
4. E.R. Murray, Opt. Engineering 17 (1978) 30.
5. H. Kildal and R.L. Byer, Proceedings of the IEEE 59 (1971) 1644.
6. Laser Focus 13 (1977) 56.
7. R.J. Allen and W.E. Evans, Rev. Sci. Instr. 43 (1972) 1422.
8. J.D. Spinhirne and J.A. Reagan, Rev. Sci. Instr. 47 (1976) 437.
9. J. Harms, W. Lahmann and C. Weitkamp, Appl. Optics 17 (1978) 1131.
10. J. Harms, submitted for publication to Appl. Optics.
11. G. Horlick, Anal. Chem. 48 (1976) 783 A.
12. C.F. Dewey, Jr., Opt. Engineering 13 (1974) 483.
13. M.C. Teich in R.J. Keyes, ed., Optical and Infrared Detectors, Springer-Verlag Berlin - Heidelberg - New York, 1977, p. 229.
14. T.R. Gurski, H.W. Epps and S.P. Maran, Appl. Optics 17 (1978) 1238.
15. D.W. Fradin, P.P. Chenausky and R.J. Freiberg, IEEE J. Quant. Electr. 11 (1975) 631.
16. K.O. Tan, D.E. Rothe, J.A. Nilson and D.J. James, Conference on Laser Engineering and Applications, June 1 - 3, 1977, Washington, D.C.; paper 6.11.

SESSION 4

ELECTRO-OPTICS IN MEDICINE

Session Chairman
Dr. F. W. Hofmann
Siemens AG
Erlangen, West Germany

Session Co-Chairman
Dr. L. H. J. F. Beckman
NV Optische Industrie de Oude Delft
Delft, The Netherlands

PERFORMANCE CHARACTERISTICS, IMAGE QUALITY AND DEVELOPMENT TRENDS IN X-RAY IMAGE INTENSIFIERS

Dr. W. Kuhl, Ir. J. E. Schrijvers and Ir. B. van der Eijk
N. V. Philips, Development Image Intensifiers
Eindhoven, The Netherlands

Introduction

In medical practice X-rays are used for diagnostic purposes because of their penetrative power.
Interaction with matter which converts the X-ray energy into a more usable form

X-RAY DETECTION

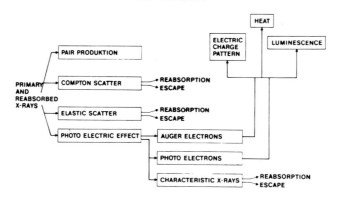

Fig. 1. Interactions of X-rays with matter

is the only possible way of detection.

The smallest dose that still yields the necessary diagnostic information must be used because of the well-known hazards of ionizing radiation. This requires efficient detectors. The film-screen combination used in full size radiography is rather suitable but real-time X-ray imaging with fluoroscopic screen is not. In fluoroscopy the demagnifying X-ray image intensifier has been applied for over 25 years now.

In todays image intensifier systems the X-rays are absorbed in a CsI detection screen, generating visible light that is used to emit electrons from the photocathode. The electrons gain energy while being accelerated and focussed to form a demagnified output image simultaneously. The visible light from the output phosphor is directed, via multi-channel optics, to TV, cine- and spotfilm cameras.

Modern X-ray image intensifier systems are as well suited for quantum-noise limited TV-fluoroscopy at low X-ray dose rate as for spotfilm fluorography made from the output screen instead of full size radiographs with adequate image quality in both applications.

Image quality

A description of the quality of displayed images requires a detailed discussion of those performance characteristics that are part of the imaging function.

Gain

The bare minimum for fluoroscopic perception limited by X-ray quantum noise for the dark adapted human eye at normal viewing distance is about 10^5 visible photons for every X-ray quantum. Thus even ideal conversion of all X-ray energy into visible light would not be sufficient.

After the recommendation of ICRU and standardization by IEC gain is defined as conversion factor expressed in (output) luminance in $cd \cdot m^{-2}$ per (input) exposure rate in $mR \cdot s^{-1}$ at 7 mm HVL-Al. For image intensifiers with unity magnification a conversion factor of 1 $cd \cdot m^{-2} \cdot mR^{-1} \cdot s$ corresponds with the above stated minimum gain.

The relevance of the defined conversion factor may be at stake when the devices for picking up the output image have spectral sensitivities differing from that of the human eye (TV pick-up tubes, photographic film).

Resolution

Determining limiting resolution of any imaging device is a subjective test. It seems to be easy at hand and leads to the conclusion: the higher the number of linepairs per millimeter resolved, the better the image quality. Overrating limiting resolution even leads to wrong developments.

A much more informative characteristic is the Modulation Transfer Function T. Despite the fact that it is well-known and frequently discussed it is, unfortunately, not yet standardized.

As a result MTFs can hardly ever be mutually compared and may be useless to assess image quality.

According to its definition the normalized MTF has unity value at zero spatial frequency. Most of produced MTFs are normalized to unity at a finite lower frequency, leaving out the Low Frequency Drop which is a prominent part of the genuine MTF of any image intensifier, much more so than e.g. with optical lenses.

This normalization error in many cases is caused by the fact that the MTF measuring set-up is seldom adequate to determine LFD and modulation factors at high spatial frequencies at the same time. Often then the large area contrast ratio is measured, e.g. as prescribed by NEMA, instead of LFD.

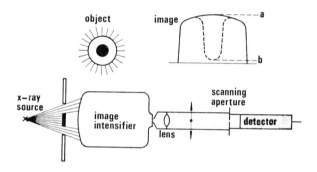

large area contrast ratio = a/b

Fig. 2.   Measurement of large area contrast ratio

But depending on the relative contributions from the various sources causing more or less far reaching "veiling glare" (halation at the output window, scattered radiation at the input window etc.) it is evident that there is no one-to-one correlation between the LFD and the large area contrast ratio for image intensifiers of different construction. And it is definitely the LFD that affects image quality.

As mentioned already limiting resolution has very little bearing on image quality. If a single number of greater relevance in this respect had to be looked for it would be, in our experience, the Spatial Noise Equivalent Bandwidth that is the integral over the square of the MTF. Taking the LFD into account or leaving it out results in SNEBs that differ by more than 50% for most contemporary electronoptical X-ray image intensifiers.

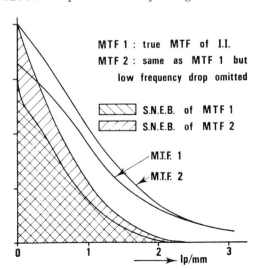

MTF 1 : true MTF of I.I.
MTF 2 : same as MTF 1 but low frequency drop omitted

S.N.E.B. of MTF 1
S.N.E.B. of MTF 2

M.T.F. 1
M.T.F. 2

Fig. 3.   A true image intensifier MTF and the SNEB derived from it compared to the same curves but with omission of the LFD.

Noise

The latent information in X-ray diagnosis is carried by X-ray quanta obeying Poisson statistics. Because of the limited admissible or available amount of X-ray energy, and the high energy per X-ray quantum needed to penetrate the human body, the number of X-ray quanta available in the radiation image is low. Hence the latent information is relatively poor compared to what we are accustomed to receive as visible information under normal viewing conditions.

According to fundamental laws of physics the signal-to-noise ratio at the output of a real detector is always poorer than that at the input. This holds also true for each component of a composite system (as X-ray quanta $\longrightarrow$ electrons $\longrightarrow$ visible photons).

On condition that the gain of an image intensifier intended for fluoroscopy exceeds a lower limit the efficiency of exploitation of the incoming X-ray flux can be determined by the noise power factor F defined as

$$F = \frac{(S/N)^2_{in}}{(S/N)^2_{out}} \tag{1}$$

The noise power factor can be measured by determining the S/N ratio at the output of the image intensifier and comparing this to the input S/N ratio as measured by an ideal detector.

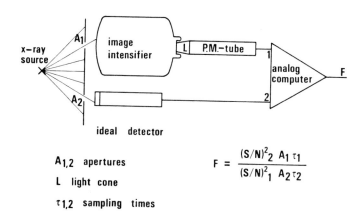

$A_{1,2}$ apertures

L light cone

$\tau_{1,2}$ sampling times

$$F = \frac{(S/N)^2_2 \; A_1 \, \tau_1}{(S/N)^2_1 \; A_2 \, \tau_2}$$

Fig. 4. Measurement of the noise power factor F.

In a properly designed image intensifier F is almost entirely determined by the X-ray transparency of the input window and by the amount $(g \cdot cm^{-2})$ and the nature (atomic numbers) of the primary detector. This first component where incoming X-rays are converted to subsequent signal generating particles (light photons) must be the quantum sink.

The thickness of the CsI layer is a compromise between the MTF of the input screen and absorption.

Incomplete absorption means a loss in signal-to-noise ratio due to mainly two effects. Firstly, X-ray quanta may penetrate the detector without exchanging energy at all – the information they are latently carrying is lost. Secondly, the smaller the absorptive power is for the primary X-ray quanta, the more of the generated eigen-radiation will escape, thereby increasing the variance of the output signal more than would be deduced from the absorbed energy of the primaries.

Perceptibility

Apart from the systems resolution a lower limit to the perceptibility of small details is set by the number of quanta in an X-radiation image. This fact has been clearly established by various researchers even working with visible light (e.g. A. Rose, Vision, 1973, Plenum Press, New York).

Because of the inherent small number of quanta in case of X-rays an image intensifier should have maximum attainable absorptive power.

Under conditions of low signal-to-noise ratio as is usually the case in medical fluoroscopy the detail-contrast perceptibility for a detail with the characteristic dimension x by means of an image intensifier with noise power factor F and modulation transfer factor $T\,(1/x)$ has been empirically established to be

$$(Cx)_{min} \sim \frac{\sqrt{F}}{T(1/x)} \tag{2}$$

where C is the radiation (input) contrast of the object detail. Reduction of F and increase of T in order to perceive smaller detail contrast at the same exposure rate are contradictory requirements because it means a thicker primary detector with non-the-less better resolution.

However, developments in the last ten years have been successful in reducing the lateral light propagation in CsI screens by controlled growth of suitable layer structures.

The radiation input contrast C is caused by local differences of density and/or atomic composition within an object that show up as differences in attenuation super-imposed upon that caused by the thickness of the transradiated object. If various influences of geometry and scatter are neglected C can be approximated by

$$C = 1/2 \; x \; (\mu_x - \bar{\mu}) \tag{3}$$

x = irradiated length of detail
μ_x = attenuation coefficient of detail
$\bar{\mu}$ = mean attenuation coefficient of surrounding matter

Consequently the Wiener-spectrum of such details is more or less hyperbolic

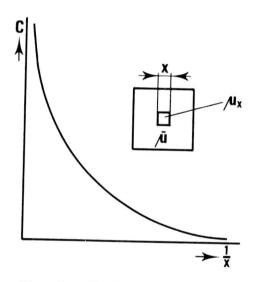

Fig. 5. The X-radiation input contrast of a detail as function of its characteristics dimension x.

in other words, the spectrum is "pink", whereas in natural visible scenes object dimensions and -contrast can be independent of each other, leading to a "white" spectrum. This may raise questions as to what a meaningful spatial bandwidth may be for X-ray imaging systems.

The new generation of multi-mode image intensifiers.

Characteristics of modern image intensifier systems

The use of 100 mm spotfilm recordings of the image intensifier output screen instead of full size radiographs has made high demands upon the image quality to be gained by the image intensifier system. These are being met in the newest generation.

To the fluoroscopic and cinematographic properties of the image intensifier are added the convenience of fast change-over from fluoroscopy to fluorography and the feature of the electron optical zooming capability.

The 14" X-ray image intensifier

A rather recent development is an image intensifier with an input diameter of 14" in order to provide anatomic coverage for most medical indications like e.g. abdominal angiography and colon examinations.

Examples of some major design problems were
a. the X-ray transparancy of the input window
b. the chromatic aberration of the electron optics, and
c. the low frequency drop in the MTF.

The solution of combining the necessary mechanical strength to withstand atmospheric pressure and high X-ray transparancy in order to achieve small X-ray scatter was attained by using a thin (0.25 mm) titanium membrane as input window. Its transmission of 85% even compares favourably with that of the usual glass windows of smaller image intensifiers.

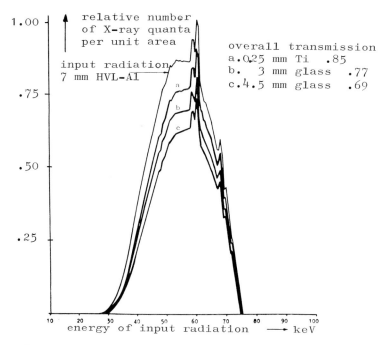

Fig. 6. The transmission of titanium and glass for 7 mm Al HVL X-rays

This results in a very small contribution to the LFD from the input section.

Chromatic aberration is a consequence of the non-zero initial energy of the photo electrons. The MTF T_{EO} of the electron-optics due to the chromatic aberration can be approximated by

$$T_{EO} = \exp\left[-0.7 \; \frac{\overline{E}_o \cdot \nu}{F_c} \right] \qquad (4)$$

$\overline{E}_o$ = mean initial energy of the photo-electrons
F_c = photocathode field strength
ν = spatial frequency at the cathode

The only parameter that can be influenced is the cathode field strength

$$F_c \sim \frac{V_A}{L.R} \qquad (5)$$

V_A = anode voltage with respect to cathode
L = electron-optical tube length
R = demagnification factor of the tube

Chromatic aberration in general is worse for larger tubes and greater demagnification factors.

The choice of a relatively short tube led to large geometrical aberrations like pin-cushion distortion and image curvature, due to the large diameter to length ratio.

In order to minimize the LFD in the MTF the use of a fiber-optics output window was required. Making it plano-concave at the same time enabled the correction of afore mentioned geometrical aberrations.

As a result the new 14" image intensifier has negligible LFD as compared to conventional tubes with glass input and output windows.

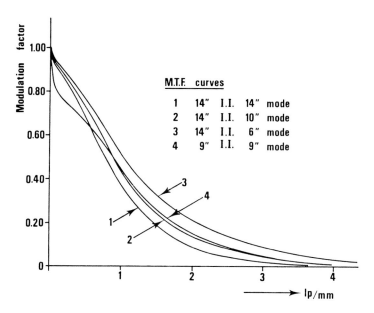

Fig. 7.   The MTFs of the 3 modi of the 14" X-ray
image intensifier compared to the MTF
of the 9" image intensifier

Trends

The first X-ray image intensifiers had an input screen of 5" diameter and were intended
for fluoroscopy and cinematography. Radiologists, especially in Europe, demanded larger
formats and this led to the 9" and 10" tubes.

Now the availability of the 14" format with its excellent image quality opens up
the field of large format image intensifier fluorography and this will support the trend
to 100 mm spotfilm fluorography instead of full size radiography.

In the past the need for better detailcontrast perceptibility has frequently been
translated into requirements for higher limiting resolution only. Recently however,
a trend can be noticed to decrease the noise power factor also.

The crucial importance of the modulation transfer factor of the spatial frequencies
under 1 lp/mm for the image quality in the medical field is better understood presently.
This will lead to extensive technological developments to decrease the LFD.

In an experimental 2" X-ray image intensifier an external CsI screen is coupled by
a fiber optics window to a photocathode in vacuum. This results in less low frequency
drop and a lower noise power factor at the same screen thickness due to the absent
absorbing vacuum envelope. The application of such a small image intensifier with very
high resolution lies in the field of micro-angiography and surgery of the extremities.

Large format video fluorography is unacceptable without a high resolution television
chain that also allows for image processing. This requires relatively cheap electronic
mass memories and hard-copy devices, probably on non-silver base. The monitor photo-
graphy that is the only available solution today will not be adequate in the future.

X-RAY IMAGE CONVERTER WITH VIDEO-SIGNAL OUTPUT

B. Driard, J. Ricodeau, H. Rougeot
Thomson-CSF, Electron Tube Division
Paris, France

Abstract

Imaging systems incorporating X-ray image-intensifier tubes are widely used in medicine and industry. The image quality given by these systems is comparable to that given by film/intensifier screen combinations.

In conventional equipment, the optical coupling system between the intensifier and the image pickup and recording system can appreciably degrade the quality of the final image. The result is reduced contrast and sensitivity, loss in resolution, particularly towards the edges of the field, and vignetting. This equipment is also bulky and heavy, and prone to change in optical alignment with time. Even using fiber-optic coupling with a special vidicon is not the perfect solution, as fiber size can limit the resolution, and dead fibers cause blemishes.

In this paper, we describe a new tube in which an electron-beam scanning section, similar in design to that of a vidicon, is incorporated into the intensifier. This beam reads out the charge-pattern image that is created on a silicon target by the photoelectrons.

With this tube, doses on the order of 1 to 4 μR per image, and 100 to 300 μR per image are required for fluoroscopic and fluorographic operation respectively. The limiting spatial resolution (5 % modulation), obtained by measurements with bar charts, is 32 lp.cm^{-1}.

Introduction

The image quality obtainable with systems using X-ray image-intensifier tubes, together with television readout or 100-mm film recording, is now comparable to that given by standard radiographic film/intensifier screen combinations. Because of improvements in tube performance, particularly so far as contrast, spatial resolution and spatial-resolution uniformity are concerned, the final image quality is limited principally by the characteristics of the other components of the system. For example, the optical image-transfer system can introduce vignetting, and resolution loss towards the edges of the image.

The operation of such systems may present certain problems. They are quite bulky, due to a large extent to the optical system and to the recording camera, and loading and unloading the film in close proximity to the patient may not always be convenient or desirable.

A system in which the intensifier tube directly produces a video output signal, this then being displayed on a cathode-ray tube for photographic recording, would have certain obvious advantages. With such a system, which would be extremely flexible to operate, an intermediate visible image would no longer be supplied because the intensifier tube would detect the X-photons and convert them into a useable electrical signal. Consequently, the bulky, image-degrading optical-transfer system would no longer be required. In addition, the output signal could be processed in various ways before recording.

Requirements of a Video-Output X-Ray Image Intensifier

The principle of operation of a system using a video-output X-ray image intensifier is shown in Figure 1. The only components near the object field are the video-output intensifier itself and a control monitor. The recording, processing, and principal display equipment are located some distance away.

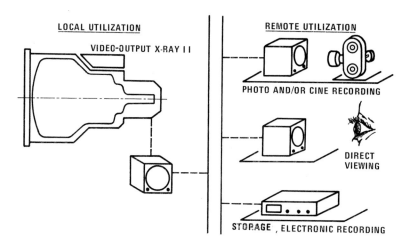

Figure 1 - Different ways in which the video-output X-ray image intensifier can be used.

In fluoroscopy, the requirements are the same as for when a standard X-ray image intensifier is used. Images should be obtained with dose rates of 20 to 100 μRs^{-1}, that is to say with 1 to 4 μR per image.

For fluorographic operation, the image recorded on the 100-mm film must be of very good quality, with a spatial resolution on the order of 40 lp.cm^{-1} (with respect to input field), and with no perceptible noise. To avoid noise problems arising from the quantum nature of the X-rays, doses per image of 100 to 300 μR are required.

It must thus be possible to vary the gain of the intensifier tube within wide limits, so as to cater to the different dose rates found in fluoroscopic, fluorographic and cineradiographic operation.

The line standard of the television system is fixed by the spatial resolution required for photographic recording (600 and 900 line pairs per diameter for input fields of 15 cm and 22 cm, respectively). The television image frequency should be kept as low as possible so as to reduce the bandwidth of the transmission system. The operating conditions are summarized in Table I.

Table I - Characteristics required of the all-video imaging system.

EXPLOITATION CHARACTERISTICS	FLUORO-SCOPY	FLUOROGRAPHY			35-mm CINE
		15-cm FIELD	22-cm FIELD	32-cm FIELD	
SPATIAL RESOLUTION NEEDED AT INPUT lp.cm^{-1}	15 to 30	40	40	40	25
CORRESPONDING LINES/IMAGE	875	1200	1800	2400	875
IMAGE RATE s^{-1}	25 to 30	1 to 10	1 to 10	1 to 10	50 to 100
DOSE PER IMAGE μR	1 to 4	100 to 400 105-mm spot film	75 to 200 105-mm spot film	50 to 150 105-mm spot film	10 to 30

Design Principles of Video-Output X-Ray Image Intensifiers

Tubes meeting the previously discussed requirements can be made in various ways. One method is to make an intensifier tube with a fiber-optics output window, this then being coupled directly to a camera pickup tube. The use of fiber optics, however, limits the spatial resolution, and dead fibers degrade the appearance of the image.

Because of this, we decided to make a different type of tube consisting of an image section and a scanning section, separated by a silicon diode array target, the whole being enclosed within the same evacuated envelope. The structure of this tube is shown in Figure 2.

The Image Section

The image section resembles a conventional X-ray image intensifier. The X photons are detected inside a high-absorption CsI screen

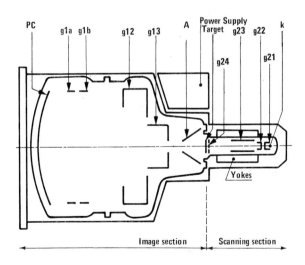

Figure 2 - Diagram of the video-output tube.

and are converted into light photons. These excite a photocathode that is deposited on the CsI screen. The photoelectrons so emitted are accelerated and focussed, after which they bombard the silicon-diode-array target.

The Silicon-Diode-Array Target

This component is of great importance, and its characteristics exert a major influence on the overall performance of the tube. Basically, it is a 25-mm diameter silicon wafer that has been thinned down to a few microns thickness. A diode array faces the scanning section, and a continuous signal electrode is deposited on the opposite side.

This assembly of reverse-biased diodes constitutes a matrix of resistive-capacitive elements connected to the signal electrode. The latter consists of a material whose thickness has been chosen such as to slow the electrons down to a suitable energy level.

The photoelectrons from the image section bombard the target and create p-type carriers that migrate towards the diodes and discharge them. The scanning electron beam recharges these diodes, and a signal current is thus generated.

Characteristics of a Video-Output X-Ray Image-Intensifier Tube

Figure 3 shows a video-output X-ray image-intensifier tube with an input field of 15 cm. It has a housing that incorporates both magnetic and X-ray shielding. The high-voltage power supply is included inside the housing and the TV-camera electronics are mounted around the scanning section. The photocathode is at a negative high voltage, and the signal electrode is grounded.

This technique results in an unitary device that is extremely compact (length 36 cm, diameter 22 cm) as compared to the bulky systems incorporating optical image-transfer sections.

Figure 3 - The TH X1413 video-output X-ray image intensifier.

Conversion Factor

The silicon target is capable of giving a much higher gain than that necessary for radiological applications. The required conversion factor depends on :

- the X-ray dose needed to form an image,
- the maximum charge that can be stored by the target.

For a 15-cm tube, and with the doses already mentioned, the conversion factor is several thousand electrons per incident X-photon in fluoroscopic applications, whereas it is only a few hundred electrons in fluorography.

The variation in conversion factor is obtained by changing the multiplication factor of the silicon target. For this reason, the input face of the target carries a barrier layer. Figure 4 shows the way in which the target gain varies with barrier-layer thickness, and with the photoelectron accelerating voltage.

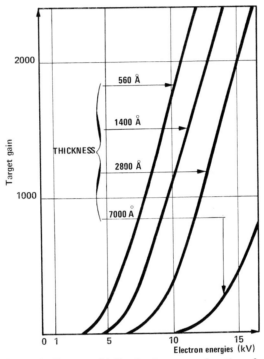

Figure 4 - Target multiplication factor as a function of accelerating voltage and barrier thickness.

Figure 5 shows how the signal current varies with dose for different values of accelerating voltage.

The changeover from fluoroscopic to fluorographic operation involves dropping the accelerating voltage by about 3 kV (from 13 to 10 kV). To obtain high spatial resolution, the electric field at the cathode must be quite high. Because of this, the electrons are first accelerated to 25 kV, and are then slowed down in a second section to give then the correct energy.

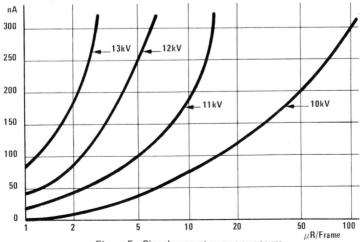

Figure 5 - Signal current versus exposure.

Because the target is grounded, the gain is adjusted by varying the photocathode potential (negative high voltage). Figure 6 gives the dose per image as a function of photocathode potential for various signal currents.

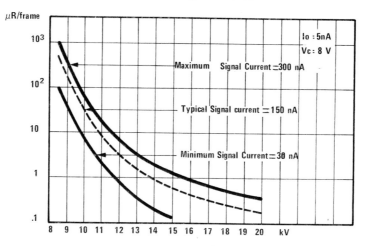

Figure 6 - Dose per frame as a function of the photocathode voltage, for a defined signal current.

Spatial Resolution

A limiting spatial resolution (5 % modulation) of 32 lp.cm^{-1} is obtained when looking at bar charts. Figure 7 shows the MTF of the tube, and of its constituent parts.

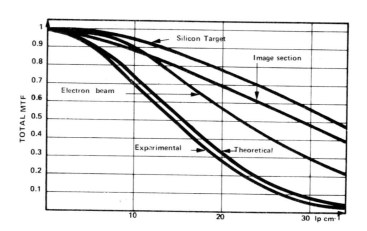

Figure 7 - MTF of the different sections of the video output X-ray image intensifier.

In the image section, which resembles a conventional intensifier without an output screen, the MTF is limited by the input screen characteristics, and by electrooptical aberrations. The MTF of the target is limited by target thickness, diode pitch, and the diffusion length of the carriers.

In the tube discussed, these various characteristics are such that for a spatial resolution on the order of 30 lp.cm^{-1} the target is not the primary cause of resolution degradation. Loss in resolution is mainly due to the electron beam.

Noise and Lag

Like conventional X-ray image-intensifier tubes, this video-output tube is designed so that tube-generated noise is very low as compared to noise due to X-photon fluctuation.

Quantum noise is of particular importance in fluoroscopy. It depends on the detection of the X-rays by the input scintillator, and also on the temporal characteristics of the tube. A high-absorption input scintillator (70 % for a HVL of 7 mmAl) is thus used. The temporal characteristics depend on the capacitance of the target, and hence on the resistivity of the target material. In this tube, target material resistivity is 10 Ω cm, and the lag is 15 % at 60 ms for a signal current of 200 nA (see Figure 8).

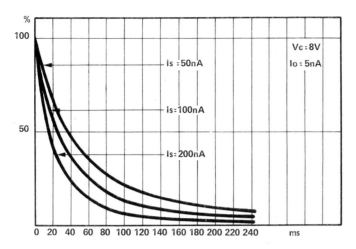

Figure 8 - Typical lag characteristics.

In fluorography, quantum noise is limited by using doses on the order of 100 μR per image, so that sufficiently high quality images can be obtained. It is thus essential that noise generated by the tube and its associated electronics be negligible. This result is obtained by scanning the target at a slow rate (5 to 10 images per second), so as to reduce the bandwidth of the signal transmission channel.

Using a Video-Output Tube

The appreciable reduction in bulk and weight with respect to conventional systems makes video-output tubes much simpler to use. This is one of the major advantages of these new tubes.

In fluoroscopy, a conventional radiological TV system is used. The output image is noteworthy for its high contrast, and lack of vignetting, due to the suppression of optical lenses.

In fluorography, displaying the final image on a TV monitor leads to operational flexibility, because no film manipulation near the patient is required. Using the new tube, 100-mm photos can be taken using low doses. The conversion factor can, if desired, reach a very high value, which permits darkening the film with only a few microroentgens. The dose to be used depends only on the acceptable noise level.

The capabilities of the tube are only fully exploited if a very high resolution, low-noise TV system is used.

One interesting feature of the video-output tube is the possibility of integrating and storing images by cutting off the scanning electron beam. The storage time is on the order of several tens of seconds, being limited by integration of the dark current. This dark current depends on the target voltage, and on the temperature. It can be made negligible by cooling the target. Figure 9 shows the variation of dark current with target voltage.

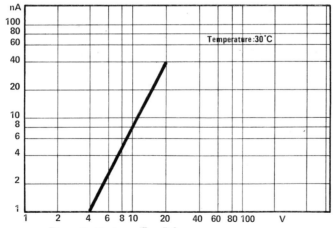

Figure 9 - Dark current with respect to target voltage.

Finally, the use of a video-output tube permits using signal processing techniques, such as gamma correction and contour enhancement, before displaying or recording the image.

Conclusion

Although the spatial resolution in recordings made using this tube is still not completely adequate, the video-output X-ray image intensifier appears to hold much promise for the future.

HIGH RESOLUTION TV FLUOROSCOPY AND IMAGE STORING

J. Haendle, H. Sklebitz
Siemens AG, Bereich Medizinische Technik
Erlangen, West Germany

Abstract

The investigation of the image-quality determining sections of the X-ray TV chain is an important matter for the optimization of the whole system. The modulation-transfer functions (MTF) of each component as the most important factor for the evaluation of the separate sections of the X-ray chain were measured. Special attention was given to gain exact measurements of MTF and other behavior of image pick-up tubes. In the second part of this paper image-quality determining factors and development tendencies of appropriate video stores for high resolution are discussed.

Image Intensifier Television Chain

The quality-determining components of the X-ray TV chain are shown in Fig. 1. First of all we will consider the sections which influence the image quality of the direct fluoroscopic image. These are the image intensifier, the tandem optics, the TV-camera tube - this is generally a vidicon - , the video amplifier and finally, as the last section of the transmission chain, the image-display tube.

When storage of the video image takes place, the storage media also have a decisive influence on the reproduced image quality. These can be magnetic memories, i.e. video-tape-recorders and video-disk-recorders, or also in certain circumstances video-storage-tubes.

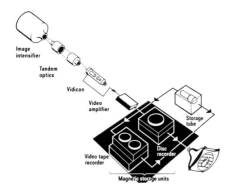

Fig. 1 Image intensifier television chain and video store units

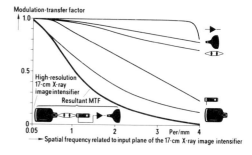

Fig. 2 Spatial modulation-transfer function (MTF) High-resolution X-ray TV system

Spatial Modulation-Transfer Functions

Since the spatial modulation-transfer function has turned out to be the most important factor of image quality, in this paper we should like to discuss the sections mentioned here above all in terms of this image-quality parameter.

Fig. 2 shows the modulation-transfer functions of the sections of a present-day high-resolution X-ray TV chain, consisting of a high-resolution 17-cm X-ray image intensifier and a high-resolution X-ray TV system.

The modulation-transfer functions have been measured separately for each individual section. The spatial frequencies are in each case related to the input of the 17-cm X-ray image intensifier.

The curve at the bottom is the resultant modulation-transfer function applying for the whole transmission chain.

If we look at the sequence of the sections - going from high to low transmission quality -, then first of all there is the video amplifier, to which a modulation-transfer function can likewise be assigned for vertically oriented structures. It can be seen that a modern, wide-band video amplifier in this case no longer has an effect on the image sharpness. Next comes the high-resolution image-display tube which, as with the new tandem optic, likewise brings hardly any loss in quality. On the other hand, more transmission losses are caused by the high-resolution vidicon and finally also the high-resolution X-ray image intensifier, even though this is a modern high-resolution image intensifier having a limiting resolution of 6 to 7 Per/mm.

The individual sections of the high-resolution X-ray TV system have to be better than the X-ray image intensifier, so that the output image of the high-resolution X-ray image intensifier can be reproduced on the TV monitor as far as possible without major losses in quality.

With this representation, the coarse contrast behaviour of the X-ray image intensifier is not included in the modulation-transfer function, because the modulation-transfer function has been normalized to 1 not at the spatial frequency zero, but at 0,05 Per/mm. This is appropriate for the comparison of the X-ray image intensifier with the sections of the TV system. That is because with the TV system is practically no such coarse contrast behaviour, and of more interest is the attainable maximum resolution which can be obtained e.g. for appropriate collimation of the X-ray source.

The advances which have been achieved for the individual sections of the high-resolution 1249-line X-ray TV chain in comparison with a 625-line standard TV system are shown in Fig. 3. The modulation-transfer functions associated with the individual sections are drawn underneath in each case. The continuous curves belong to the high-resolution X-ray TV system. The dotted curves show the modulation-transfer functions of the sections of the 625-line standard TV system.

The curves on the far right of Fig. 3 are the resultant transfer functions for the whole chain. These resultant curves show that the limiting resolution - that is the spatial frequency for which the modulation-transfer function has fallen off to about 4% - has doubled for the high-resolution X-ray TV chain compared with the standard TV system.

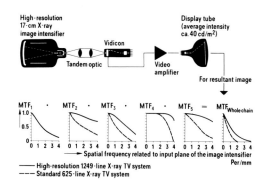

Fig. 3 Spatial modulation-transfer functions (MTF) X-ray TV system

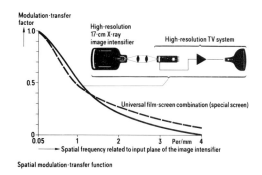

Fig. 4 High-resolution X-ray TV chain and universal film-screen combination

How far we have come with regard to the resultant modulation-transfer function for the high-resolution X-ray TV chain we see in Fig. 4 in comparison with the dotted modulation-transfer function of a universal film-screen combination with special screen. The MTF of the high-resolution X-ray TV chain, already comes quite close to the modulation-transfer function of the film-screen combination.

Investigation of the TV Image Pick-Up-System

Now it shall be demonstrated for the high-resolution vidicon, specifically using the results of measurements, how we would proceed and which problems and image-quality parameters must be investigated to get hints for optimization possibilities.

The point-spread functions of vidicon tubes is to be seen in Fig. 5. The dotted curve is the point-spread function of a high-resolution 1" vidicon, the other curve is the point-spread function of the 1" vidicon of a 625-line standard X-ray TV system. The radii for the 1/e value of the functions show that the radius for the high-resolution vidicon is almost twice as small. The same deflection focussing coil assembly was used during both measurements.

The effects of different distributions of the magnetic focussing and deflection fields on the modulation-transfer function for one and the same high-resolution vidicon is also important to investigate. We thus hope to gain some useful hints on how a deflection focussing coil assembly can be optimized for a high-resolution vidicon.

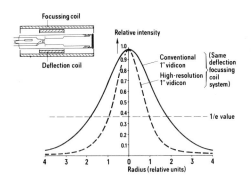

Fig. 5 Point-spread function of the vidicon including deflection focussing coil system

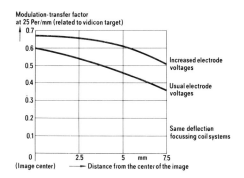

Fig. 6 Sharpness distribution for increasing the vidicon electrode voltages

For such measurements with high-resolution vidicon tubes and coil assemblies, however, we first had to construct a new special measuring unit which permitted for the first time the detection of even the slightest differences appearing with system changes. With conventional measuring units for vidicon tubes, such accurate measurements have previously not been possible, above all because of the high spatial frequencies, here involved.

Virtually a real TV camera must be developed for each deflection focussing system, since with the high bandwidth necessary for measurement e.g. video amplifier and coil system matched to one another must form a unit.

With the pick-up system here realized , frequency response, gain etc. of the video amplifier can be checked in the real operating state. Also the vidicon parameters can be measured at controlled target temperatures, even below ambient temperature.

For optimization of the high-resolution vidicon including the deflection focussing coil assembly, among other things the sharpness distribution over the whole target area must be accurately measured as an important parameter. This permits the determination of influences on the sharpness distribution, and appropriate empirical optimization attempts can be carried out. An example of measurements for the sharpness distribution over the image area of the high-resolution vidicon is shown in Fig. 6. Here, the modulation-transfer factor for 25 Per/mm is plotted as a function of the distance of the position of measurement from the center of the image. The electrode voltages of the vidicon, for example, were altered as parameters.

Considerably higher electrode voltages lead to higher image sharpness and better sharpness distribution.

With the measuring unit developed specifically for measurements on high-resolution vidicons, as already mentioned, the slightest improvements can be detected, which would not have been possible earlier because of the standard measuring units unsuitable for measurements at the highest spatial frequencies.

Similar problems have had to be investigated and optimization problems solved for the image-display tube and the associated deflection yokes, for which a new specific measuring unit was likewise to be constructed.

Video Stores

The use of video memories in X-ray diagnosis is increasing strongly. The video memory has a considerable effect on the quality of the reproduced image. This applies all the more so to the high-resolution X-ray TV image.

Hitherto existing magnetic video-memory systems, whether video-tape-recorders or video-disk-recorders,were essentially designed for the 5 MHz bandwidth usual with broadcast television. For the storage requirements of high-resolution 1249-line X-ray TV system, however, a considerably higher bandwidth is necessary.

Storage Tubes

If only the storage of an individual X-ray TV image is required, also a memory tube can be used, for which an increase in the electrical bandwidth has been relatively simply achieved by improving the electronic circuitry.

However, the frequency characteristics, the modulation transfer functions and signal-to-noise ratios of appropriate storage tubes measured by us contradict the use of these storage tubes for high-resolution X-ray TV systems.

In Fig. 7 the relationships are displayed in schematic form.

With the magnetic video memory the signal-to-noise ratio is higher than with the storage tube. With the storage tube, the relatively poor signal-to-noise ratio, i.e. a low number of resolvable grey steps, is determined by the target structures, vignetting etc.

The frequency characteristics plotted underneath show quite clearly that practically no decrease is present up to the frequency limit of the magnetic memory; only then a steep fall off takes place.

In contrast to this, with the storage tube a decrease of the transmission characteristic appears even with relatively low frequencies. The doubly effective influence of the electron optics with the writing and reading process is responsible for this.

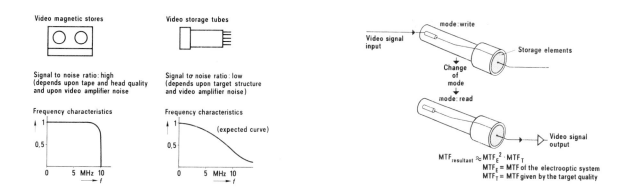

Fig. 7 Image quality of video magnetic stores and video storage tubes

Fig. 8 Modulation transfer function of
video storage tubes

These relationships with respect to the resultant modulation-transfer function for the memory tube can be seen in Fig. 8. The modulation-transfer function of the electron optics and the modulation-transfer function of the target structure are effective for the charge distribution obtained at the target as result of the write mode. The modulation-transfer function of the electron-optical system is again effective to a first approximation in the read mode.

On account of these disadvantages with storage tubes, we have turned intensively to the development of high-resolution magnetic stores.

Magnetic Stores

The development of a laboratory sample of a high-resolution video-disk-memory showed that, above all with improved electronics and also an increased rotational speed, a doubling of the electrical bandwidth from 5 to 10 MHz can be attained with relatively high signal-to-noise ratio.

Since, however, the video-disk-memory has only restricted possibilities of application in radiology and, possibly would only be used in a relatively small number, on account of the high price, we saw no chance for the high-resolution video-disk-recorder.

We have, however, transferred the electronic experience gained to the video-tape-recorder, without changing the mechanics, and have been able to double the video band-width from 5 to 10 MHz, without alteration of the relative speed of video head to video tape. Despite the high bandwidth, it was even possible to improve the signal-to-noise ratio further. Here, the technological advances with video tapes and video-head ferrites came to our aid.

The electrical frequency responses of the video memories can be reproduced as modulation-transfer function for the horizontal direction of the image. Fig. 9:you see as a dotted curve the modulation-transfer function of the hitherto existing 5-MHz video-tape-recorder. On the right of this is the curve for the high-resolution 10-MHz video-tape-recorder. Additionally, for comparison, is shown the resultant modulation-transfer function of the high-resolution X-ray image intensifier TV chain.

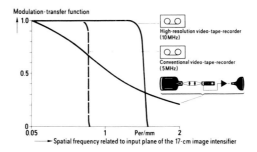

Fig. 9 Spatial modulation-transfer function (MTF)
High-resolution X-ray TV system and high-
resolution video-tape-recorder

The spatial frequencies are again related to the input screen of the 17-cm X-ray image intensifier.

With the increased image quality of the video store, it pays for the first time to use as high an image dose as with X-ray cinematography. Then with the same image dose the image quality of the stored video images is no longer mentionably different from that of the X-ray cinematography image. Therefore even today X-ray cinematography could be replaced by video-tape recording.

When it is a question of only storing and replaying a single image with X-ray TV, digital semiconductor memories could prevail with high-resolution X-ray TV some time in the more distant future.

At present, however, with the high data rate of the high-resolution X-ray TV, lying in the order of magnitude of about 100 to 200 Megabit/sec, an economic product solution is not yet possible.

Summary

This paper presented a brief insight into advances and work in the field of high-resolution X-ray TV technology including image memories. Therapeutic vascular catheterizations, particularly those developed recently, which are being extended specifically to the smallest vessels, have on the side of the user led increasingly to a requirement for increased resolution with fluoroscopy. It can be foreseen that in addition high-resolution X-ray TV systems with their high image quality will be a necessary prerequisite for future image evaluation or image processing.

References

1.   Pohl, D., "Die Grenzen der Detailwiedergabe bei Bildverstärker-Fernsehanlagen". Fernseh-Kino-Technik 30 (1976),9, 301 - 304
2.   Meyer-Ebrecht, D.W. Spiesberger, "Neue Verfahren in der medizinischen Röntgen-diagnostik - eine Herausforderung für die industrielle Röntgenprüftechnik." Materialprüfung 19 (1977), 10
3.   Delgambe, R., "La troisième génération des amplificateurs de brillance. J.Belge de Radiol. 58(1975), 101 - 105

4.   Pfeiler, M., J. Haendle, "State of the Art and Development - Tendencies of X-Ray Image Intensifier Television Systems".Electromedica 5 (1975), 148 - 157

5.   Haendle, J., H. Horbaschek, M. Alexandrescu, " Das hochauflösende Röntgenfernsehen und die hochauflösenden Videospeicher". Electromedica 3-4/1977), 83 - 91

PHOTOELECTRONIC IMAGING FOR DIAGNOSTIC RADIOLOGY AND THE DIGITAL COMPUTER

S. Nudelman, M. P. Capp, H. D. Fisher, M. M. Frost, and H. Roehrig

University of Arizona Health Sciences Center

Tucson, Arizona 85724

Abstract

A digital image processing facility is being installed at the University of Arizona to serve the needs of clinical practice and research for improved imaging and diagnosis from radiology. In addition, this facility will support similar needs from other image generating services throughout diagnostic medicine that use ultrasonic devices, gamma cameras, and thermography. Attention is drawn to the structure of this system and its design for dealing with the most severe problems facing diagnostic radiology. These include the input of radiological images to a digital computer, storage, and display. Possible solutions are discussed in terms of utility, performance, economics, and acceptance by the diagnostic radiologist.

Introduction

Our program began during the summer of 1973 with a study of the status of imaging for radiology. It included an examination of film, computerized axial tomography (CAT), photoelectronic imaging devices and systems, and the role of the digital computer for image processing. The study was pursued over a period of about 18 months and came to the following conclusions:

(a) The most likely significant new advances for improved imaging in radiology are likely to come from direct reproduction methods, utilizing photoelectronic imaging devices, systems, and techniques.

(b) Image processing for feature and contrast enhancement is a fertile area for new research directions and offers substantial promise for improved diagnostic performance.

(c) Indirect imaging methods, such as computerized axial tomography (CAT) and coded apertures, will become more important because of the concentrated research and development programs emphasized by industry.

(d) Film-based radiography is clearly the state of the art, and an abundance of research prevails both in industry and university. However, on the negative side is the feeling that further significant improvements are not likely, and that the cost of film-based radiology is rapidly becoming very expensive.

From the above conclusions and influenced by the strength of our experimental experience, the decision was made to pursue a program of research in radiology making optimum utilization of photoelectronic imaging devices and digital image processing. This combination appeared to offer the likelihood of being able to extract and present to the viewer, any information present in the x-ray image. Accordingly, it should do as well or better than film. The main limitation to be faced was in not being able to provide axial tomography in any simple, inexpensive manner. On the other hand, we anticipated resolving objects with contrast as small as that discernible from CAT, and stereo systems could be improved to the point where they would provide attractive new applications to radiology.

Medical Imaging

We soon realized that this Department of Radiology was becoming in practice a department of medical imaging. In addition to the imaging devices used in diagnostic radiology (x-rays) and nuclear medicine (gamma rays), we were actively involved in ultrasonic imaging, thermography, and video. These were all being used in our examination and therapy rooms as well as in research. Ultrasonics in particular has expanded dramatically within the past year to include three different types of imaging devices and has become a separate division within the department.

Examination of imaging requirements and procedures utilized in other departments throughout the Health Sciences Center has revealed a wide variety of applications with images acquired in the ultraviolet, visible, and near infrared spectral regions. They involve studies in such diverse departments as ophthalmology, internal medicine, dermatology, and surgery. Analysis ranges from examining features of cancer to observing and quantifying the movement of red blood cells in microcirculation.

Thus it is clear that in planning for the establishment of our research laboratories and the selection of our digital computer, it is necessary to allow for expansion to accommodate imaging needs and services throughout our Health Sciences Center.

Goals

The immediate result of that early study was to set three short-term goals, and one long-term goal. They were the following:

(a) Establish laboratories for research in (1) photoelectronic imaging devices and systems devoted to improved imaging for medicine, and (2) medicine that uses state of the art imaging devices and systems.

(b) Acquire a digital computer system able to (1) accommodate research on medical images acquired throughout the Medical College, and (2) service clinical imaging activities throughout the University Hospital.

(c) Undertake programs of research and development in support of practice at the earliest possible time.

(d) After achieving short-term goals, undertake to expand the program through establishment of a center for medical imaging to include a multi-faceted education program directed toward meeting the needs of medical students, physicians in practice, and graduate students.

Facilities

The research laboratories are outlined in Figures 1 and 2, and are essentially divided into a laboratory for research in medicine and biomedicine (Figure 1) and another for research in photoelectronic devices and systems applied to medicine (Figure 2). They are the result of an ongoing expansion in which all activities were housed formerly in the laboratory of Figure 1.

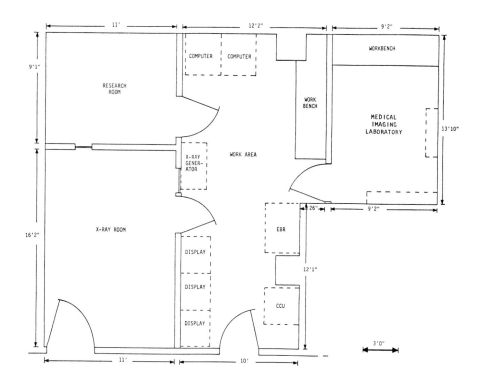

Fig. 1. X-ray digital video laboratory.

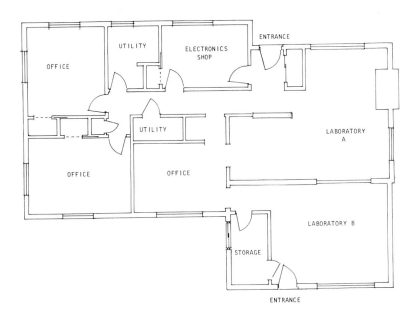

Fig. 2.   Research laboratory for photoelectronic imaging in diagnostic medicine

Imaging for Radiology

Activities in x-ray imaging are particularly exciting at this time. They can be summarized with reference to Figure 3, where four possible x-ray sensors are shown, each in combination with some video capability. We have seen, for example, within very recent years, the emergence of new rare earth phosphors, [1], and the production of the proximity focus x-ray intensifier, [2] the 14" diameter Philips x-ray intensifier, [3], and the Thompson-CSF x-ray intensifier SIT camera tube. [4] On the video side, Siemens is offering a high-resolution TV camera for coupling to an intensifier, [5], and for large screen optics there is the new Electrodelca lens. [6].

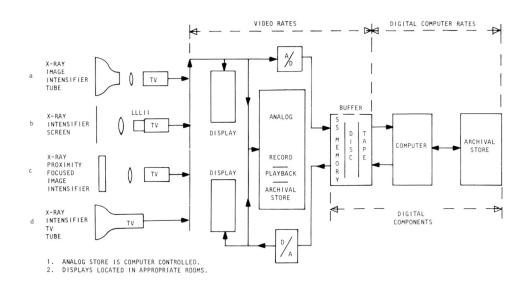

Fig. 3.   Diagram of four x-ray video digital camera systems.

The time is ripe for innovation, and there is a rare opportunity to explore these new devices for both old and new applications to medicine. In general, there is the feeling that one cannot predict with any certainty ultimate performance nor the winners versus the losers from the viewpoint of dominance in the field. In a large part, our research is directed toward: (a) monitoring and evaluating device performance, (b) determining their best areas of application, and (c) undertaking to demonstrate their capability by assembling and utilizing systems appropriate for important procedures in medical imaging.

The main features of the x-ray laboratory are: (a) a selection of state-of-the-art x-ray intensifiers permitting trade-offs between contrast factor, resolution, and image diameter, (b) a selection of x-ray tubes with spot sizes ranging from as small as 50 μm focal spot to as large as 1.2 mm with the latter able to accept a maximum heat loading of 1.8 million heat units, (c) three video cameras with associated optics having combined features which make available video electronic bandwidths up to 65 MHz selective raster rates, raster line number and aspect ratio, and a dynamic range expressed in peak signal to RMS noise in excess of 500:1. The latter is particularly important because the largest dynamic range anticipated from clinical x-ray images is expected to approach 500:1. We are fortunate in having achieved a measured performance of 800:1 and anticipate being able to demonstrate in excess of 1000:1 with a new low noise preamplifier.

The three cameras are paired with x-ray intensifiers to provide different features including: (a) snap-shot operation for static studies needing to deliver the best spatial resolution possible (in excess of 2000 TVL), (b) dynamic studies requiring continuous imaging with best contrast rendition (maximum dynamic range) and with modest spatial resolution (e.g. 512 x 512 TVL), (c) either snap-shot or dynamic capability with the versatility to trade off between spatial resolution and contrast resolution as needed for any particular procedure. A list of the current intensifier-camera tubes that can be accommodated is presented in the table below. These systems can be rearranged with appropriate optics to accommodate any medical requirement forseeable.

Intensifier-Video Tube Combinations

SYSTEM	X-RAY INTENSIFIERS		VIDEO		MAIN PURPOSE
	TYPE	DIAMETER (inches)	CAMERA & BANDWIDTH	TUBES[d]	
1	Siemens 17H		SRL #1[a] 60 MHz	2" Vidicon[a] 1.5" FPS	1. General duties 2. Snap-shot & dynamic studies 3. Combined resolution and contrast matched to medical requirements
2	Thomson-CSF TH-9428D	9-6-4	SRL #2[b] 60 MHz	2" Selenium Vidicon West. WX-32620	1. Static studies 2. Snap-shots 3. Maximum spatial resolution
3.	Philips 14"	14-10-6	Sierra[c] 30 MHz	Plumbicon Amperex 45XQ	1. Dynamic studies 2. Continuous video 3. Best contrast resolution

a) SRL camera #1 provides a capability to set the raster rate, TV line number, and aspect ratio to match imaging requirements. The camera is set up to accept vidicon type tubes and coils whose maximum dimensions do not exceed those of the Westinghouse 2" vidicon WX-5140.

b) SRL camera #2 provides the same capability as #1, with the additional features of being able to accept return beam tubes, and the very large high resolution tubes such as the Westinghouse SEC 75 mm photocathode tube, model WX-32193, and Schade's return beam vidicon [7]

c) The Sierra camera can be set up to accept any 1.5" tube, such as the Chalnicon, Newvicon, Saticon, and Vidicon.

d) The tubes listed have been acquired for current research programs in radiology.

The research room in Figure 1, adjacent to the x-ray laboratory, has an interconnecting window which is transparent to x-rays and is opaque to light. The room is light tight and is designed to accommodate research on photoelectronic systems by allowing x-rays to pass through the window while maintaining independent lighting controls. This arrangement facilitates setting up video cameras with x-ray intensifiers and screens for arrangements optimized to a particular medical task. It is also intended to suit our research on psychophysical testing of radiologists performing from video displays, compared to that from film.

Research for Medical Imaging

The imaging laboratory in Figure 1 is intended to satisfy our needs for research suited to medicine, other than radiology. Its principal imaging devices at this time are two Sierra 30 MHz video cameras with one extra head, and an AGA thermographic camera operating in the 8 to 14 μm region. The Sierra cameras are being set up to accommodate three kinds of vidicon type tubes, to take advantage of the different sensitivities, spectral response and lag characteristics available from chalnicons, newvicons, plumbicons, saticons, and vidicons. These cameras are being set-up to interface with optical imaging devices including those used in ophthalmology (e.g. the funduscope), internal medicine (e.g. endo-scopes), and pathology (e.g. microscopes).

The work area in Figure 1 is occupied by instrumentation including the computer satel-lite, a three-color display with memory, A/D converters, any interfacing apparatus between the different cameras and computer, 16 mm electron beam recorder and an analog video disc able to record for 20 sec with a minimal dynamic range of 50 db. These instruments combine to offer a variety of desirable features, such as the ability to snap-shot video frames into memory and display; to record continuously analog for 20 sec with wide dynamic range; to transmit a continuous video run of much longer duration to a Bell and Howell high density digital tape machine located in the main image processing computer facility (see Figure 4); to manage simple image processing (e.g. subtraction) in the imaging laboratory and complex processing at the main computer; and to offer an interactive display for image manipulation.

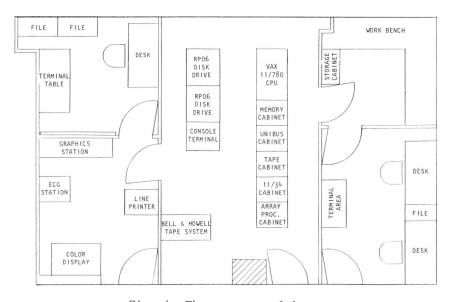

Fig. 4. The computer laboratory.

Photoelectronic Imaging Devices and Systems for Medicine

The arrangement of rooms shown in Figure 2 provides two main laboratories and four offices devoted to this research. Laboratory A is used for research in photoelectronic imaging devices which includes intensifiers, signal generating tubes and displays, and their combination. Laboratory B is devoted to evaluation of the performance of these devices and systems. An ongoing program, for example, is to evaluate the performance of x-ray screens and intensifiers on an absolute basis. Notable results to date on x-ray screens utilizing photon counting techniques provide an actual measure of screen gain and avoid estimates based on using the concept of conversion efficiency. The combination of laboratories A and B establishes the research environment necessary to meet the wide variety of needs in imaging emerging from medicine, in part by creating a working facility wherein physicians, scientists, and engineers can meet to discuss problems and work toward their solution.

Image Processing

The central computer laboratory for image processing is shown in Figure 4 and is built around a new DEC VAX 11/780. Major attributes include a floating point fast array processor, an RGB (color) interactive display, and a Bell and Howell high-density digital tape machine. The latter is particularly important for real-time studies since it is able to operate at video rates (30 frames/s) for a raster of 512 x 512 pixels, and 14-bit words per pixel. It has a speed reduction of 64:1 on playback to facilitate input into the computer. Furthermore it can acquire data of 7.5 frames/s, for a raster of 1024 x 1024 pixels, or 1.875 frames/s at even higher resolution of 2048 x 2048 pixels.

The complete system includes a group of satellites located near the imaging sources. Figure 5 illustrates the arrangement by a network between the VAX 11/780 and the different kinds of imaging laboratories and Figure 6 shows a typical satellite. The layout between the VAX 11/780 system and five of the satellites is shown in Figure 7.

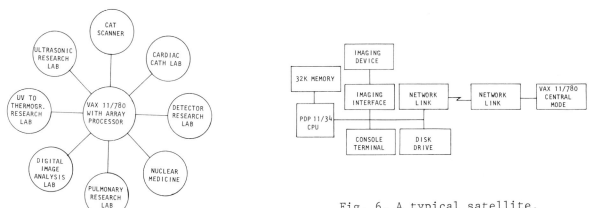

Fig. 5. The network

Fig. 6. A typical satellite.

Specifics of the Network

The core of the network will consist of a DEC VAX 11/780 computer. The computer system has 350 million bytes of disc storage, a 1600/800 BPI magnetic tape system, 256,000 bytes of core memory, and a floating point processor. In addition, the central facility has a directly viewed, video-refreshed 512 x 512 pixel x 8 bit color display, a vector graphics display, a line printer/plotter, low speed A/D and D/A converters, plus video and hard copy terminals. An array processor is included to provide the processing capability necessary for research in various image processing techniques (in particular, the nonlinear approaches now being investigated). The VAX 11/780 system will then function as a "network scheduler," queuing tasks for the array processor, providing a timeshare environment for general image analysis, and providing mass storage capability and hence a common data base for the acquisition nodes in the network.

Of special importance is the balance in capabilities between the general-purpose processor in the center of the network (the VAX 11/780 computer) and the array processor. The VAX 11/780 is the most advanced of the minicomputers, featuring cache and bus architecture that gives it the internal computation speed of a medium-scale computer.

The critical characteristics of the array processor are asynchronous operation in conjunction with the queuing or scheduling processor (the VAX 11/780 computer); pipelined arithmetic in a variety of data formats (e.g., integer, floating point, etc); bulk high-speed memory for buffer storage of image data supplied by the queuing and scheduling processors; and hard-wired or microprogrammed instructions for the most common image processing operations, e.g., convolution, fast Fourier transform, correlation, etc. An array processor with features such as this can be utilized as a special computational resource by the central computer. For example, a large block of image data can be retrieved from disk and transferred to the array processor by DMA channel. The array processor is then initiated, e.g. a series of Fourier transforms in an asynchronous fashion freeing the central computer to perform a number of tasks such as communications with the other nodes or scheduling other data or programs for the array processor. If the network is active, the central processor always has a certain background of system management to supervise, with array processor service requests being given the highest priority of interrupt. It is important to note that a lower capability computer in the center of the network probably is not adequate to manage the network and the array processor. Instead, capabilities rise from the periphery of the network moving inward, with the computer having sufficient resources to manage the network and to provide network access to the most powerful resource, the array processor. It is in this sense that it is believed that the network is balanced.

There will be eight real-time acquisition nodes in the final network providing research support for five different imaging modalities. These are nuclear medicine, diagnostic radiology (x-ray), thermography, ultrasonics, transaxial tomography, and radiation oncology.

The block diagram of a typical acquisition node is given in Figure 6. The major components include: (a) a central processor, (b) mass storage device, (c) real-time imaging interface, (d) control console, (e) network link. In general, components have been chosen to provide a mobile satellite able to accommodate present as well as anticipated needs for new research. PDP 11/34 processors will be used in all new nodes in the system. Features of this processor include hardware integer math, memory mapping hardware, floating point hardware option, compact physical size, and low cost. Each imaging modality has different requirements on mass storage speed and size. At present two different mass storage systems are used. These are (a) DEC RK06 disk system (7 million words with 250,000 words/sec transfer rate), and (b) DEC RL01 disc system (2.5 million words with 80,000 words/sec transfer rate).

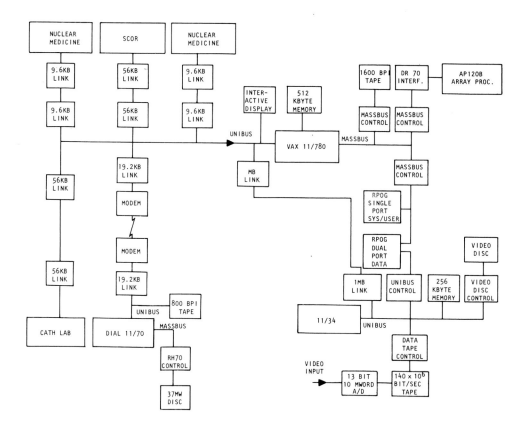

Fig. 7. Image processing computer system.

The imaging interface is unique to each application, as is the choice of control console. The network data transmission hardware consists of DEC DMC11 network links operating at 56,000 bits/sec rate. These provide DMA data transfer with internal error checking to reduce system overhead. They also allow for downline loading of programs and data to remote nodes that have no mass storage devices.

Benefits of having similar hardware throughout the network are (a) applications software written for any node can be used at any other node, (b) interfaces and other system components can be exchanged between nodes in the system without modification, (c) maintenance will be simplified by having redundant hardware throughout the system.

Research Programs

Non-Invasive Imaging of the Coronary Arteries for Detection of Atherosclerosis

We have assembled a system similar to that illustrated in Figure 3a, whose main function is to provide image enhancement by simple subtraction. It is designed to permit avoiding angiography, by a technique for obtaining radiographic images of a quality suitable for detection of atherosclerosis through intraveneous (rather than arterial) injection of contrast media. The procedure is to obtain first an x-ray video recording before injecting any contrast media, second another recording after injecting contrast media, and third a difference image by subtracting the first picture from the second. The resultant picture then depicts only the location of the contrast media, since all other features appear in both the first and second pictures, and are subtracted out. An example of the power in this technique is illustrated by Figures 8a, 8b, and 8c. They reveal an early result of injecting iodine (1 cc/kg of dog) into a vein located in the foreleg, and recording images of the heart. The difference picture clearly reveals the method's potential power for imaging, as shown by the major cardiac chambers and blood vessels. We have since succeeded in improving the system to obtain difference images obtained in real time, either by subtracting a first image (without contrast), for all those that follow, or by subtracting successive video frames from one another. The first run follows the contrast media as it appears, builds up and then disappears, while the second presents a video run, recording the change of iodine concentration between successive frames. Note that the difference image in Figure 8c is simply from subtraction and does not involve any digital processing planned for extracting and displaying specific features and/or edges.

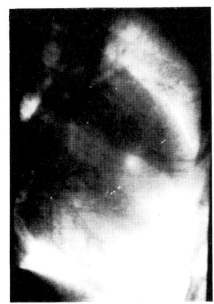 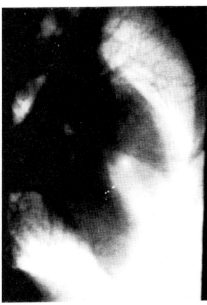 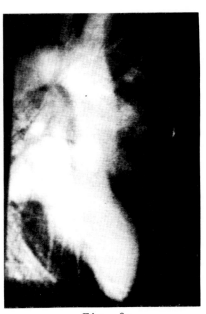

Fig. 8a	Fig. 8b	Fig. 8c
Dog's chest before contrast injection	Dog's chest after contrast injection	Subtracted image

A 16-kg anesthesized dog was used for this procedure. An 18-gauge needle was inserted into the external jugular vein of the dog. A scout image (8a) was obtained of the dog's chest using 75 kV, 50 mA, 1/30 sec. Then 16 cc (1 cc/kg) Renografin 76 was injected into the vein as a bolus. Six sec later a second image (8b) was obtained of the heart. The images were then subtracted to form the final image (8c). This was our first attempt to image the heart.

Photoelectronic (Video) Radiology

Radiology carried out with photoelectronic imaging devices (PEID) and digital processing of images is being given special attention in our active research planned for the near future.

The stimulation for this research is in three parts. First, an examination of the signal and noise levels in the diagnostic radiographic image was determined and found to be within the capabilities of pertinent PEID-s. In addition, resolution and motion requirements had to be estimated for each type of radiographic examination, but again it was found that in general, PEID-s could meet these needs. Second, a basic experimental study was made of performance from five of our radiologists to compare their reading of film with that from a display. The tests were set up to ascertain (1) the extent to which small contrast objects discernible in a film radiograph could also be read from a high-quality TV camera display system and (2) the effect of TV lines in a raster on their performance. Results indicate that indeed the radiologist could perform as well with one system as with the other. Third, a cost analysis of our diagnostic film based radiology facility was carried out and compared to that expected from a PEID-based system. The results indicate that the PEID-based system should prove substantially less expensive to set up and operate. Thus, on the basis of positive results from preliminary engineering, psychophysics, and cost studies, we decided to pursue an in-depth research program in this area.

Reference to Figure 3 provides an overview of the PEID-digital system envisioned. At the sensor end, it is clear that the Philips 14" tube operating in the 6" mode with a contrast factor of 36, causes a contrast loss of less than 3%. When one considers that just a few years ago intensifier contrast factors were only in the range of 10-12, and the intensive research activity prevalent on all sensors (i.e. a to d in Figure 3), one can anticipate substantial improvement. Video cameras and tubes are available today with sufficient dynamic range to permit discernment of objects with contrasts less than 1%. Furthermore, state of the art low noise preamplifiers should ensure that one system will serve the needs of both fluoroscopy and diagnosis. Spatial resolution of these tubes appears able to meet the widely different requirements of radiological imaging directed toward procedures such as for the chest, abdomen, bone, brain, and heart.

We are in the process of acquiring most of the instruments necessary to undertake complete psychophysical test program of radiologist performance from PEID-digital systems compared to film. All the digital equipment identified in Figure 3 is on hand, excepting for satisfactory digital archival storage. At the moment, the most likely candidate to fill the need is a laser beam disc recorder, such as that under intensive development by the Philips Corp [8]. Our displays need improvement for quantitative studies. However, we will be able to test with all imaging sensors excepting that in Figure 3d beginning in January 1979.

References

1. A.L.N. Stevels, New Phosphors for X-Ray Screens, Medicamundi, 20, 12 (1975).
2. S.P. Wang, C.D. Robbins, and C.W. Bates, Jr., A Novel Proximity X-Ray Image Intensifier Tube, Proc. SPIE, 127, 188, 1977.
3. W. Kuhl and J.E. Schryvers, Performance Characteristics, Image Quality and Development Trends in X-Ray Image Intensifiers, Proc. Fourth European Electro-Optics Conf., Utrecht, The Netherlands, Oct. 1978 (To be published by SPIE, 1979)
4. M. Driard, M. Ricodeau, and M. Rougeot, A Vidicon Output Integrated Vidicon X-Ray Image Converter, Proc. of 7th Symposium on Photoelectronic Imaging Devices, Imperial Col., London, 1978 (To be published by Academic Press, London, 1979.
5. J. Haendle and H. Sklebitz, High Resolution TV Flouroscopy and Image Storing, Proc. Fourth European Electro-Optics Conference, Utrecht, The Netherlands, Oct. 1978 (To be published by SPIE, 1979).
6. L.H.J.F. Beckman, The Electrodelca, A New Photofluorographic Camera with Image Intensification, Proc. Fourth European Electro-Optics Conference, Utrecht, The Netherlands, Oct. 1978 (To be published by SPIE, 1979).
7. O.H. Schade, Sr., Theory and Performance of High Resolution Return-Beam Vidicon Cameras, in Photoelectronic Imaging Devices edited by L.M. Biberman and S. Nudelman, Plenum Press, N.Y., Chapt 17, 1971.
8. G. Kenney, D. Lou, P. Janssen, J. Wagner, F. Zernike, R. McFarlane, A. Chan, An Optical Disc Data Recorder, see Proc. SID, Los Angeles, 1978 and Proc. CLEOS, 1978.

Acknowledgments

This work was supported by NIH Grant 5-P50-HL 14136/07, by NIH Contract 02 N01-HV-7-2931, and by FDA Grant 5 R01 FD00804.

THE ELECTRODELCA, A NEW PHOTOFLUOROGRAPHIC CAMERA WITH IMAGE INTENSIFICATION

L. H. J. F. Beckmann, A. J. Vermeulen
Oldelft, The Netherlands

Abstract

Photofluorographic cameras of the Odelca type have been widely used for more than 25 years in mass chest x-ray examination. Such cameras provide full diagnostic detail on a reduced size low cost photograph, but although they are equipped with extremely wide aperture optics, the required x-ray dose is somewhat higher than that for a full size radiograph. This disadvantage, which has limited the range of application of photofluorography in the past, is eliminated in the new Oldelft "Electrodelca". This camera includes an electro-optical image intensifier of the proximity focussed type, the output of which is fiberoptically coupled to the photographic film.
The design criteria for this approach are discussed and performance data are presented.

Introduction

In x-ray imaging for diagnostic purposes, it is important that the dosage to which the patient is exposed, is kept at such low levels as is compatible with the required information content of the image. It can be generally stated, that this goal has been achieved whereever modern x-ray image intensifiers can be applied to medical diagnosis. However, the range of sizes for which x-ray image intensifier tubes are presently manufactured, is limited by complexity and therefore cost. As is well known, sizes between 150 and 250 mm diameter are widespread and tubes of up to 350 mm diameter have been produced. The even larger sizes that would be needed to cover a complete chest x-ray, typically 400 mm x 400 mm, have so far been outside this technique, and chest x-ray images are presently taken either as full size photographs between x-ray screens at dose levels around 1 mR, or as reduced size photofluorographs, which, while saving photographic film, require higher dose levels by a factor of about 5.

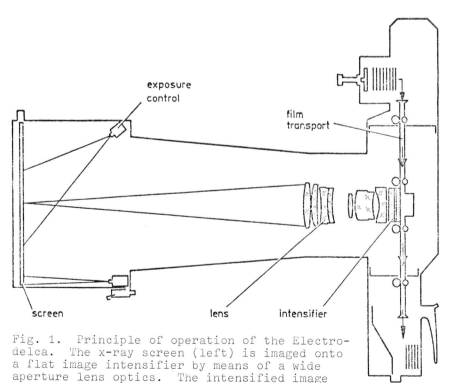

The subject of this paper, an instrument called the Electrodelca, is a new approach to chest x-ray photofluorography with image intensification and allows a substantial dose reduction for a reduced size chest x-ray. The basic principle is shown in fig. 1.
The Electrodelca has a flat external x-ray screen. The low brightness visible image obtained on this screen is transferred to the input of a single stage image intensifier by means of a wide aperture lens. The output from the image intensifier is fiberoptically coupled to the photographic film which stores the image.

Fig. 1. Principle of operation of the Electrodelca. The x-ray screen (left) is imaged onto a flat image intensifier by means of a wide aperture lens optics. The intensified image is fiberoptically transferred to a photographic film.

Electrodelca System Components

Screen. An external x-ray screen as used in this system gives us some extra flexibility, since it could be easily adapted for other than the chest x-ray application for which it was primarily conceived. This aspect is dealt with in the last paragraph.

Optics. The optics which demagnifies the screen image upon the image intensifier presented a particular design challenge. In this application, as in the classical photofluorography, a wide aperture is mandatory and essentially determines the design approach. The actual lens has a numerical aperture of 0.4, roughly corresponding to an f number of one. It is designed for a fixed optical demagnification of 4.6. Its resolution, or more exactly the modulation transfer function, could be kept high enough throughout the field, so that the lens does not essentially determine the overall system MTF.

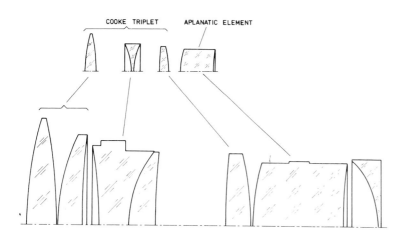

Fig. 2. The Electrodelca lens design. The basic triplet plus aplanatic magnifier, from which the design has been derived, is shown at the top.

The design, shown in fig. 2, has a total of 7 lens elements. It can be considered an extension of the well known triplet approach in which a negative lens element between two positive elements create a flat field while keeping positive power. A rather thick extra lens, the sixth element, further increases the optical power. It can be looked at as an aplanatic magnifier and the last element compensates for the residual image curvature, which is caused by the aplanatic magnifier.

The measured performance is summarized in table 1, which lists MTF values for different positions in the field referred to the object plane.

Intensifier tube. The image intensifier tube for the Electrodelca is a proximity focussed diode shown in fig. 3.
The tube has a useful field of 90 mm x 90 mm with rounded-off corners. The glass entrance window is 8 mm thick.
The output window is a fiberoptic plate 12 mm thick, so that the total thickness of the tube with 4.5 mm spacing between cathode and anode is only 24.5 mm. Image intensifier tubes of this type so far have found surprisingly little attention.

Table 1. Measured MTF of the Electrodelca lens referred to object plane

spatial frequency :	1	2	3	4 lp/mm
MTF on axis :	93	82	68	56 %
at 160 mm diameter:	79	63	50	38 %
at 290 mm diameter:	63	45	31	21 %

This may be due to the fact, that a processing chamber is needed, which allows for putting together the front and the rear parts of the tube in vacuum, and only after the cathode has been processed in an appropriate location while the output phosphor is shielded.

We use copper pinch-off techniques for sealing the two parts of the tube together while it is still in the processing chamber. The proper design of the processing chamber was, of course, a key factor in this development.

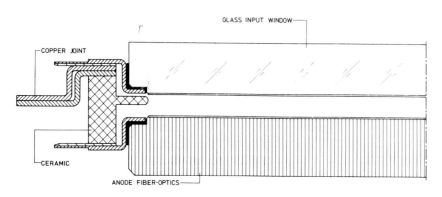

Fig. 3. Cross section of the proximity focussed image intensifier tube for the Electrodelca. The image field is 90 mm square.

The resolution capability of the proximity focussed diode is known to be governed by two parameters: the distance d and the voltage difference V between cathode and anode.[1,2] For a given spatial frequency f, the factor

$$\frac{fd}{\sqrt{V}}$$

is then a constant, which means that within certain limits we can trade off between spatial resolution and gain. The gain, of course, increases with the voltage across the tube. A limit is set by the field strength E = V/d, which must be kept below such values at which field emission causes excessive background brightness. For a background brightness under $3 \cdot 10^{-6}$ lm m^{-2} sterad^{-1}, the field strength could be raised to about 2.2 kV mm^{-1}. Combined with a cathode to anode distance of 4.5 mm, the voltage across the tube is approx. 10 kV which results in a tube gain of 15.

Resolution. For the above parameters the resulting MTF curves are shown in fig. 4 which summarizes the contributions of the various components to the overall system MTF.

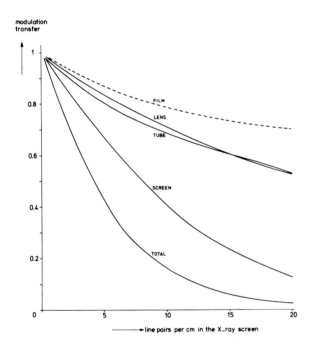

Fig. 4. Electrodelca system MTF and contributions of its components (measured values). All values are referred to the input plane.

The limiting resolution of the tube alone is close to 30 lp/mm which translates to 6 lp/mm in the input screen of the Electrodelca. These numbers are all quite conservative and apply to early pre-production models. It can be expected, that with more production experience the tube voltage may be raised and that both gain and resolution will be improved somewhat. The limiting resolution of the present system is about 2.5 lp/mm. It can be clearly seen from fig. 4, that this is largely determined by the properties of the ZnCdS x-ray screen.

Film. To record the image from the intensifier tube, a photographic film in the popular 100 mm x 100 mm sheet film format is pressed against the fiberoptic output of the tube.

Photon balance

Fig. 5 summarizes the photon balance of the Electrodelca. On a logarithmic vertical scale we have the number of paricles per image element to produce a density of one on the film. The size of the image element has been chosen to be 0.25 mm x 0.25 mm at the screen, which corresponds to the limiting resolution of the system. From left to right, we have the number of x-ray quanta which impinge on the patient, on the grid that filters scattered radiation, and on the x-ray screen. Next we have the visible light photons emitted by the screen, and those falling on the photocathode of the intensifier tube.

Finally, there are the photoelectrons in the intensifier tube and the photons which reach the photographic film. The largest number of visible photons occurs at the x-ray screen: about 2.10^6. At the final image, however, the photons are concentrated on a much smaller area, so that the luminosity is higher, although there are less total photons at the corresponding image point.

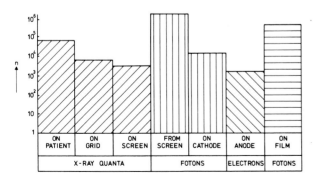

Fig. 5. Photon balance of the Electrodelca for an image element of 0.25 mm x 0.25 mm and for a film density of one (see text). Film: RP1S.

Results

As an indication of the image quality which is achieved in the Electrodelca, fig. 6 shows an x-ray picture of a skull.

This picture represents only a portion of 27 x 23 cm² in the x-ray screen. On the original, the x-ray grid (22 lines per cm) is discernable.

Fig. 7 shows an example of a chest x-ray taken with the Electrodelca. The screen dosis is 0.3 mR, which compares favourably with the 0.7 to 1 mR needed for a full size x-ray.

Although dose reduction in chest x-ray photofluorography has been the principal goal of the Electrodelca program, the gain provided by the image intensifier could be utilized for a different compromise between x-ray dose and image resolution.
As we saw in figure 4, the ZnCdS x-ray screen has by far the largest effect on the resolution of this photofluorographic system. This is because we would normally want the highest possible light output, which requires relatively thick screens.
If one puts resolution at a premium, however, one might use a finer screen with higher resolution but somewhat lower light output, which would be allowed due to the image intensification of the system. This then brings us to dose levels approaching those for full size chest x-rays, and thus still substantially lower than in conventionale photofluorography.

Experiments are currently under way to select x-ray screens for various compromises between dose level and image resolution.

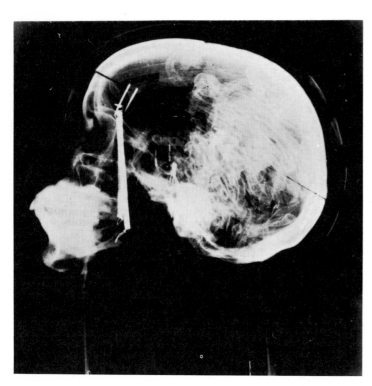

Fig. 6. X-ray picture of a skull taken with the Electrodelca. Detail of 23 x 27 cm from the x-ray screen.

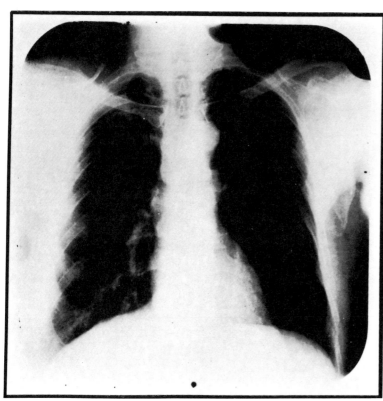

Fig. 7. Example of chest x-ray taken with the Electrodelca. Input format 40 x 40 cm.

References

1. Csorba, I.P., Applied Optics 16 (1977), 2647.
2. Eberhardt, E.H., Applied Optics 16 (1977), 2127.

LASER VIDEORECORDER FOR MEDICINE APPLICATIONS

Christian Slezak

Soro Electro Optics, Arcueil, France

Abstract

This new video recorder - the Videographe - provides the capability for transferring television images directly onto a photosensitive dry-processed paper. This laser recorder produces, with standard composite video signal, immediately visible recording with both high contrast and resolution (sixteen grey shades, 750 lines, 625 spots per line) in less than 35 seconds and in large dimensions (21 cm x 30 cm). This recorder will be very useful for medical applications (ultrasonic, X rays, thermal and nuclear imaging) giving immediate high quality and large dimensions images.

Introduction

The laser videorecorder Videographe is a new development in laser imaging technology. This apparatus is derived from the Visor line, presently on the market and oriented towards sonar, teledetection and meteorology applications. The Videographe is a completely independant equipment and can be directly connected to a standard video signal.

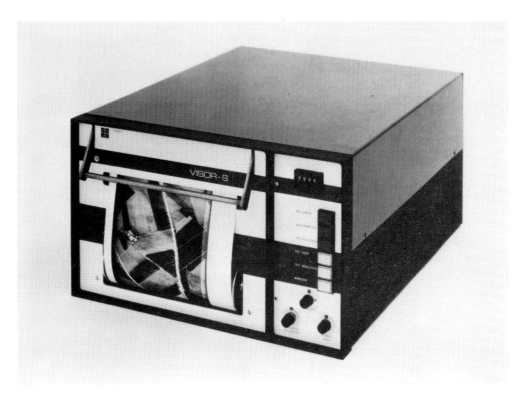

Fig. 1.   Videorecorder

Principle of operation

The principal application of this image recorder is to obtain from a TV screen a good quality image on paper or film in near real time. The slow scan television signal is delivered by a video sampler : conventional "real-time" television signals are sampled in order to achieve a large reduction in bandwidth. The recorder utilizes a low power laser beam, modulated by an acousto optic modulator and focused on the recording medium through a galvanometer scanning mirror. The recording spot produces a latent image record of processed input signal on the recording paper. Then this paper is driven into the thermal processor and the final image is then available in about 5 seconds.

Optical System

The major functions that have been realized in the optical system are :
- establishement of a basic recording energy source
- modulation of this energy source by the signals to be recorded
- focussing of the modulated energy source into a high energy density recording spot
- scanning of a recording paper by the modulated spot.

The laser is used as an energy source because it is extremely bright and is compatible with wide band intensity modulation requirements. The monochromaticity and spatial coherence of the laser beam make it possible to focus the wavefront into a high efficiency diffraction limited point which becomes the recording spot. In our recorder we use a low power, industrial type and reliable He-Ne laser. Wide band intensity modulation is accomplished by acousto-optic techniques. The Soro acousto optic modulator (model IM 20) and its amplifier control the exposure with respect to the intensity modulation response and gamma curve of the recording paper, producing continuous tones over a wide density range. More than sixteen equally spaced grey shades (from 0,1 to 1,5 D) can be recorded on the paper. Laser beam deflection systems requiring more than 1000 spots resolution employs galvanometer deflectors. Their low cost and optical simplicity together with a high resolution capability allow them to be used in a such application. The scanner we use is a servo-actuated moving iron galvanometer with position transducer. With the good accuracy of this scanner (linearity and repeatability better than 1 % of peak-to-peak excursion) and its performances in resolution (1500 spots per line of scan) and speed (200 lines per second), one can produce high quality images.

Electronics

The video sampler is a device that can extract pictorial information from a standard television signal at a rate compatible with the date rate of the scanner. The video sampler utilized a sampling technique to obtain a reduced bandwidth signal. With the 625 lines signal, the sampling rate is about 16,000 samples/second. The readout use an internal clock which is synchronized to the television horizontal rate and data is taken from both the two interlaced TV fields. As a result of the sampling technique used, the output data comes from a vertical sampling line that scans from the top to the bottom of the original TV raster. This sampling line is moved from left to right across the raster to complete a single frame.

Paper transport and processing

The paper transport is achieved by a low cost step motor, which is driven by the sampler. After the recording of one image, the paper is cut and developped in the thermal processor. The dry-silver paper is a high resolution, negative acting photorecording medium specially designed for systems where red sensitivity is desired ; the elimination of wet chemical development, the near real time heat processing and low cost are the keys reasons for the interest in this paper. This dry silver paper can achieve an optical density of 1,5 and can be processed in less than 5 seconds. The paper roll is 150 m long and one can obtain 500 pictures (21 x 30 cm) with one roll. In a new development we provide the possibility of getting smaller size images : 10 x 15 cm.

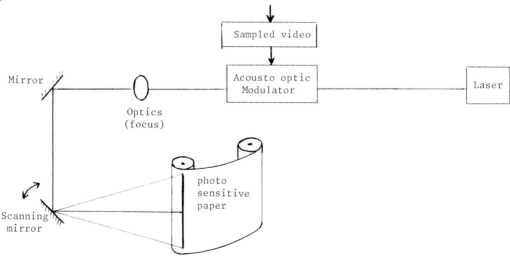

In conclusion, the necessary time to scan the entire video frame is 30 seconds. The development process lasts 5 seconds. One can obtain the picture in 35 seconds. The quality of the image is impressive, not only through its resolution but also by the shapes contrast and the gloss of the paper.

Applications

The range of medical applications for the Videographe is quite large.

. In the field of ultrasonic scanners, the Videographe is a very useful instrument. Due to its 16 grey levels the Videographe provides high quality images and can be connected to scan converter type equipment or to digital memory type equipment. In order to obtain an image, the operator may apply pressure on a pedal and thus keep his hands free. The cost of a 21 cm x 30 cm image is close to 20 american cents and the 10 cm x 15 cm image (in a new development available in February 1979) will cost less than 10 american cents. These costs are quite competitive with other imaging processes.

. In the field of X Ray scanners (head or body) the Videographe is also quite useful since one can obtain many hard-copies from the same medical examination with high quality and low cost.

. In the fields of thermography and nuclear medecine, the Videographe can also be used as well as in radioscopy (in orthopedic surgery for instance).

Thus, the Videographe is a suitable instrument for the modern radiologist who can connect this equipment to almost any other video equipment.

THE RETINAL RESOLVING POWER MEASURED BY LASER INTERFERENCE FRINGES

Bernhard Rassow

Department of Medical Optics
University Eye Hospital Hamburg
Martinistrasse 52, 2000 Hamburg 20, Germany

The use of lasers in clinical routine work of ophthalmology is known up to now only for retinal coagulation. I would like to present another application of the laser in Ophthalmology as a diagnostic tool.

Vision is an extremely complex process. The formation of an image on the retina is only a very simple part of the whole phenomenon. The optical information is coded by the retina and transferred to the brain, but we don't understand these processes in detail. The assesment of visual acuity usually includes the evaluation of all three parts of the optical pathway: optical media, retina and brain (fig.1). Since the days of Maxwell we know another way of optical observation. In the so called Maxwellian view (fig.2) the object is focussed in the nodal plaine of the eye. Therefore no clear image of the object comes to the retina. MAXWELL used this method to compaire the luminosity of stars.

LE GRAND [1] proposed to use the Maxwellian view with two beams, getting interference fringes on the retina if the two beams have the same wavelength and a stable phase relation, or in other words, if we use coherent light (fig.3). This idea of LE GRAND is nothing else than the old Young-double-slit experiment. What is the advantage for the measurement of visual acuity? The influence of the optical media decreases and everybody, the emmetropic as well as the hyperopic or myopic will have the same subjective impression.

But it took another thirty years since CAMPBELL and GREEN [2] realized the idea of LE GRAND. Using a helium neon laser they had a good coherent light source. Meanwhile several groups worked with this technic and LOTMAR [3] showed, that such pattern on the retina also could be produced by Moire-technic. But the methods were limited to small groups of devoted subjects as a result of the complex apparatus and the time-consuming experiments. We tried to construct a small and stable instrument for an easy applicable daily routine measurement in an eye-hospital.

The optical pathway of our device is shown in figure 4. The laser beam of an 1 mw helium-neon-laser is reflected at both surfaces of a glass plate. The both bundles created by this reflection are focussed in the nodal plane of the subjects eye. The interference fringes in the overlapping area are, as mentioned above, independent of the optics of the eye. Seeing the fringes means that the resolving power of the retina-brain system is sufficient for this frequency. A variation of the line-frequency is done by changing the glass plates by a revolving system. Glass plates of different thickness give different distances of the two light points in the pupil of the subjects eye. A small distance corresponds to low and a great distance to high frequency.

The method is not only useful for scientific research of physiological optics, as mentioned above, but also for two practical problems of the routine ophthalmological work.

1. This apparatus allows for quick and easy applicable screening test of the eye of an unknown patient. The threshold of frequency visible to the patient gives an information of the maximal capability of visiual acuity of this eye. We measured nearly two-hundred patients with clear ocular media and different visual acuity, to find the correlation of normal visual acuity and the so called "retinal visual acuity". The results are shown in figure 5.

About thirty-three lines per degree of vision angle do correlate with a normal vision.

2. The starting point for our activites in this field was another question, however. The two small bundles are able to penetrate even opaque media, particularly if there are microscopically small clear areas. This is just the situation we often find in the case of opaque ocular media, especially in the case of lens turbidity or cataract. The penetrating coherent light creates the interference fringes behind these opacities. Now we are able to measure the function of the retina and the brain in those cases, where the vision as well as the direct inspection of the fundus is disturbed.

The results of the examination is a diagnostic and prognostic aid for the ophthalmologist. But for a patient with cataractous eye the fringes are disturbed by granulation, created by the scattered coherent light of the laser beam. This may be the reason of an error of the subject. To avoid mistakes there is a device to change the direction of the lines by a dove prism. The patient will be asked which direction the lines have, rather than wether or not he is able to see the lines. Figure 6 shows the results of the measurement of the retinal visual acuity for nearly 300 patients, with an opacity of the optical media. The values of retinal acuity are in most eyes higher than the ones of optical acuity. This means that in all these cases the reduction of the visual acuity is caused most likely by the opacity. A smaller group shows a reduced retinal as well as optical acuity. In these cases either the opacity is to thick. so that neither optical nor retinal visual acuity is possible. or a macular disease is the reason for the poor vision.

With a smaller collective(about 150 patients)we took the retinal visual acuity before cataract-extraction and compared this with the normal visual acuity before and after surgery. The results are shown in figure 7 and 8. At the time of the first examination the visual acity in most cases is less than 0,2. After cataract-extraction most of these eyes show the predicted visual acuity of more than 0,7. All the points on the lefthand side of the diagonal are eyes with high optical density. The luminous level of the scattered light therefore was also high and the patients did not have the impression of fringes. It is not easy to explain the group of nearly thirteen per cent of points lying on the righthand side of the diagonal. In these cases the pre-operative measurement predicted a visual acuity which wasn't reached post-operative.

There seem to be three reasons for such a deviation.
1. A macular desease with progessive loss of visual function.
2. A retinal desease remaining only small spots of high resolving power, which are too small for normal visual tests.
3. The high contrast of the lines sometimes seems to generate an inadequate high visual acuity.

We are about to reduce the rate of errors by refining the method. We just now added a contrast-sensitivity unit. It allows for the variation of the contrast of lines. The threshold contrast for all given line frequencies may be measured. By this modification we might get a new kind of test of visual acuity which is more adequate to the daily use of human eyes. Testing vision generally means to find out the spatial resolution of our visual system with small optotypes of high contrast. But the objects we normally have to see are greater structures with smaller contrast. The interference fringes apparatus with the possibility of line contrast-variation will just test this quality of the ocular visual system.

Finally I want to mention two possibilities to improve the apparatus in another way. Up to now our test is a subjective one. that means we need the cooperation and the answer of the patient. We now try to change the test to an objective one in two ways.

The first one is to move the interference lines in a steady way on the retina of the subject. This generates an instinctive eye movement called "optokinetic nystagm", which may be indicated objectively by an infrared detector.

Our first results with the optokinetic machine are promising. but the method is rather delicate.

The second way to get better results is the derivation of cortical potential created by the interference fringes. To get an alternating potential – which is easier to amplify – the interference lines move with a given frequency. The amplitude is half of the line width. The measured potentials are disturbed by a high amount of electrical noise. We tried to avoid this difficulty by a fourier-analysis of the signal. By this way not the amplitude of the signal itself but the amplitude of the fourier-component of the exitating frequency is measured. If this component is significantly greater than the noise. we may be sure that the lines were recognized by the subject.

Summarising we may state, that retinal acuity measurement by laser interference fringes is just becoming a clinical routine method.

We think that the contrast sensitivity unit and – in future – the objectifying of the patients answer by visual evoked potential will give a higher propagation of the method.

References

1. Le Grand, I.. "Sur la mesure de l'acuité oculaire au moyen de franges d'interfé-rence," C. r. hebs. Séanc. Acad. Sci., Paris, Vol. 200. pp. 490-491. 1935.
2. Campbell, F. W. and Green. D. G.. "Optical and retinal factors affecting visual resolution," J. Physiol., London, Vol. 181. pp. 576-593. 1965.
3. Lotmar, W.,"Use of Moiré fringes for testing visual acuity of the retina." Appl. Optics. Vol. 11. pp. 1266-1268. 1972.

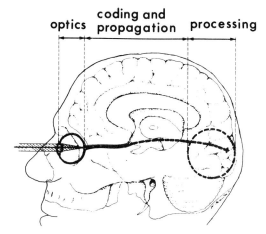

Fig. 1: Pathway of optical information in human visual system

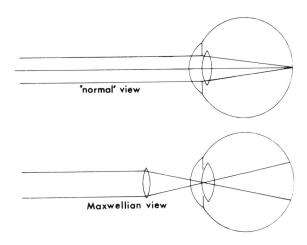

Fig. 2: Optical pathway of "normal"
and Maxwellian view

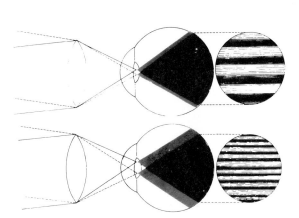

Fig. 3: Principle of interference-
fringes method

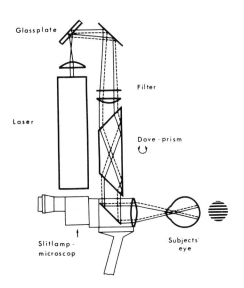

Fig. 4: Optical pathway of the
"Retinometer"

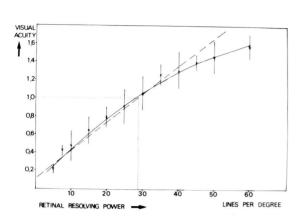

Fig. 5: Visual acuity vs retinal
visual acuity for subjects
with clear ocular media

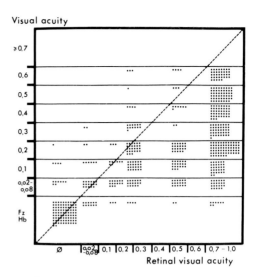

Fig. 6: Visual acuity vs retinal
visual acuity for subjects
with opaque ocular media

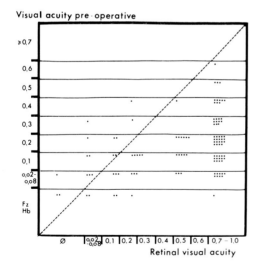

Fig. 7: Visual acuity vs retinal
acuity for subjects with
cataract as measured before
treatment

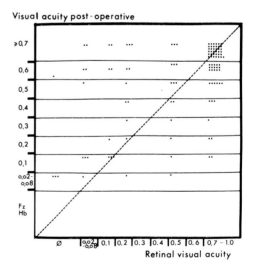

Fig. 8: Visual acuity post cataract
extraction vs retinal visual
acuity before treatment

FOURTH EUROPEAN ELECTRO-OPTICS CONFERENCE

Volume 164

SESSION 5

ELECTRO-OPTICS IN MEASUREMENT AND CONTROL

Session Chairman
Dr. Lionel Baker
Sira Institute Ltd.
Kent, United Kingdom

Session Co-Chairman
Dr. Hans Tiziani
Wild Heerbrugg Ltd.
Heerbrugg, Switzerland

SHAPE MEASUREMENT FOR HOT ROLLED STEEL STRIPS

P. Petit, G. Tourscher, M. Machet (IRSID, Maizieres-les-Metz, France)
M. Kassel (Delta, Strasbourg, France)

SUMMARY

A device that measures the hot rolled strips or plates shape was developed by the Irsid, in collaboration with the firm Delta, with the Ministry of Industry financial help.

It uses new technics : photodiodes areas for the sensor, a microprocessor unit for the signal processing.

An industrial prototype is set up on the hot rolling mill of the firm SOLMER at Fos-sur-Mer. It will allow to take up a new stage in the flat products rolling leading.

I. INTRODUCTION

The constant improvements carried out during those latest years in the opto-electronic components technology, bring an important evolution in the measure systems conception.

In particular case of the iron and steel industry, problems henceforth classical, as : the sizes measurements, the aspect inspection, the position control, etc... can be taken up in a new way with the appearing of new opto-electronic components, such as the photodiodes areas.

In addition, the association of those components to the means of data processing of a high level of integration, such as the microprocessors, gives the facility of treating more intricate problems, with equipments which cost remains reasonable.

So, the Irsid, in association with Delta, undertook the realization of a device intended to the hot rolled strips and plates shape checking "on line". The first prototype, described in this article, was realized with the State help, within the framework of a predevelopment agreement. It is now undergoing industrial tests on the hot rolling mill of FOS-sur-MER, with the SOLMER rolling mills and measure control departments help.

II. SHAPE NOTION

For the good understanding of the subject, it is useful to specify what you mean by a strip or plate shape. More exactly, you will refer to a good or a bad shape, or you will say the product is good-shaped, or on the contrary, it shows shape defects.

A strip or a plate is good-shaped when its surface is in contact everywhere with a reference plane ; if it is not, the product is affected by a shape defect.

In this article, we will only deal with the main defects appearing on the hot rolled products. They are of two kinds :

— long side (s) : one of the edge (or the two) shows an approximately sinusoidal wave ;

— long centre : the strip shows a series of wide buckles in the middle.

The appearance of those defects is bound to a thickness reduction, which is not uniform on all the product width ; the most loaded fibres lengthen more and undergo the shortest fibres cohesion strength, which causes a buckling phenomenon.

The longest fibres take then a wavy appearance.

That bad distribution of the reduction is mostly bound to an inadequate camber or, on the contrary, to a too sharp camber of the rolling mill cylinders, whence the main defects appearance at the edges or in the centre of the product.

III. SHAPE MEASUREMENT PRINCIPLE

Consequently, in order to check the strip or plate shape, you can be satisfied with measuring the length of three fibres in the rolling direction respectively located in the centre and at each edge.

A central fibre, longer than the edges average, will be typical of a "long centre defect" ; the opposite situation will correspond to a "long sides defect".

Moreover, a length variation between the edges fibres will reveal an asymetry of the rolling mill setting up (levelling).

The results of those measurements will then allow to undertake corrective actions, by acting on the cylinders camber (bending, thermal action, stands load modification) or (and) screwing modification of one stand side.

Thus, the shape gauge which was realized, is constituted of three sensors which function is to measure the strip surface vertical position on three points (edges and centre), in an almost continuous way. Through the moving, at a high speed, of the strip under rolling, the electric signals coming from the sensors are the corresponding three fibres images. Those images are shown to the mill operator on a video display unit for a shape first valuation.

At the same time, a microprocessor treats those signals. It computes the length of three fibres, taking the reduction caused by a coiler tension into account.

The computation results, in form of elongation and shape index, can be used either in monitoring, or in a closed loop control.

IV. DEVICE TECHNICAL DESCRIPTION

We shall especially describe the properly so-called gauge, specifying however the data processing is ensured by a microprocessor system INTEL 80/20.

A. *Sensors principle*

The measurement principle is described on FIGURE 1. A light source P is used in order to create a split-shaped light S on the product.

The optical system of that light source is suitably chosen in order that the light spot S is sharp in a field of some tens of centimetres.

A lens C, which axis is located in the specular reflexion direction of the pencil of rays issued by P, forms the S image S' on a photosensitive detector.

The point S position depends on the product vertical position. Particularly, when the plate is no plane (long side in the case on Figure 1), the spot will be in S1 and, consequently, the image of that spot will be transferred in S'1 on the photodetector.

The angle α being small, you can easily show the segment S' S'1 is directly proportional to the S1 height h, with regard to a reference plane (for instance : plane tangent to the rollers).

The detector D is constituted of a line comprising 512 photodiodes. To that line is associated a shift register which allows to scan the diodes state (lighted or not) from a clock. So, by computing and determining the detector lit zone, you can know the segment S' S'1 size and, consenquently, by knowing the apparatus geometry, deduct from it the S1 height with regard to the reference plane.

With the optical systems used on the device, the measurement field is 300 mm high, with a resolution that is consequently about 0,6 mm.

B. *The shape gauge (figures 2 and 3).*

So, it is constituted of identical three sensors, cameras and projectors are kept in water-cooled cases made of moulded alloy. A compressed air blowing avoids the optical system cloging. In order to fit the strip width (strip from 600 to 2 100 mm wide), the edges sensors are mobile. They move on highly linear slides and are moved through a motor-screw device. A logical unit ensures an automatic position from the data given by the rolling mill width gauge.

All those units are assembled in a movable box, embedded in the floor of a cabin that spans the product. That box is pressurized thanks to a turbine.

V. SAMPLE RESULTS

The figure 4 shows the recording of the signals issued by the sensors during the rolling of a strip on which shape defects were voluntarily created.

You easily discern the experience three stages on that recording :
- *creation of a long centre at the beginning of the strip ;*
- *good-shaped area ;*
- *creation of long sides successively at each edge.*

From those signals, we computed the parameters allowing to quantify the defects, i.e. the relative elongations of each measured fibre :

A.1   Operator side $A = \dfrac{\Delta 1}{1}$ $\Delta 1 = 1 - 1o$
A.2   Centre
A.3   Motor side   1 : measured fibre length
                        1o   length of a right fibre

And the shape and level indexes :

$$P = \frac{A1 + A3}{2} - A2 \qquad\qquad N = A1 - A3$$

The P signal indicates the defect nature, and its absolute value characterizes the amplitude.

P > 0 = long side
P < 0 = long centre

N expresses a "long side defect" asymmetry and, consequently, a stand lack of balance.

N < 0   Too tight on the operator side
N > 0   Too tight stand on the motor side

The Figure 5 shows the evolution of those parameters for the test taken into consideration.

VI. CONCLUSION

After a six months test period, the equipment reliability revealed itself satisfying. Minor mechanical modifications were brought on the mechanical and pneumatic servitudes in order to improve those results.

The device industrial exploitation will begin with the gauge definitive connection to the microprocessor which ensures the signals processing in real time. So, we shall have an equipment that should allow to achieve a further step in the strips and plates rolling mills conducting.

Downwards :

· operator side

. centre

· motor side

Product quick moving direction from the right to the left.

Scales :

. vertical scale : one division = amplitude of 14 mm

. horizontal scale : 1 cm = 5 m on the strip

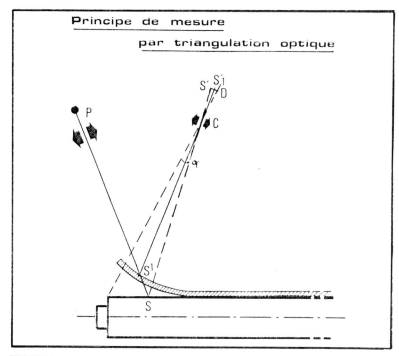

FIGURE 1 — MEASUREMENT PRINCIPLE BY OPTICAL TRIANGULATION

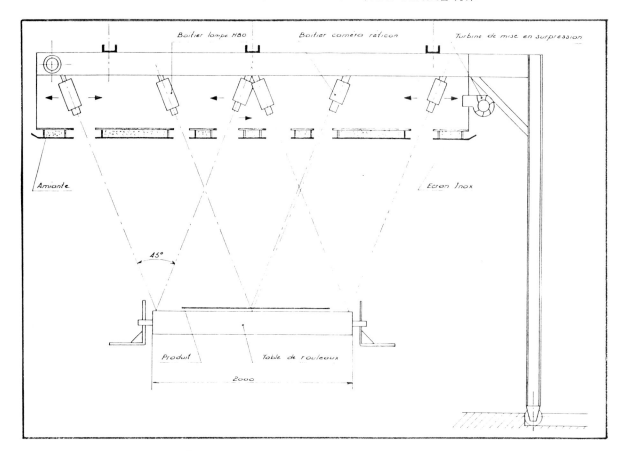

FIGURE 2 — GAUGE CROSS SECTION VIEW

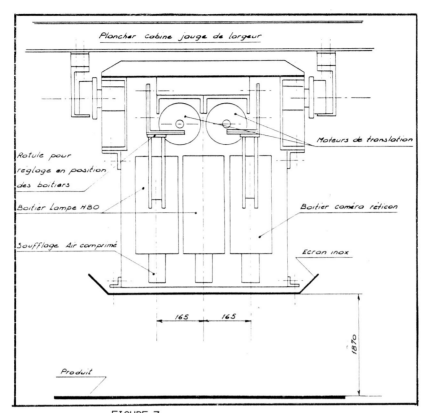

FIGURE 3 - GAUGE CROSS-SECTION VIEW

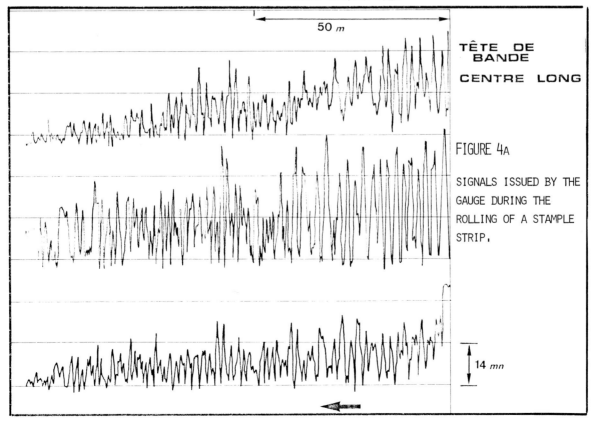

50 m

TÊTE DE BANDE

CENTRE LONG

FIGURE 4A

SIGNALS ISSUED BY THE GAUGE DURING THE ROLLING OF A STAMPLE STRIP.

14 mn

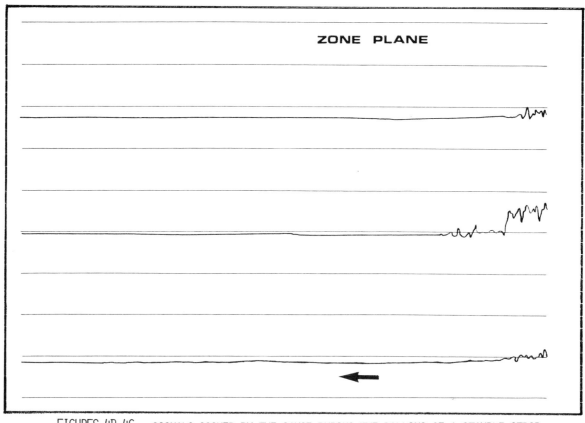

FIGURES 4B-4C - SIGNALS ISSUED BY THE GAUGE DURING THE ROLLING OF A STAMPLE STPIP

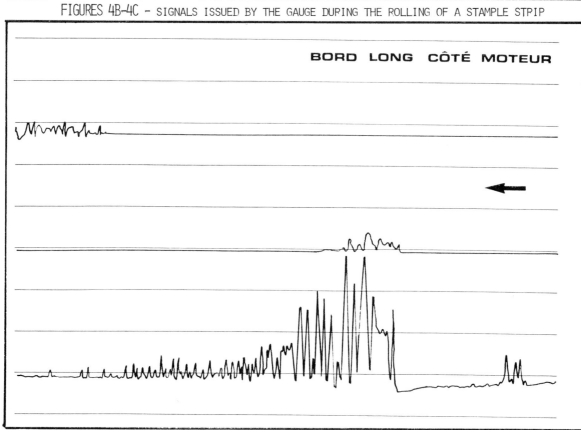

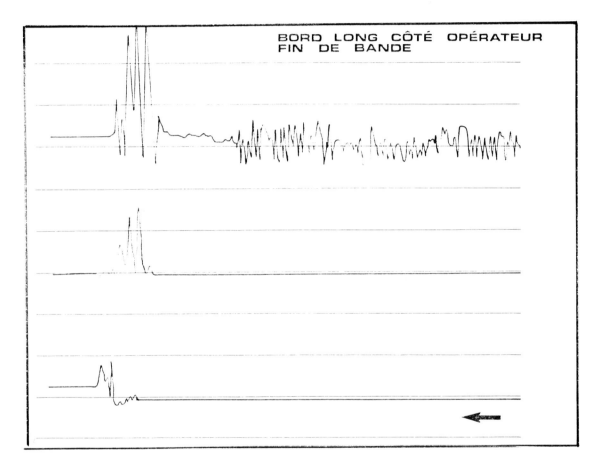

BORD LONG CÔTÉ OPÉRATEUR
FIN DE BANDE

FIGURE 4D – SIGNALS ISSUED BY THE GAUGE DURING
THE ROLLING OF A SAMPLE STRIP

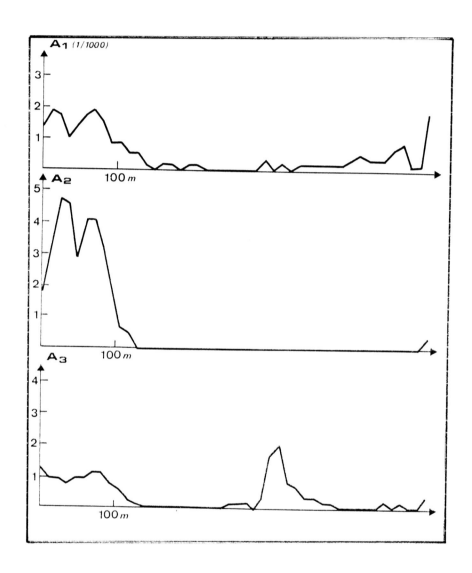

FIGURE 5A - RELATIVE ELONGATIONS

A1 OPERATOR SIDE

A2 CENTRE

A3 MOTOR SIDE

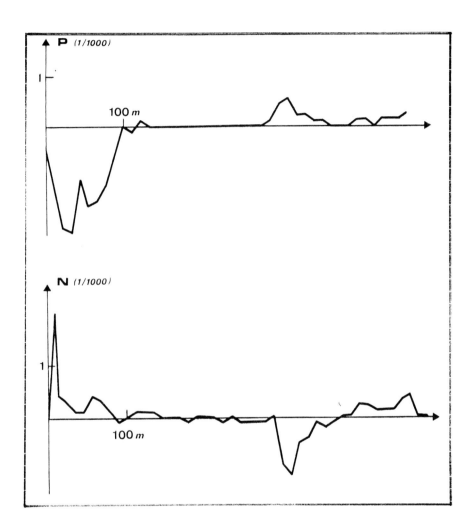

FIGURE 5 B — SHAPE INDEX P
LEVEL INDEX N

COMPUTER-CONTROLLED LASER BEAMS IN OBJECT DIMENSION MEASUREMENTS

J. Johansson, C.-O. Falt, and S. T. Eng
Department of Electrical Measurements
Chalmers University of Technology
Fack; S-402 20 Goteborg, Sweden

Abstract

This paper describes some general features of a computer-controlled laser system for noncontact measurements of dimensions of objects. The laser beams are scanned over the object and a system of detectors mounted behind the object detects whether the beams are shadowed by the object or not. The detector signals are converted to digital values and transferred to a minicomputer, where they are processed. This may require special software to allow the dimensional data to be presented in real time. An experimental system with a HeNe laser and controlled by an HP 21MX minicomputer has been built and tested. At a rate of 100 measurements per second the systematic error is in the order of 1 mm for absolute measurements and 0.1 mm for relative measurements. The imprecision is about 0.1 mm for both cases.

Introduction

The use of lasers in noncontact measurements of dimensions of objects has been tried in different applications. One very interesting is automatic log scaling in sawmill industries (1) - (6). It is essential to measure the dimensions of each log so that the yield from it can be maximized. The work leading to this paper was done in order to investigate the requirements on and performance of a computer-controlled laser system for noncontact measurements of objects. The technique that has been used is, to scan laser beams over the object and measure the time interval during which each laser beam is shadowed from a detector, mounted behind the object.

General aspects

A general computer-controlled laser measurement system is shown in Fig. 1. The system is

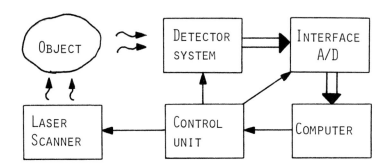

Fig. 1. A general computer-controlled laser
measurement system.

controlled by a minicomputer and a control unit. The control unit relieves the computer of the most trivial control tasks, such as timing and synchronizing. There is always an act of balance in what should be built as hardware in the control unit or written as software in the computer. For experimental setups the software way is often the most convenient, since it is easier to change the software than the hardware. However, in a final measurement system it is essential to have a highly developed control unit and to use the computer for more complex signal processing in real time. The light source contains one or several lasers, that can be scanned over the object e.g. by means of rotating mirrors. The light is detected by a detector system and the detector signals are converted to digital values by the interface and transferred to the computer.

The scanning technique is very important for minimizing the possible measurement error. Fig. 2 shows two different setups, the angular and the parallel scanning, both containing three laser beams. The diameter of the object, e.g. the cross section of a log, is measured in different directions. The references, that are shown in Fig. 2 for each laser beam,

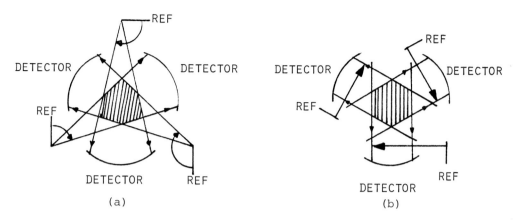

Fig. 2. Different scanning techniques; (a) angular scanning; (b) parallel scanning.

are used when the dimensions of the object are calculated. The angular case requires some-what more of the computer software, since all calculations must include trigonometric func-tions. However, this is no important disadvantage, since trigonometric functions can be effectively programmed by means of microprogramming features, that are commonly available on computer systems. The parallel case requires an optical arrangement, e.g. a parabolic mirror, that converts the rotation of a mirror, placed in the focus of the parabola, into a parallel scanning of the beam. In many applications for instance those in the sawmill industry, it is essential to minimize the number of optical components due to the rugged conditions they must work under. In those cases the angular scanning is a good choice, since it can be done by a small rotating mirror as the only optical component. Both arrangements in Fig. 2 will give as a measurement result six lines that defines the object. As long as one edge of the object touches each line, the object can have any shape. Care must be taken to arrange the scanning of the laser beams so that the error due to invisible details of the object is minimized. In a general case this can only be done by increasing the number of measurement directions or if possible, rotate the object during the measurement.

The experimental system

The experimental work described in this paper was done with the system shown in Fig. 3. For a more detailed description see (7).

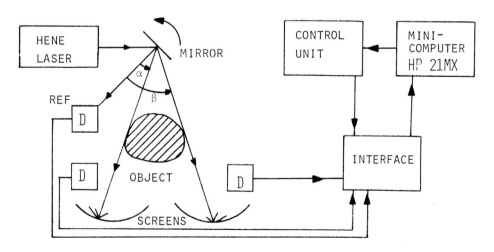

Fig. 3. The experimental system.

The experimental system has only one HeNe laser and uses angular scanning of the beam by means of a rotating mirror. The detectors are PIN diodes each followed by an amplifier. The three detector signals are used by the interface to create three digital numbers corresponding to the time of a complete revolution and the angles α and β, respectively. This is done by counting 10 MHz pulses during the corresponding time intervals. The detectors for the α- and β-angles detect the total light intensity reflected from the reflector screens. This has an advantage, namely that the diffraction pattern, created when the beam passes over the edge of the object, will be integrated to the total light intensity and, therefore, not cause any false detector pulses. The angles α and β are calculated by the computer using the corresponding time intervals and the time of a complete revolution. By using the setup shown in Fig. 2 it is possible to calculate the dimension of the object. In order to simplify the test equipment, we rotated the measurement object round a known axis instead of building several laser scanners for different measurement directions.

The experimental system has been tested with measurements on objects with known dimensions. Taking 100 measurements per second, the systematic error was always below 1 mm for object sizes from 0.1 m to 0.5 m. However, relative changes could be detected with a systematic error of below 0.1 mm. The imprecision calculated as a 95 % confidence interval from 10 values with the method described below, was 0.1 mm for both cases.

Measurement errors

Using the terminology of (8), we divide the measurement errors into the systematic error and the impression. The systematic error is calculated as the worst case error and the imprecission is estimated by statistical means. The error analysis is made for a general case of parallel scanning of the laser beam as shown in Fig. 4. The analysis for the angular case is similar (7).

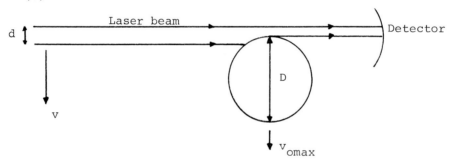

Fig. 4. Error analysis for a general case of parallel
scanning of the laser beam.

Systematic error

The rise time of the detectors is caused both by the gradual shading away of the laser beam and the electrical features of the detector element and the amplifier. Adding the squares of these two contributions and taking the root of the sum, we calculate the total rise time τ as

$$\tau = \sqrt{\left(\frac{d}{v}\right)^2 + \tau_e^2} \tag{1}$$

where d is the diameter of the laser beam enclosing 80 % of the light intensity; v is the velocity of the beam; and τ_e is the rise time of the electrical components. The contribution ΔD_1 to the error in the object diameter D can be considered to be proportional to τ and v and thus we get

$$\Delta D_1 = k \cdot \tau \cdot v \tag{2}$$

where k is a constant that depends on how well the light intensity from the reflector screen can be correlated to a specific position of the laser beam, at the edge of the object. In our experimental setup $|k| < 0.2$.

The possible movement of the object in the scanning direction may cause an error ΔD_2 calculated as

$$\Delta D_2 = \frac{v_{omax} \cdot D}{v - v_{omax}} \tag{3}$$

where v_{omax} is the velocity of the measurement object in the scanning direction. As $v_{omax} \ll v$ for all practical cases, we obtain

$$\Delta D_2 \approx \frac{v_{omax}}{v} \cdot D \tag{4}$$

As the actual measurement values are converted to digital numbers, a digitizing error ΔD_3 must be added to the systematic error. We obtain

$$\Delta D_3 = \frac{v}{f} \tag{5}$$

where f is the frequency of the pulses, that are counted during the time for passage of the laser beam over the object.

By summing the contributions to the total systematic error ΔD, we obtain

$$|\Delta D| = |\Delta D_1| + |\Delta D_2| + |\Delta D_3| \tag{6}$$

or using (1), (2), (4), and (5)

$$|\Delta D| = k \cdot \sqrt{d^2 + \tau_e^2 v^2} + |\frac{v_{omax}}{v} \cdot D| + |\frac{v}{f}| \tag{7}$$

The systematic error as function of the scanning velocity is shown in Fig. 5.

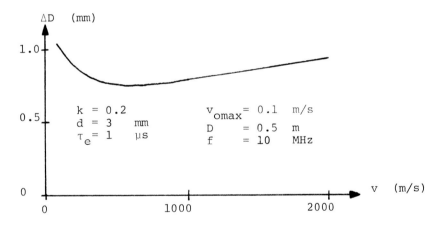

Fig. 5. The systematic error as function of the
scanning velocity.

Imprecision

The imprecision of the measurement method is estimated by taking a number N of measurement values from the same object. We can then calculate the standard deviation of the mean value as (9)

$$\sigma_{\overline{D}} = \frac{1}{\sqrt{N}} \cdot \sqrt{\sum_{n=1}^{N} \frac{(D_n - \overline{D})^2}{N-1}} \tag{8}$$

where $\overline{D}$ is the mean value. This equation can only be used if the imprecision is purely stochastic. An investigation of the statistical properties of D should be carried out using e.g. autocorrelation or fourier transform. D could for instance contain low frequency components, cause by vibrations in the mechanical setup of the system. The imprecision can now be described by a confidence interval

$$[\overline{D} - t_p \cdot \sigma_{\overline{D}}, \overline{D} + t_p \cdot \sigma_{\overline{D}}] \tag{9}$$

where t_p is taken from the t-distribution and depends on the number of measurements N and the desired confidence level.

Conclusions

An experimental computer - controlled laser system for noncontact measurements of dimensions has been built and tested. Taking 100 measurements per second the systematic error is in the order of 1 mm for absolute measurements and 0.1 mm for relative measurements. The imprecision is about 0.1 mm for both cases, calculated as a 95 % confidence interval from 10 measurement values using the t-distribution. Care must be taken to eliminate disturbing vibrations from the mechanical setup of the system. Otherwise, they may completely destroy the precision of the measurements.

References

1. J. Wiens and H. Quilitzsch, "Computerized log scaling, breakdown: sawmill impact and implementation," *Forest Industries*, vol. 102, pp. 61-63, May, 1975.
2. H. Reiter. "Photon-detectors used for measuring planks and round timber," *6-th International IMEKO Symposium on Photon Detectors*, Siofok, Hungary, pp. 349-357, Sept., 1974.
3. V.S. White. "Laser scanner gives St. Regis pine mill accurate handle on recovery," *Forest Industries*, March, 1976.
4. A.W. Porter and P.Eng, "Innovations in soft wood lumber production methods," *Canadian Forest Industries*, vol. 92, pp. 37-41, Sept., 1972.
5. P.A. Mueller and R.T. Herbert, "Increasing production yields with a laser scanner lumber inspection system," *Conference on Laser and Electro-Optical Systems*, QE Council OSA, San Diego, May, 1976.
6. D.M. Swing, "Dimensional gauging using a scanning laser," *Proceedings of the Society of Photo-Optical Instrumentation Engineers*, Quality Control & Reliability Problems with Optics, vol. 60, pp. 118-123.
7. J. Johansson, C-O Fält, and S.T. Eng, "Limitations of computer-controlled laser beams in dimension measurements", submitted for publication.
8. M.G. Natrella, *Experimental Statistics* (National Bureau of Standards Handbook 91). Washington, D.C.: U.S. Government Printing Office, 1966, chap. 23.
9. M.G. Natrella, *Experimental Statistics* (National Bureau of Standards Handbook 91). Washington, D.C.: U.S. Government Printing Office, 1966, pp. 2-6--2-7.

AN AUTOMATIC FRINGE ANALYSIS INTERFEROMETER
FOR RAPID MOIRE STRESS ANALYSIS

Dr. J. McKelvie, D. Pritty, Dr. C. A. Walker

University of Strathclyde, Glasgow, Scotland, U. K.

Abstract

In optical methods of stress analysis, the reduction of raw data to engineering quantities has proved to be a difficult problem.   A portable instrument has been developed which interrogates a specimen grid point-wise using a narrow beam of coherent light, and electro-optically measures interference-fringe-spacing, in three directions simultaneously.   A microprocessor is incorporated to convert the strain component values to convenient quantities such as principal stress and shear, and these are displayed.   The read-rate is adjustable up to 5 kHz, the discrimination level is 4 $\mu\varepsilon$, and a facility for temperature compensation is included.   The logic behind the design of the major features of the unit are described.

1. Introduction

Over a number of years, the moire fringe method has been developed into a sensitive tool for the experimental stress analyst.   Its use, although widespread, is inhibited by the magnitude of the effort involved in converting deformation component information into the stress contours which are the desired end point of a whole-field analysis.   Since each moire picture contains information about one deformation component only, three pictures must be analysed to provide enough information for a rosette calculation at each point of interest.   Several approaches to a solution of this problem, have been described;   mechanical differentiation of the basic moire photographs, which produced strain component contours (1);   computer programmes have been written to accept moire data in digitised form, and to calculate stress contours automatically (2) ;   and automatic fringe analysis machines have been built and operated which would measure the fringe spacing at any point in a field of moire fringes (3) .   The drawbacks with all of these methods, however, is that they are indirect and post factum; there is no way in which an immediate set of readings can be made from a specimen;   by and large they are time consuming, and prone to operator fatigue and error.   It is the object of this paper to describe an instrument designed to give an instantaneous analysis of the strains in a specimen grid;   the instrument is portable and self aligning;   apart from the need for electrical power, it is self contained, with all the required electronics for control and computation built in.   The output may be chosen to be in the form of principal strains or principal stresses (and direction), or strain components.

2. Moire Technology

The basis of the moire system is the comparison of a grating that deforms with the surface of a specimen, with a standard grating.   The sensitivity of the moire system depends directly upon the density of the gratings used;   for engineering stress analysis, gratings with densities in the range 100-1000 line/mm are required.   At such densities diffraction effects make it impractical to "beat" one grating directly with another, in reflection, as can be done at lower line densities, and so one has to resort to interrogating a high frequency specimen grating with an interferometer designed for the purpose(4) .

Such interferometric methods, however, lay great demands upon the stability of the complete system;   and while it may be that in certain instances, the specimen can be loaded on a vibration-free table which also carries the optics (4) , in general this is not a possibility.   Two alternative approaches, which have been used successfully, have been the recording of the strained specimen grid photographically (5) or by making a cast replica of a relief grating (6) .   These recordings can then be analysed separately in a suitable moire interferometer.   Analysis of the effects of vibration upon high-frequency moire fringes has shown that the main problem is one of fringe visibility, i.e., the moire fringes are present, and their accuracy is undiminished, but they are sweeping across the field of view too fast for the eye to follow.   A suitable grid, and an optical system have been described (7) which enables these fringes to be photographed, and which are, therefore, capable of producing pictures of deformation fields in situations which are not vibration free; we know, then, that with a sufficiently short data sampling time (less than 2×10^{-3} sec) vibration effects are not troublesome.

The essential step that is required to provide an instantaneous strain read-out is the measurement of the moire fringe spacing.   On a whole-field basis this is a problem of some magnitude in view of the quantity of data produced.   The device described in this paper provides a strain readout from one point at a time.   In essence this is achieved by interrogating the specimen grating along three axes simultaneously with an interferometer. The resultant three fringe patterns are read and analysed electronically and from the resultant fringe spacings, the strain tensor is calculated by an integral microprocessor.

3. System Description

3.1 Optics

A schematic description of the optical system is shown in Fig. 1a & 1b which show one of the three axes of the interferometer. The complete optical system is shown in Fig. 1c.

The raw laser beam is expanded to 3 mm diameter and split in two; one beam traverses the electro-optic modulator and a 90° polarisation rotator; the other beam passes through a path compensation block and the two beams are recombined. The combined beam is directed on to the grating, which splits the light into a total of eight diffracted beams, (see Fig. 2). There is also one reflected return beam. Six of the beams are used in pairs to provide information in the directions, 0° and 90°. The two extra beams are used for position control (section 3.2.4).

Each pair of measurement beams is treated in a similar manner: both beams pass first a polarisation-selective filter and then one of them passes a 90° polarisation rotator. The two beams are recombined and directed onto the photodiode system. By this means, one unmodulated beam is interfered with one modulated beam, in the correct polarisation orientation, and together they form a fringe pattern on the photodiodes.

The design is dominated by the requirements of characterising the fringe spacings (since strain α fringe frequency). In fact, a system has been adopted that is somewhat similar to that used by Dandliker (8) for measuring holographic fringe spacings. In essence, the fringes are viewed by two stationary photodiodes, and the fringe pattern is swept past them by modulating the phase of one arm of the interferometer relative to the other, (Fig. 2). A comparison of the phase of the signals from the two photodiodes yields the fringe spacing. The advantage of the phase comparison method over, for example, the use of a self-scanned photodiode or CCD array to characterise the profile of the optical density, is that with phase comparison one is examining a moving fringe pattern against a stationary background of "noise" which inevitably arises from dust on the optics. The signal/noise ratio is thereby much improved, since noise affects only the intensity and not the phase of the signal.

In theory, a separate modulator could be used for each arm of the interferometer. The advantages to be gained from using one modulator for all 3-axes include a reduction in weight, and volume, at the expense of a slight increase in the optical complexity.

3.2 Electronics

The main task of the electronic and microprocessor system is to perform the signal processing and computation necessary to transform phase differences of the waveforms produced by the three photocells in the interferometer head into the principal components of stress.

3.2.1 General System Considerations

An electro-optic modulator sweeps the fringe pattern back and forth across an array of individually accessible, linearly responding, photodiodes. The wave forms produced by the second and subsequent photocells in the array are thus phase displaced from the first by an amount proportional both to the spacing between the photocells and to the strain. The drive to the modulator is a linear ramp of voltage (Fig. 3). The position displacement of the fringe is proportional to the voltage applied to the ramp. Thus for a given maximum drive voltage, the received frequency from the photocells is proportional to the frequency of the ramp. A ramp frequency of 5 kHz was chosen as providing a high enough working frequency to allow vibrating strains to be measured but not sufficiently high to cause circuit design problems for the phase measurement or the ramp drive circuitry.

3.2.2 Fringe Analysis by Phase Comparison

When consideration is given to the fringe analysis circuitry, one is faced with the opposing demands of speed and accuracy.

The relationships governing a phase-measurement system are as follows:-

If fringe spacing is P mm. and the grid pitch is p mm, the strain is given by

$$\varepsilon = \frac{p}{P} \tag{1}$$

If we cause the fringes to "sweep" across the field at constant velocity, a single linearly-responding photodiode at a point will give a sinusoidal output as the illumination at the point alternatively increases and decreases. Two diodes separated by a distance D mm (which must be less than P/2 to avoid ambiguity), will give similar outputs, and phase-shifted one relative to the other by an amount

$$\phi = \frac{360 D}{P} \quad \text{degrees} \tag{2}$$

Therefore, by measuring we know the strain

$$\varepsilon = \frac{p}{360} \cdot \frac{\phi}{D} \tag{3}$$

For the greatest accuracy an analogue comparison, employing for example two phase-lock-loops, averaging over very many cycles, can give discrimination to a fraction of a degree of phase. For our purposes, however, a strain resolution much better than 10 $\mu\varepsilon$ is not likely to be useful but good time-resolution is important, and so it was convenient to choose an alternative, digital, method of phase comparison, capable of giving a strain reading for every modulator sweep, i.e., every 150 μ sec. (see Fig.3). This system counts 11 MHz clock-pulses between zero-crossings on both signals. At a 5 kHz sweep-rate the system has been found to have a repeatability better than $\pm$ 2° phase shift, which, with D = 1.25 mm and p = .001 mm, corresponds to a strain repeatability of $\pm$ 4 μϵ.

It will be plain that the closer the two phase comparison photodiodes are, the higher is the maximum fringe density which can be accommodated; likewise, the wider the photodiodes spacing, the more accurately can widely spaced fringes be measured. By using an array of individually accessible photodiodes, at spacings of 0.125mm, 0.375 mm. and 1.25 mm (e.g., Integrated Photomatrix type 1412) along with the digital fringe spacing analysis, three ranges are available, viz. O $\rightarrow$ $\pm$ 400 μϵ, O $\rightarrow$ $\pm$ 1300 μϵ, O $\rightarrow$ $\pm$ 4000 μϵ, when the grid frequency is 1000 line/mm.

In order to measure the phase difference between the two waveforms sufficient drive voltage has to be applied to produce a displacement of at least one full cycle i.e., 2π. Various factors which are discussed below mean that in fact 3π of displacement has to be provided. One of the most accurate ways of measuring the phase of a signal particularly where noise is present is to use a phase locked loop, where the phase of a clean square wave or sinusoid is locked to the incoming waveform from the photocell. This averages out the effects of any noise and distortions on the incoming waveform. Such systems however tend to work over many cycles. Our requirements was the measurement of the phase in only one cycle of the ramp. The alternative of operating the whole system at, say, ten times the present frequency (i.e. 50 kHz) was unattractive for a number of reasons.

As the phase displacement has to be converted into a digital form for the microprocessor it is obviously attractive to use a digital method of measuring phase displacement. Referring again to Fig.3., if V_{S1} and V_{S2} are perfect sinusoids then the phase shift between V_{S1} and V_{S2} is given by $\phi = 360° \; t_1 / t_\lambda$. If there are any distortion or noise present then t_1 may not equal t_2, and a more accurate value for the phase shift is :

$$\phi = 360° \; (t_1 + t_2) / 2t_\lambda$$

The signal processing to evaluate such a relation is straightforward. One can establish the times of the zero crossings by providing sufficient amplification of the waveforms before a limiting stage. The various times can be measured using suitable logic and two counters, one for $t_1 + t_2$, the other for t_λ.

With a constant ramp frequency , and a known maximum voltage applied to the modulator, t_λ will not vary much from its predefined value. Thus simple look-up tables provide the necessary corrections to apply to $t_1 + t_2$ when t_λ varies slightly from the standard value.

3.2.3   Modulator Drive Circuits

The electro-optic modulator requires a drive voltage of 360 volts to produce a 2π displacement of the fringes. As the initial position of the fringe with respect to the photodiodes is indeterminate we require to cause a displacement of just over 3π radians in the fringe position in order to guarantee that the displacement provided by the modulator gives a complete cycle of the fringe starting from a zero crossing of the photocell waveform (see Fig.3).

The ramp is therefore driven by a sawtooth waveform with a peak to peak amplitude of 540 volts.

Fig.4 shows the essentials of the ramp drive circuit.

The upward part of the sawtooth is made linear by providing a constant current drive to the virtual earth of a two stage amplifier which has a 100pF feedback capacitance connected between the output and the virtual earth. A high voltage power transistor used in grounded base mode provides the output stage of the amplifier. The upward part of the sawtooth takes 150μS. The ramp is returned to zero in about 50 μS., by turning on the transistors in the drive circuits. An output from the microprocessor drives the monostable which generates the 'ramp-up' pulse (of 150 μS).

Logic circuitry exists in the phase counter section to ensure that phase and wavelength measurement only start from a zero crossing of the photocell waveform and not just the beginning of the ramp as discussed above.

3.2.4   Video Amplifiers and Limiters

These circuits are located at the interferometer head. An outline diagram of the circuits is shown in Fig. 5. The amplifiers take the signal from the photocells and raise it to a suitable level to feed the limiter circuits. The SN72733 integrated wideband video amplifier circuits, which are used, have a gain bandwidth product of 400 MHz, and a maximum voltage gain of 400. They provide a push-pull output which is capacitatively coupled to an integrated circuit comparator (the SN72720 which is the dual form of the SN72710) which forms the limiter circuit.

To allow for changes of scale in the strains being measured one of three photocells spaced at multiples of 0.1 mm from the reference photocell can be selected. This is achieved by having an MOS analogue multiplexing switch route the output from the required photocell to the input of the second video amplifier.

As the differential phase of the signals must not be affected by the amplifier and limiter, the time constants in the amplifier chain are chosen such that they provide less than 0.1 radian of phase shift at the 10kHz signal frequency. The signal current out of the photocells is of the order of 50nA r. m.s.

This develops approximately 200μV r. m.s. across the input to the video amplifier which gives a reasonable signal-to-noise ratio for phase measurement purposes. (The position of the zero crossings of the signal will be altered by the noise and can thus cause a random fluctuation of the phase shift. The microprocessor averages the phase shifts over at least 10 cycles unless the most rapid read repitition rate of 5 kHz is selected). The noise bandwidth of the system is limited to 150 kHz with a simple filter on the output of the video amplifiers.

As mentioned above, the video amplifier has a typical gain of 400 raising the signal level to 80mV r.m.s. at the input to the 72720 comparator. The comparator is automatically biased in the linear region and a signal change of $\pm$ 1mV. at its input will cause full scale output i.e. transmission from a logical "1" to a logical "O" or vice versa. The response time of the comparator is approximately 50ns. Limiting of the signal is therefore better than 1% and phase errors due to the rise time of the limiter are of the order of $\frac{1}{4}°$. The TTL compatible digital signal produced by the comparator is fed down a 75Ω coax line to the main unit using a SN75121 integrated line driver.

3.2.5 Phase Counter Circuitry

The digital signal from the head is received by SN75122 dual line receiver circuits and passed through further logic to the phase counters. Fig.3 shows the required output (V_ϕ) to the phase counters when V_1 leads V_2. Fig.6 shows the logic required to produce the waveform V_ϕ. D type flip-flops detect the delay between the positive going edges of V_1 and V_2 and the negative going edges of V_1 and V_2 and the outputs are ORed to provide V_ϕ .

The case of V_1 lagging V_2 is catered for by a further two D type flip-flops. The 'lead' output is fed to the 'up' input of a 12 bit up/down counter clocked at 11MHz and the 'lag' input fed to 'down' input of the counter. The 11MHz clock rate gives a phase discrimination of $\pm$ $^1/6°$.

The 'lead' and 'lag' outputs are ORed, fed through logic to determine one period and into an 8bit wavelength counter clocked at $2\frac{3}{4}$MHz. The $2\frac{3}{4}$MHz clock gives an accuracy of $\pm$0.2% in the measurement of the wavelength.

Two 203 x 203 mm boards contain the three phase counter circuits and the input logic to the microprocessor board. The dual-in-line logic packages are interconnected by wire-wrapping.

3.2.6 Microprocessor System Hardware

This follows modern practice with maximum use of readily available MSI circuits for the interface electronics.

An M6800 microprocessor was chosen for the project as this processor had been successfully used on past projects and therefore the hardware and software design requirements were familiar to the authors.

The microprocessor board which measures 203 x 203 mm incorporates 1K bytes of static RAM (8 off 210's) and has the capacity for 4-8K bytes of fixed program held in UV erasable PROM's.

3.2.7 Input/Output

The M6800 has no I/O instructions as such and therefore all input output must be memory mapped. This implies that each I/O register is assigned a memory location and data is then merely read from or written to that location depending on whether an input or output transfer is required.

In order to simplify timing requirements, information from the phase counters is transferred to 8 bit D type flip-flop storage registers at the top of the ramp. The information in these memory mapped registers has therefore to be transferred to the microprocessor before new data is fed in on the next cycle of the ramp i.e., within 200 μs. The back edge of the ramp up pulse is fed to the maskable interrupt line of the processor. The interrupt routine invoked then transfers all data from the registers to processor memory.

The only other major input/output area provided in the prototype design is the display and front panel controls. Three five digit L.E.D. displays provide the main read-out from the instrument. These are pulse driven on a time multiplexing basis from a 16 x 4 bit RAM carried on the display board. A 4bit counter addresses the RAM and each LED in turn through a 4 to 16 line decoder. A two line to one line multiplexer allows either the addresses generated by the counter or the address generated by the processor to be selected. This allows the RAM contents to be updated by the processor as required. A 150 x 100 mm board contains the logic required to drive the LED displays and to input the signals generated from the push buttons on the front panel to the processor. The final output drive transistors to the LED's are contained on a board on which the LED's are also mounted.

3.2.8 Microprocessor Software

The entire operation of the calculation and measurement section of the instrument is controlled by the microprocessor software. The main program senses the required operation of the instrument as given by the state of push-buttons on the front panel. Appropriate routines are then called to perform the various functions and once these have been completed the instrument returns to the main routine and awaits further commands.

The instrument imposes no particular demands in terms of speed of computation. Almost all the software for the instrument is therefore being written in a real-time high-level language for microprocessors known as PL/F. This language was previously developed in the University of Strathclyde, Department of Computer Science (10).

3.2.9 Alignment

There are two aspects to consider in the alignment of the interferometer with the specimen grating (see Fig.7) - firstly those motions which have a bearing on the accuracy of measurement, and secondly those motions which must be controlled due to the limited aperture of the optics. It has been shown (9) that rotations about the 0° and 90° grid axes have only a second order effect on the measurement, but for accurate work, alignment should be maintained within 0.1°, or due allowance made for misalignment.

Rotation or translation of the system about the input beam axis cause the final pairs of beams to lose registration and so these motions must be controlled, although less accuracy will suffice. All of these control functions were accommodated by using the spare pair of diffracted beams to activate a servo control system via a pair of position sensitive photodiodes (Fig.7). While alignment can be carried out manually, this procedure proved tedious when a scan of a specimen surface is desired.

The output of the position-sensing photodiodes is also used to correct the measurement for the slight misalignments which are inevitable with a simple control system.

In operation, the system is aligned approximately by eye. The servo system then takes over and aligns the interferometer and holds it in registration so that a scan of the complete grid can be made without the need for manual alignment at each point.

3.3 Specimen Grids

The requirements, placed upon the specimen grids, are that they should diffract light efficiently and be capable of application to a wide variety of substrates. Accordingly, a crossed holographic relief grating (10) was used as the master from which replicas of the profile were cast in epoxy resin and transferred to the specimen surface (9) - see Fig.8. The special properties of the master grating ensured that a high proportion of the light was diffracted into the desired orders, with only a small proportion being lost into unwanted orders. Such performance could not be obtained from a bar-and-space amplitude grating of the type commonly used in moire work; in addition, the accuracy of the grating should be such that nowhere is there an error greater than 5-8 microstrain, and a fidelity of this order is hard to achieve by means other than the replica casting. Gratings of this type have been found to be stable in use over a number of years.

3.4 Auxiliary Features

With the computational power of the microprocessor available, it has been a relatively straightforward matter to incorporate useful additional capabilities, viz:-

1. Allowance can be made for Young's modulus, Poisson's ratio, temperature coefficient of expansion and specimen temperature, by entering these parameters through an integral keyboard.

2. Automatic range selection.

3. The output display can be selected as either strain or stress, and as components or principal values.

4. Displayed values can be expressed in metric or imperial units.

5. The memory can be programmed to display only the largest value encountered during a scan over an area of grid.

6. Triggering of the start of modulator sweep can be synchronised with an external event.

4. Further Work

The development of grids suitable for high temperature operation is proceeding. These will have lower line densities (e.g., 40 line/mm) to allow of more remote operation, but at the corresponding sacrifice in accuracy.

5. Conclusion

An instrument has been designed and built, capable of direct rapid interrogation of a high-frequency specimen grid, in a non-contacting point-wise manner and of displaying the deformation information in a selection of modes, thus overcoming some of the previous disadvantages of grid methods.

List of References

1. Post, D., and MacLaughlin., T.F. Strain analysis by moire fringe multiplication. Experimental Mechanics Vol.28. No.2. 1971, pp 408-413.

2.   Bossaert, W., Dechaine, R., and Vinckier, A., Computation of finite strains from moire displacement patterns. Jnl. of Strain Analysis, Vol.3, No.1., 1968. pp 65-75.

3.   Hitchings, D.J., and Luxmoore, A.R., An automatic high-speed scanner for moire fringe techniques. Jnl. of Strain Analysis, Vol.7. No.2., 1972, pp 151-156.

4.   Marchant, M., and Bishop, S.M., An interference technique for the measurement of in-plane displacements of opaque surfaces, Jnl. of Strain Analysis, Vol.9. No.1. 1974. pp 36-43.

5.   Daniel, I.M., Rowlands, R.E., and Post, D., Strain Analysis of Composites by Moire Methods Experimental Mechanics, Vol.30 June 1973, pp.246-252.

6.   McKelvie, J. and Walker, C.A., Analysis of Strain Histories in Structures USAF. Report No.AFOSR. 73-2566A-0002. (1975).

7.   McDonach, A., McKelvie, J., and Walker, C.A. Fatigue Behaviour of Carbon Fibre Reinforced Plastic visualised by a moire fringe method. Presented at BSSM Annual Meeting, Bradford, Sept. 12, 1978.

8.   Dandliker, R., Ineichen, B. and Mottier, F.M. High resolution interferometry by electronic phase measurement. Opt. Commun. Vol.9. (4), 1973, pp 412-416.

9.   McKelvie, J. and Walker, C.A., A practical multiplied moire fringe technique. Experimental Mechanics, Vol.18, No.8. 316-320, Aug. 1978.

10.   Pritty, D., A continuous path microprocessor NC system.   Advances in Computer Aided Manufacture edited by D. McPherson, North-Holland Publishing Co. and also published in Proceedings of the Third International IFIP/IFAC Conference on Programming Languages for Machine Tools, "PROLAMAT" , June 1976.

FIGURE 1a.

GENERATION OF INTERROGATION BEAM.

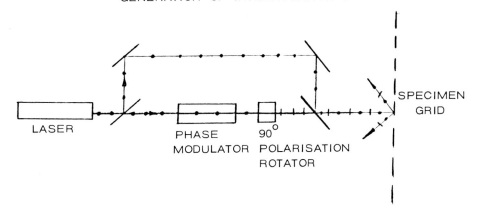

FIGURE 1b.

INTERFEROMETRIC OPTICS.

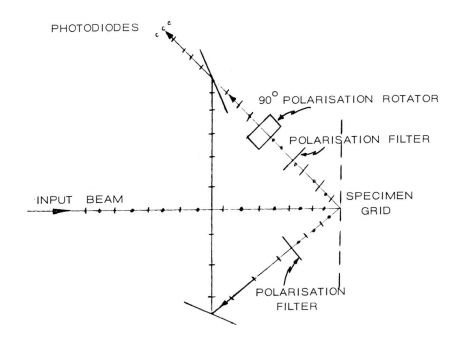

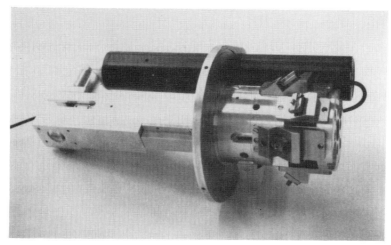

Side view of Interferometer head.

Front view of Interferometer Head.

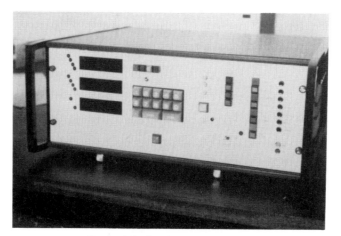

Control & display electronics.

Figure 1C. Interferometer head & control
electronics.

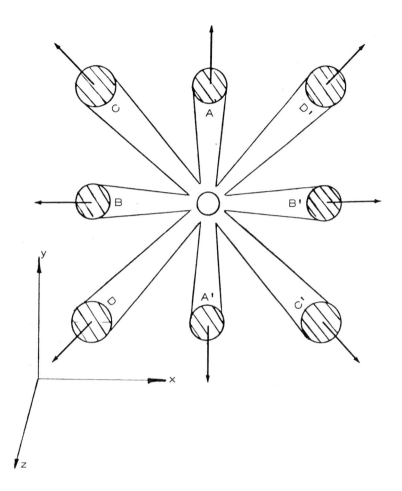

Fig.2.   Diffraction of laser beam by a crossed (90°) blazed
         Specimen grating.

         The specimen grid is in plane x-y (the plane of the paper
         and the input beam is directed along the z-axis towards
         the paper.

         Beams A, A' are combined interferometrically to yield
         a fringe pattern which is characteristic of deformation
         of the specimen grid in the y-direction.   Likewise,
         beams B, B' and C, C' are combined to measure
         deformation in the x- and 45°- directions.

         Beams D, D' are used for position control purposes.

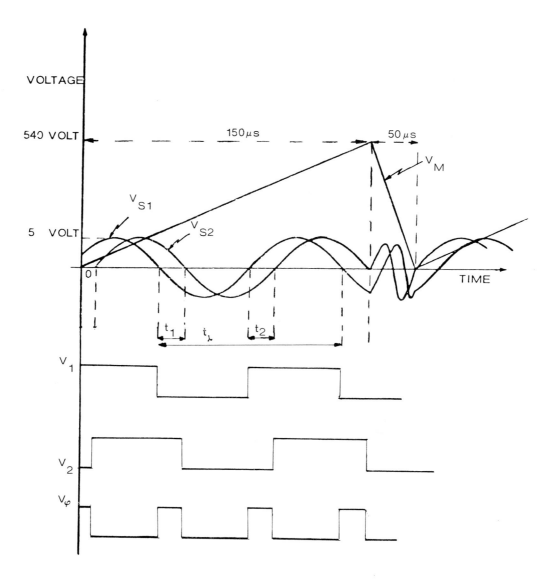

Fig. 3   Modulator, Photodiode, Limiter and Counter voltages.

V_M – Modulator drive voltage

V_{S1}, V_{S2} – Voltages from a pair of photodiodes.

V_1, V_2, – Limiter output voltages.

V_φ – Phase counter voltage.

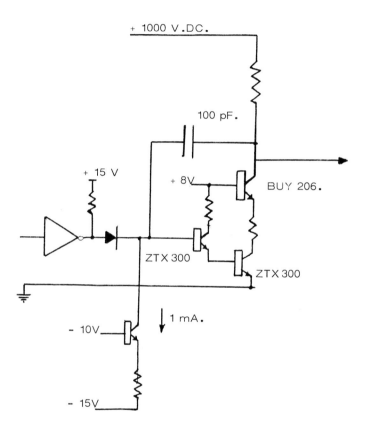

Fig.4.   Essentials of ramp drive circuit.

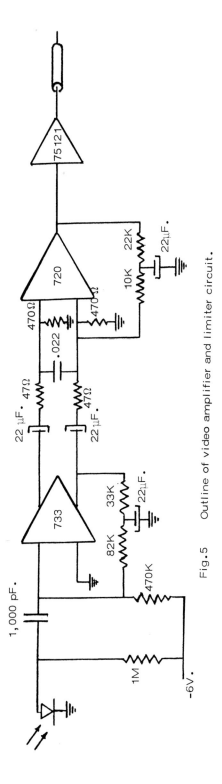

Fig.5   Outline of video amplifier and limiter circuit.

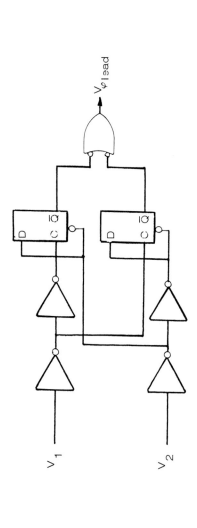

Fig.6   Phase counter logic circuit.

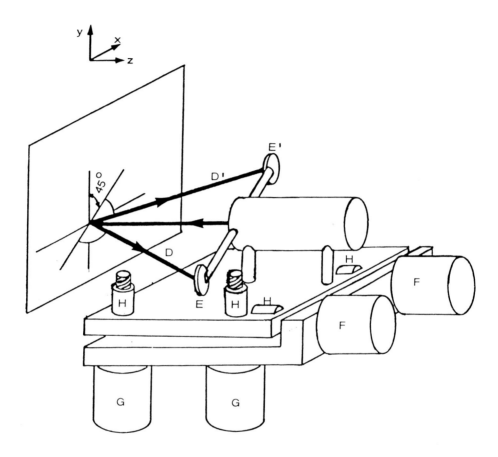

Fig.7. Schematic of Alignment Mechanism.

Diffracted beams D, D' strike position-sensing
photodiodes E, E'.   Duly processed signals drive
motors F & G.   Motors F provide rotation about y-axis
and motors G provide rotations about x-and z- axes,
through nuts and drive-screws H.

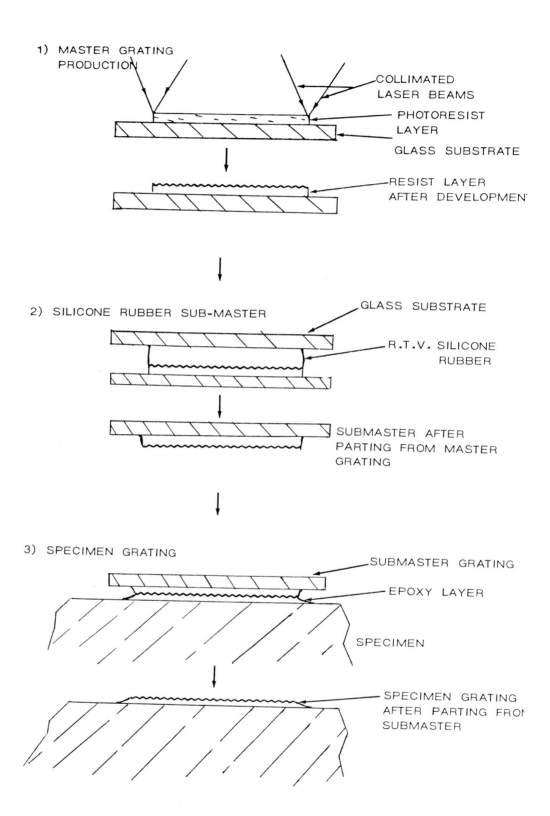

Fig. 8 . GRATING REPLICATION PROCESS.

STRESS PATTERN ANALYSIS BY THERMAL EMISSION (SPATE)

D. S. Mountain
Admiralty Surface Weapons Establishment
Portsdown, Cosham, Portsmouth, Hants, England

J. M. B. Webber
Sira Institute Ltd.
South Hill, Chislehurst, Kent, England

Abstract

This paper describes a novel method of stress determination in structures under dynamic loading conditions. The principle is based on the thermodynamic property of a material in which, under adiabatic conditions, heat is produced or absorbed in direct proportion to the magnitude of the stresses to which it is being subjected. An instrument is described which uses infra-red radiation to measure, remotely, the resulting minute local temperature changes in the material, with oscillating mirrors to provide a raster type scan of the item under test. Spatial resolution down to 1 mm and a temperature discrimination of about 0.002 °C have been achieved, this latter representing a stress change of less than 300 lb/in^2 in steel. Results are given which demonstrate the ease with which the equipment can be used to assist in the design of a structure and an indication is given of its potential value in the field of Non-Destructive Testing, for example in the location of fatigue cracks.

Introduction

Adiabatic compression or expansion of a gas produces a heating or a cooling of it, but what may not be so generally appreciated is that solids display the same property although to a much less noticeable extent. Any pressure in a gas however is automatically equalised throughout, whereas pressures arising in a solid show up as local higher stressed regions where the heat is generated or absorbed. Stress Pattern Analysis by Thermal Emission or SPATE is a technique which uses infra-red radiation to measure the resulting local temperature changes remotely, and thus provides detailed information on the associated stresses.

Objective

The original objective of a research contract, placed by the Admiralty Surface Weapons Establishment with the Sira Institute, was to try to see whether the basic idea could be developed into a practical instrument to provide an easier, quicker and cheaper method of determining stress distributions, particularly in a complex structure such as radar antenna pedestal, of which Figure 1 is a typical

example. The structures concerned are designed primarily for shipborne use and are often mounted high on a mast where the considerable distance above the ship's roll centre means that they must be as light as possible if not to have a serious effect on the overturning moment of the ship. In any structure designed for minimum weight, the utilisation factor of the material used must be high, ie it should be designed, ideally, to have equal stress levels throughout since it would be dangerous if any parts were overstressed and wasteful if they were understressed. To achieve such an optimised situation it is essential that detailed information on the whole of the stress distribution be readily available during the design stage, and it was with that requirement in mind that the idea for SPATE was originally conceived.

Theory

The equation which can be derived (1) to link temperature rise to stress under adiabatic conditions, from which it can be seen that there is a linear relationship between them, is as follows:

$$T = - \frac{S}{\dfrac{9398.4c\rho}{R a} + \dfrac{3aE}{1 - 2\sigma}} \qquad \ldots\ldots (1)$$

Where T = Temperature Rise
S = Stress
R = Reference Temperature
c = Specific Heat
ρ = Density
a = Coefficient of Thermal Expansion
E = Young's Modulus
σ = Poisson's Ratio

Fig. 1. Typical radar antenna pedestal.

If typical values for the physical constants are inserted into that equation, it is found that the temperature rise is very small, only about 0.2 or 0.3 °C at the ultimate strength of the material. It is also very transient, since being an adiabatic situation the heat is rapidly dissipated by conduction, for example between the tension and compression parts of the body, and by convection and radiation.

Early Experiments

A number of tests were carried out to check the practicability of the proposals. An early experiment repeated one described by Lord Kelvin in 1855, in which a rubber band under tension, if heated, will contract and do work, for example by lifting a weight. In another a thermocouple was attached to a steel wire which, when put under tension, produced a marked drop in temperature. A series of thermocouples attached along a cantilevered bar with an artificial high stress point were used to demonstrate that the temperature change does occur locally and that a region of higher stress can readily be located by the discontinuity in the temperature pattern. Finally a TGS detector was used to record the temperature changes remotely by measuring the IR emission from a local high stress point.

Equipment

Performance

New electro-optical and signal processing techniques have had to be specially developed by the Sira Institute in order to achieve the required performance, but an equipment has now been produced which can measure the IR emission remotely and can register temperature changes with a discrimination of about two thousandths of a degree centigrade and with a spatial resolution of better than 1 mm. This temperature discrimination will allow a stress change of something less than 300 lb/in^2 in steel or about 100 lb/in^2 in aluminium to be distinguished.

Description

The Sira Institute is at present re-engineering the equipment to make it neater and more compact, but Figure 2 shows it in the experimental form in which more control facilities were provided than has subsequently been found necessary. On the right is a box housing the infra-red detector and optical and scanning mechanisms with the control and display unit on the left. A schematic diagram of the detector unit is shown in Figure 3. The infra-red detector is a Lead/Tin/Telluride cell approximately 100 microns across cooled

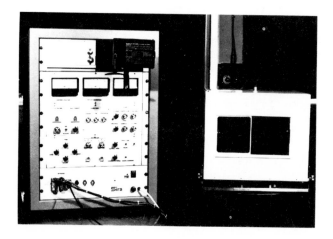

Fig. 2. SPATE equipment.

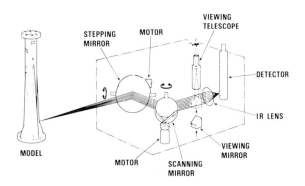

Fig. 3. Schematic diagram of optical head.

with liquid nitrogen. Immediately adjacent is a germanium f0.8 lens which focuses the incoming radiation onto the detector, and a two mirror scanning system is used to provide a TV type raster. A telescope is also provided so that by inserting the angled plain mirror into the IR path the scanned area can be checked visually and adjusted as required. The unit on the left of Figure 2 contains the processing and display equipment, and the control facilities. For example, the boundaries of the scan can be varied by four of the controls and its position is shown vertically and horizontally by the two left-hand meters. The meter on the right shows the processed output from the detector and this is also made available on sockets for external recording. On the top right-hand corner is a built-in oscilloscope, shown with a recording camera in place, on which the X and Y axes are used to display the extent of the scan, and the spot brightness is used as the third dimensions to record the temperature and therefore stress.

Experimental Model

In its role as a design tool SPATE would normally be used on a model of a proposed structure and one designed and made specifically for the tests is shown in Figure 4. It can be loaded as a cantilever in either of two directions at right angles. In one direction the design is poor. High stress points have been intentionally introduced by discontinuities in the webs, by joining the webs to the flange and body with rapid transitions and by making some holes with sharp right angled corners. In the other direction, 90° away, some of the anticipated high stress points have been eliminated, for example by creating smooth transitions, by slightly moving the positions and size of some holes, and by putting small radii on the corners of the rectangular ones. A number of these models were made identically by machining out of solid aluminium alloy on an NC machine, and were finally anodized with a matt black finish to give a fairly high and constant emissivity over the whole surface.

Fig. 4. Experimental model structure.

Method of Loading

One of the ways in which the temperature discrimination has been achieved has been by increasing the signal to noise ratio by moving the item under test backwards and forwards through its neutral position and synchronising the detector to sample the +ve and -ve temperatures at the peaks of the oscillations. Figure 5 shows the servo controlled hydraulic ram which is being used during the experimental work to provide the necessary reciprocating stresses. This arrangement has the added advantage that the results are obtained under the more representative condition of a dynamic load rather than statically as would normally be the case.

Fig. 5. Vibrator head and model structure.

Typical Results

Oscilloscope

Figure 6 shows a picture taken directly from the oscilloscope of the stress distribution over a portion of the model together with a physical outline showing the direction of the load and, between the dotted lines, the limits of the scan in the vertical dimension. For this test the vibrator was set to give a stressing frequency of 20 Hz and the detector was at a range of about 2 ft from the model. The more highly stressed regions around the dog leg in the web, can be clearly seen by the increase in brightness of the traces.

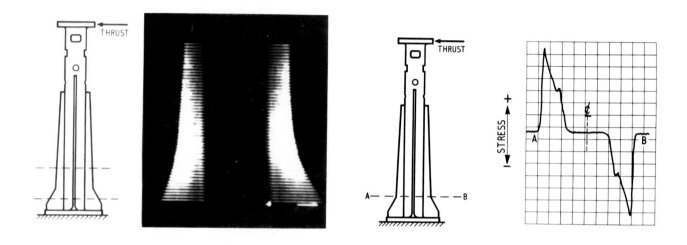

Fig. 6. Stress pattern on webs — side view.

Fig. 7. Stress levels of single scan across webs.

Pen Recordings

The exact extent of the area of highest stress is not easy to determine however despite a built-in thresholding control since it is difficult to distinguish more than perhaps 10 levels of brightness and the various reproduction processes lose some gradation as well. A more accurate method is to use the output socket to feed an X-Y recorder and so produce an "A" scan for any given sweep, with the amplitude being proportional to the stress level. One such scan is shown on the right of Figure 7, with its position defined by the dotted line on the adjacent sketch. Since at any given moment one side of the model is in compression and the other side is in tension half the trace is above and half below the zero line. The ease with which the stresses across a section can be obtained is obvious since, having got the equipment set up, a sweep like that only takes some 15 seconds. The pen recording was redrawn for Figure 7 to give a clearer picture, but Figure 8 is an enlargement of one half of an actual record. This gives an indication of the accuracy and lack of noise in the system. With the deflecting force being used in that particular test, the peak of the curve represents some 20 000 lb/in^2 of stress and a readable minimum of say 1/20th of an inch on the chart thus represents approximately 300 lb/in^2. A facility is also provided by which the scan angle can be varied as required from the horizontal one normally used for producing the raster, to the vertical, so that a "stress cut" of the type shown in Figure 8 can be taken in any desired direction.

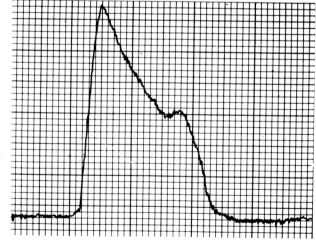

Fig. 8. Pen recording of stress across web.

Use In Design

Method

In analysing a structure, a designer would first take a relatively rapid overall picture of the whole model at comparatively long range, thereby of course sacrificing some stress discrimination and spatial resolution. Figure 9 shows such a test at a range of 4 ft and at a loading frequency of 20 Hz and took some 8 minutes. It is of the same model, being loaded in the same direction ie at right angles to the plane of the paper, but now viewed end-on with the indented web down the centre of the picture, and enables a rapid identification to be made of the main areas of interest. These appear to be at the inflexion of the web, around the small circular hole, where the top of the web runs into the central tube and also, right at the top, round the larger rectangular hole. The equipment can now be moved in to carry out a closer and more detailed examination of each of these regions in turn. Figure 10 is a stress picture of

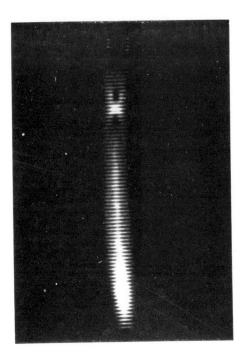

Fig. 9. Overall stress pattern on model.

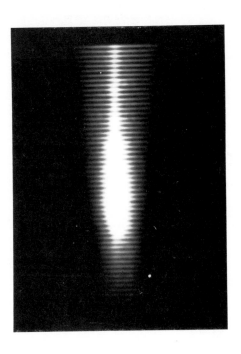

Fig. 10. Detailed stress pattern of web — end view.

the web around the inflexion and covers a similar area to that shown in Figure 6, but this time end-on instead of a side view. The bright area, indicating the high stress at the dip in the web, is clearly visible and the stress can be seen to be falling away rapidly below and more gradually above. Figure 11 is a pen recording obtained by rotating one of the mirrors to provide a single scan down the outside of the web and clearly shows the peak of the stress. Figure 12 is a close up of the area surrounding the rectangular hole with

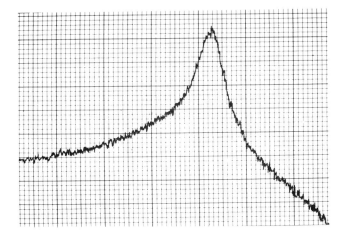

Fig. 11. Pen recording of stress down web.

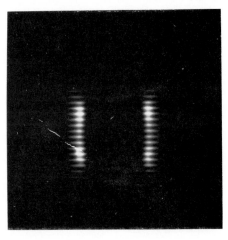

Fig. 12. Detailed stress pattern around rectangular hole.

the stress concentrations at the corners clearly visible. Figure 13 is another close up at the top of the web and around the small hole again showing the associated stress concentrations and Figure 14 is a pen record of a single scan across the specimen immediately below the top of the web.

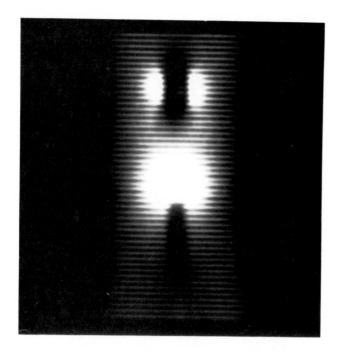

Fig. 13. Detailed stress pattern at top of web and around circular hole.

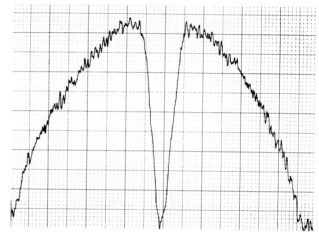

Fig. 14. Pen recording of stress across top of web.

Comments on Results

In the context of designing for minimum weight mentioned above, Figures 12 and 13 highlight a very interesting point. Considering the rectangular holes at the top of the model, it is obvious that in the bending mode there can be no stress in the material immediately above or below them since the holes cannot provide an anchorage or a buttress for any tensile or compressive forces. Thus to reduce weight, material both above and below the hole could be removed and this is clearly brought out in the stress picture of Figure 12 where those two areas, having no stress, both behave and look exactly like the hole themselves. If therefore there is no valid reason for the holes to be rectangular, one could elongate them as shown in Figure 15. By designing in this way not only is the weight reduced by the elimination of unnecessary material at Points A and B, but there is the additional advantage that the stress concentrations at the corners have been reduced since there is no longer any rapid change of section. In an exactly similar way it is obvious from Figure 13 that the small circular hole would be better elongated and that the last inch or so of the web is doing nothing at all. In addition, because there is a step where it enters the tube there is a high and quite unnecessary stress concentration at that point. The pen record of Figure 14, which took some 15 seconds to produce, shows clearly with the sudden drop in the stress level at the centre, how little stress is actually being carried by the web at this point and therefore how useless it is. In these instances therefore the structure would actually be strengthened by removing material. This is of course by no means an unheard of situation but is one which is immediately obvious from the SPATE pictures.

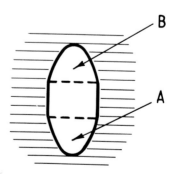

Fig. 15. Elongated rectangular hole.

Comparison with Strain Gauges

One of the model structures of Figure 4 was given to a firm specialising in stress measurements with a contract to carry out a strain gauge investigation at the dip in the web. The results obtained were comparable but were over a much more restricted area than those produced by SPATE. A tiny multiple gauge comprising 10 elements, each 31 thou long at a pitch of 80 thou were used, thus covering about 20 mm along the extreme fibre of the web. Figure 16 shows the results from these gauges and should be compared with the pen record of a single sweep by SPATE over a considerably greater distance shown in Figure 11, and which took only 15 seconds to produce. The curves are clearly similar although since SPATE gives about twice the resolution of even these minute gauges, the peak of the curve has been rather better defined. In fact one gauge was apparently faulty and one or more of the others gave readings which are obviously suspect.

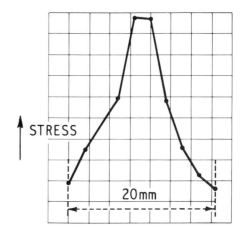

Fig. 16. Strain gauge readings at web inflexion.

Conclusion

Only a limited amount of experimental work has so far been carried out but the equipment has considerable potential value to the mechanical designer since working with models and using SPATE, he can make a comprehensive analysis of a structure himself, modify it as necessary and re-analyse any number of times at a fraction of the cost and the time needed by present methods.

Use In Non-Destructive Testing (NDT)

The equipment also has a capability in the field of fault detection which originally came to light fortuitously before any work had been done in this connection. During the course of some of the early experiments, an oscilloscope picture taken looking at the web inflexion of the model in the end-on direction whilst loading it in the high stress aspect, gave the unexpected pattern shown in Figure 17. This should be compared with Figure 10 which is under similar conditions and is of the same part of the web, Now however, instead of the bright area down the centre, there are two high stressed regions with a definite low stress area between them. The reason for this anomaly only became apparent after the black coating had been removed and a fatigue crack was discovered running from the inflexion in the web down to the central tube. This can just be seen in Figure 18. During the early experimental work to

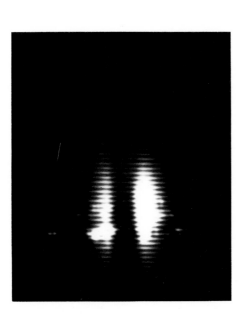

Fig. 17. Detailed stress pattern of web with fatigue crack — end view.

Fig. 18. Fatigue crack at web inflexion.

find out how to use the equipment it had been run a good deal at widely varying frequencies and displacements and as a result the model had been overstressed. The reason for the dark central area on the stress picture is now clear. The vibrator equipment is built to provide whatever thrust is needed to produce a required preset amplitude, and the cracks in the webs would immediately reduce the force at the deflection being used by a factor of perhaps 10, so although some compressive stresses would still be carried during half of each cycle, the stress in the material of the web would be only a fraction that in an undamaged model. A side view of one half of another model with a similar fatigue crack just developing is shown in Figure 19 and should be compared with Figure 6 which is a similar view of an undamaged specimen. The effect of the crack has been to move the point of highest stress away from the extreme fibre at the dip of the web to a point which, when the picture was taken, was the origin of the crack. A subsequent picture, Figure 20, shows the highest stress region moving inward toward the tube as the crack propagated across the web. A different method involving a check on the wave shape of temperature produced from a perfect and imperfect structure, is at present showing promise of greater accuracy and reliability. There has not yet been time to do much investigation of this particular application, but in a similar way it is expected that faulty welding can be shown up by comparison with the overall pattern obtained on a perfect specimen. A rapid check using SPATE can therefore be used to reveal any change in the stress distribution which immediately indicates a fault.

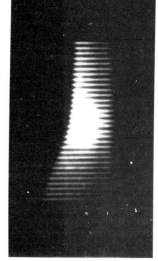 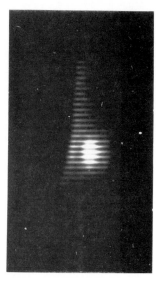

Fig. 19   Fig. 20
Stress patterns across web showing
fatigue crack developing.

Future Work

A number of areas are still being investigated by the Sira Institute. The results which have been obtained so far for example have nearly all been in a qualitative rather than a quantitative form, although calibration can be made in a number of ways. A stress calculation can for example be made at some simple point and the SPATE reading correlated with it; a single strain gauge placed in a position of reasonably constant stress can be used, or a known stress can be derived from a simple cantilever and used to establish the signal level. It is also felt that although most of the necessary information can be obtained by the instrument, full use is not yet being made of it. One possibility is to use colour to distinguish stress levels as some thermographic systems do, or the information can be stored and a processor used to output it in a variety of ways — to draw a stress contour map for example. A number of possibilities are being looked into. In the field of NDT, the results discussed above give a preliminary indication of the equipment's potential value and a more detailed assessment is at present being carried out. The time involved is also important. Although it is relatively rapid, the fact that scanning has to be provided in two dimensions since only a single detector is at present in use, makes it slower than might be desired. Considerable advances have however been made recently in the development of IR detector arrays and by using a suitable one the scan time could be enormously reduced. This might also help with another point. In many cases it is an advantage to be testing under dynamic conditions, but it means that equipment must be provided to vibrate the specimen, and there are occasions when knowledge of the stress arising from a single shock load would be desirable. If an array capable of viewing a large enough area became available, results could for example be obtained instantaneously. A considerable amount of investigation work is still being carried out since this is a completely new system using a quite novel principle and its full capabilities are not yet known. However, the above gives an indication of the considerable future potential of the instrument.

Conclusion

It has been shown that comprehensive information on the distribution of stresses over an engineering component under dynamic loads can be obtained remotely by means of an instrument recently developed by the Sira Institute. The information is obtained both rapidly and cheaply and is particularly easy to interpret. The method is also shown to have considerable potential in the field of Non-Destructive Testing.

References

1. Belgen, M. H., Structural Stress Measurements with an Infra-read Radiometer, ISA Transactions, Vol. 6, pp. 49-53. 1967.

A NOVEL ELECTRO-OPTICAL TECHNIQUE IN METROLOGY

A. H. Falkner
Department of Electrical and Electronic Engineering
Lanchester Polytechnic
Coventry, U. K.

Abstract

A new electro-optical technique has been exploited in devices of high performance for the comparative metrology of certain features of small mechanical components. The technique and the devices are described and typical performance figures given.

Introduction

New instruments have been developed for the measurement of certain features of small mechanical components exploiting a novel electro-optical technique. This technique allows the precise comparison of the features of nominally identical components, and the instruments are therefore suited to the metrology of runs of nominally identical components by comparison with a master component. High accuracy is obtainable using very simple optical systems because the comparison takes place at the real size of the component rather than with magnified images. For the accuracy obtained the speed of the devices is high. The technique lends itself to unskilled or automatic operation. The test programme has shown the instruments to be robust and suggests that they form attractive alternatives to conventional instruments in some applications. The technique has been applied to the measurement of hole position, hole direction and profile measurement at discrete points. In the following sections the technique is first described, then the instruments exploiting it with details of experimental results.

Principle of the technique

Fig.1 shows a parallel beam of light passing through circular holes in two plates onto a photo-cell. The technique is required to detect very precisely the existence of any misalignment of the holes on the beam. To do this one of the plates, the lower one for example, is vibrated with a small circular motion constant orientation being maintained. Typically this motion has a diameter of 0.1 mm. for holes of up to 2mm. diameter. Then, if the holes are aligned, the total light falling on the photo-cell does not vary during the cycle of vibration. If, however, there is misalignment then the total light varies during the cycle generating an alternating electrical signal. The amplitude measures the magnitude of the misalignment although the relationship is not linear. The phase measures the direction of misalignment. It is also possible to detect misalignment in the x and y directions using electronic processing of the signal with suitable reference phases. Again these signals are not linear functions of the misalignment in these axes but a nulling technique is used which overcomes this problem. The upper plate, the one whose hole position is to be measured, is moved in the x and y axes until alignment occurs. The necessary servo systems are operated from the x and y error signals. The x and y motions required measure the relative positions of the holes in the two plates. It is necessary that these motions can be measured precisely and the way this is achieved is described below. The use of nulling has the added advantage of making the device output independent of light intensity, photo-cell sensitivity, eccentricity in the vibrating motion, and many of the parameters of the associated electronic system. More strictly it is possible to generate second order errors if two or more of these parameters drift. In this system any symmetrical lack of parallelism in the light beam has no effect. Similarly diffraction effects are symmetrically disposed and are of no consequence. Thus it is not necessary to use a monochromatic light source, nor need it generate a beam of high parallelism. A very simple system comprising a lamp, a pin-hole and a single convex lens is adequate. If the beam is slightly convergent or divergent then there will be errors off the beam axis, but these are eliminated by fully comparative use in which the lower plate is a reference and plates in the upper position are compared among themselves. The next section shows how the technique is exploited practically.

The Measurement of Hole Position

Fig.2 is a block diagram of the device[1]. The single photo-cell is replaced by an array which corresponds in lay-out to the pattern of holes to be gauged. The reference component is fixed in the instrument and each gauged component is loaded onto a fixture. The former is vibrated in the way described above by an electro-magnetic force coil system. To measure the comparative position of a particular hole in the gauged component the corresponding photo-cell is selected electrically and the signal from this processed to detect the misalignment in the x and y axes of this hole in the two plates. Second order servo systems are used and the feedback signals move the gauged plate, again using force coils. The x and y motions required for this nulling are proportional to the currents in these coils and are available directly as electrical signals measuring the comparative position of the hole. The design of the suspension systems for the fixtures holding the plates is critical and must be such that the motions depend linearly on the electro-magnetic forces without hysteresis. This is achieved using flat springs, a robust system using no sliding or rolling parts. Fig.3 is a photograph of the device and in the centre is the platform A on which the gauged component is mounted. The reference component is concealed beneath this. B and C are one of the magnets and its associated force coil. The light source D comprises a 50 W. quartz halogen lamp illuminating a 0.5 mm. pin-hole which is at the focus of the lens E. The small angular spread in the light at any point of the emergent beam is acceptable for the reasons given above. The range of the instrument is ±0.5 mm. in each axis and the settling time 1 second for each reading. In the test programme small bearing plates with holes ranging in diameter from 0.3 to 1.0 mm. were used. In this case the loading of the component is much less repeatable than the potential accuracy of the device, and therefore one of the holes being measured is regarded as the datum and another as a twist reference, corresponding to the component specification. Then the final results require some calculation which can be carried out manually or by computer. In the tests a BASIC programme was used. The overall accuracy is of the order of ± 2 μm ±2% of reading in each axis. This is based partly on the repeatability obtained and on knowledge of the device operation. Other conventional devices which were available proved not to be accurate enough to afford a useful comparison.

Profile Measurement at Discrete Points

The above device has also been used, with some small modification, to measure a profile at discrete points[2]. In this case the fixture holding the gauged component has dummy edges opposite these discrete points leaving small gaps. A special 'reference component' is used, a plate having a hole under each gap. The servoing is then perpendicular to the edge in each case and the null occurs when the gap is symmetrically disposed over the hole. The motion required again measures the dimension comparatively, but this time there is a factor of 2 relating the motion to the dimension. The component may still have holes which are gauged, in the way already described, in the same loading of the machine. The accuracy in this extended use has been found to be similar to that of the original use.

The Measurement of Hole Direction

The technique has been extended into three dimensions in a device which measures comparatively the direction of a cylindrical hole in a small solid[3]. The latter is vibrated about two mutually perpendicular axes in such a way that the axis of the hole describes a cone of small angle. This is shown in fig.4. Any misalignment between the light beam and the hole direction then generates an alternating signal from the photo-cell. As with the hole position device, signals indicating x and y errors are derived and these are used to twist the component about the axes until alignment occurs. The magnitudes of the twists required measure the error in the direction of the hole. Again, force coils are used and the currents required for the nulling form the required outputs. The suspension comprises springs which are arranged to allow twisting but not displacement. These are clearly visible (A and B) in fig.5, a photograph from above of the part of the device holding the component C. The light beam passes down through the hole onto the photo-cell. The force coils D pass through the gaps of the magnets E.

The tests have used a hole of diameter 3 mm. and length 10 mm. The measured performance is

Range	±2° in each axis
Accuracy	±7' of arc ±2% of reading in each axis
Speed	1 second

This represents much higher speed but much worse accuracy than are available with alternative methods. The disappointing accuracy is probably associated with problems of reflections from the internal walls of the hole.

Current Developments and Conclusion

Currently work is proceeding with devices to measure other aspects of small components and to extend the ideas to larger components. The results achieved to date suggest that the techniques will result in a range of instruments of high performance and relatively low cost which will be useful in the production situation for long or even short runs. There may be applications in toolrooms and metrology laboratories.

Acknowledgments

The author is grateful to the United Kingdom Research Council for supporting the programme and to the Lanchester Polytechnic for making available the facilities used. Some of the devices are subjects of patent applications.

References

1. Falkner, A. H., 'A novel comparator for the location of holes in plates', J.Phys.E. Vol.7 pp.798-800. 1974.
2. Falkner, A. H., 'A novel comparator for the location of holes and edges in plates.' Ibid, Vol.8, pp.81-2, 1975.
3. Falkner, A. H., 'Novel electro-optical alignment technique.' Proc.I.E.E., Vol.125, pp.33-7, 1977.

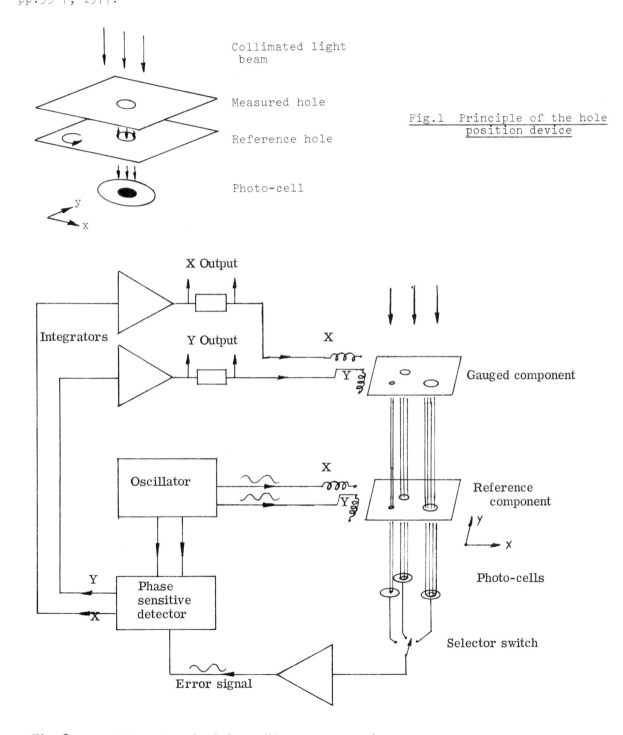

Fig.1 Principle of the hole position device

Fig. 2   The system for hole position measurement.

Fig.3 The hole position device

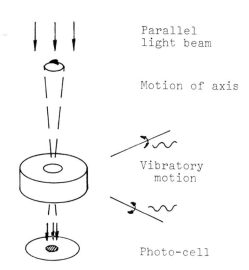

Parallel
light beam

Motion of axis

Vibratory
motion

Photo-cell

Fig.4 Principle of the
hole direction device

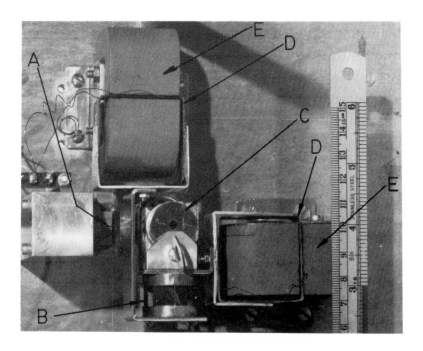

Fig.5 Detail of the hole
direction device

USE OF ELECTRO-OPTICS IN LASER VELOCIMETRY

Alain Boutier

Office National d'Etudes et de Recherches Aerospatiales (ONERA)
92320 Chatillon, France

Abstract

In laser velocimetry, flow velocities are deduced from the frequency of the light scattered by microparticles transported by the flow and seen through a fringe pattern.   This frequency being independent of the sign of the velocity, this sign can be determined by moving the fringe pattern : this operation makes use of acousto-optical modulators which change the frequency of the laser beams interfering in the probe volume. A general discussion of this method compared to other possibilities is given in this paper.

1. Introduction

In aerodynamic flows, the local velocity and its fluctuations are very important parameters. Classical techniques use probes (such as pitot tubes or hot-wires) which disturb locally the flow and whose output signal is not only a function of the velocity. For ten years, laser velocimetry has been widely developed ; this optical technique does not introduce any mechanical probe into the flow, and the instrument signal is a linear function of the velocity parameter only.

As a basic principle, the laser velocimeter uses two coherent laser beams, issued from the same laser source, which cross each other ; inside the small crossing volume, a fringe pattern appears : the fringes are equidistant parallel planes. Tiny particles, carried by the flow, cross this fringe pattern so that the light they scatter is modulated : the modulation frequency is proportional to the flow velocity component perpendicular to the fringe planes.

In highly turbulent flows, in recirculation or separation zones, which are of peculiar interest in aerodynamic studies, particles may cross the fringe pattern along any direction, so that the velocity component sign may be either positive or negative ; in order to remove this ambiguity, a solution is to make the fringe pattern move at a constant velocity, along the direction of the measured velocity component. To obtain this effect, various mechanical and electro-optical means have been put into operation, that are reviewed in this paper. Their purpose consists in shifting the laser light frequency.

To study periodic flows (in turbomachines for instance) or large scale turbulence phenomena, it is necessary to take into account the velocity informations at given times ; this may be achieved by an electro-optical shutter placed at the laser output (cf. §5).

2. Laser velocimetry principles — 2.1. Fringe velocimeter

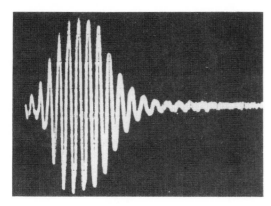

Fig. 1. Principle of laser velocimetry.

D : beam splitter,
O_1 : emitting lens,
O_2 : receiving optics,
v : probe volume,
$P.M.$: photomultiplier,
i : fringe spacing,
λ_0 : laser wavelength.

Two coherent laser beams are crossed and focussed in a small probe volume where an interference fringe pattern is created, with a fringe spacing i (fig. 1). A submicron particle carried by the flow, passing through this fringe pattern, meets alternatively dark and bright fringes, so that it scatters a light energy modulated at a frequency ν_D proportional to the velocity component u perpendicular to the fringes. The basic relationship of laser velocimetry is [1] :

$$u = \nu_D\, i \qquad (1)$$

This modulated scattered light is collected by an optical system which focusses it on a photomultiplier transforming it into an electric signal having the shape drawn on figure 2. This signal is composed of a series of quasi-Gaussian bursts containing the modulation : each burst corresponds to a particle crossing the probe volume ; its shape is due to the quasi-Gaussian intensity distribution of the laser beam.

These signals are filtered to be symmetrical relative to the zero level (ground) (see fig. 3) and are often processed by counting techniques [2] ; the period of the signal in each burst is measured with a great precision ; validation criteria are involved in the processor in order to eliminate erroneous measurements and measurements due to large particles, which do not follow the flow and provide signals of great amplitude.

Then a mini-computer performs statistics on these instantaneous values in order to calculate the mean value and the r. m. s. value of the velocity distribution (i. e. turbulence intensity) ; it is also possible to obtain the velocity turbulence spectrum.

Fig. 2. General shape of the signals.

Fig. 3. Typical filtered signal.

2.2. Velocity sign determination

Two particles crossing the fringe pattern along the same axis but in opposite directions provide the same signals. In order to determine the velocity sign, a possibility is to make the fringe pattern move at a constant velocity u_d, which is algebraically composed with the flow velocity u, so that the resulting measured velocity u_r is always positive :

$$u_r = u_d + u > 0 \qquad (2)$$

To make the fringe pattern move, the basic idea consists in slightly shifting the optical frequency of one of the interfering beams relative to the other. Then if ν_1 and ν_2 are the optical frequencies of the two beams, the fringe pattern velocity is :

$$u_d = (\nu_1 - \nu_2)\, i \qquad (3)$$

i being the fringe spacing.

A constant frequency shift of laser light is equivalent to a phase shift varying linearly with time. This can be achieved by various means described in §4.

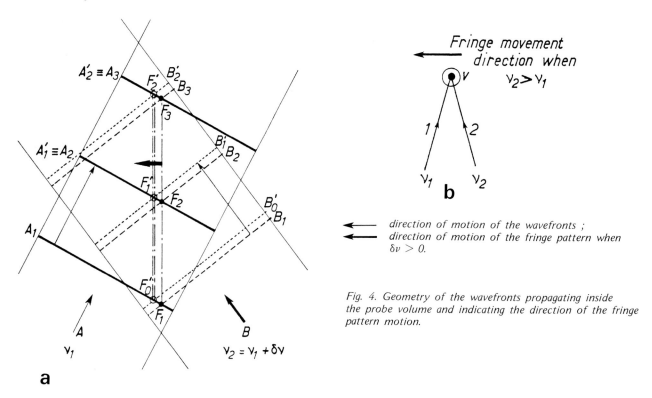

Fig. 4. Geometry of the wavefronts propagating inside the probe volume and indicating the direction of the fringe pattern motion.

The direction of motion of the fringe pattern can be easily determined by the simple argument developed below. Figure 4 shows the crossing of two beams, beam A of frequency ν_1 and beam B of frequency $\nu_2 = \nu_1 + \delta\nu$. If we look at this pattern at the time t_0, the amplitude maxima of the wave propagated in beam A are located at the positions shown by the solid lines of figure 4 (A_1, A_2, A_3, ...) separated by $\lambda_A = c/\nu_1$ (c being the light velocity), and the amplitude maxima of the wave propagated in beam B are located at the positions shown by the dotted lines separated by $\lambda_B = c/(\nu_1 + \delta\nu)$ (B_1, B_2, B_3, ...). At the time $t' = t_0 + T$, with $T = 1/\nu_1$, the solid line A_1 has moved to A'_1, superimposed to A_2, with $A_1A'_1 = cT = c/\nu_1 = \lambda_A = A_1A_2$, and the dotted line B_1 has moved to B'_1, with $B_1B'_1 = \lambda_A$. If $\delta\nu$ is positive, the displacement of a dotted line $B_1B'_1(\lambda_A)$ is greater than the line separation $B_1B_2(\lambda_B)$.

The fringe pattern is determined by the successive crossings of a solid line and a dotted line, i.e. at time t_0 : F_1, F_2, F_3, ... (see fig. 4). When $\delta\nu$ is positive, at time t', these crossings are located at F'_1, F'_2, F'_3 ... , which are on the left of F_1, F_2, F_3.... . Then it appears that the fringe determined at time t_0 by $F_1F_2F_3$ has moved to $F'_0F'_1F'_2$ at time t' towards the left. It is clear that if $\delta\nu$ is negative, the fringe is moving towards the right.

In conclusion, the fringes are moving from the laser beam of higher optical frequency towards the laser beam of lower frequency before these beams cross other (see fig. 4 b).

Another method has been proposed in the literature to determine the velocity sign : it consists in creating two parallel fringe patterns having distinct characteristics, very close to each other ; the velocity sign is determined by looking which fringe pattern the particle crossed first. To distinguish the two fringe patterns, two methods have been investigated : two perpendicularly polarized fringe patterns[3] and two different color fringe patterns (using the green and blue wavelengths of an argon laser)[4]. As in highly turbulent flows the velocity vector can have any direction, these arrangements have the following drawback : they are unable to perform measurements on particles with trajectories nearly parallel to the fringes ; actually, such methods cannot measure the zero velocity. This is why, in laser velocimetry, operational systems use mainly frequency shifting devices.

3. Two-dimensional laser velocimeter

In §2 we indicated the laser velocimetry principles which allow the measurement of one velocity component with its sign. In two-dimensional flows and for detailed turbulence studies, it is necessary to know the velocity vector fluctuations. This is obtained by measuring the instantaneous velocity vector, i.e. by measuring simultaneously two velocity components.

Two fringe patterns having different characteristics and distinct orientations are superimposed in the probe volume. The detection subsystem delivers two electrical signals in parallel, each one being representative of a fringe pattern : orthogonal polarisations[5], two colours (blue and green lines of an argon laser)[6], two different fringe motions[7]. In this last case, the two fringe patterns, having the same wavelength and the same polarization state, move at two different velocities ; there is only one photodetector ; the signal separation is achieved by means of two electronic bandpass filters set in parallel (system studied at AEDC Tullahoma - USA). Two crossed Bragg cells are used, creating four beams set in the configuration of figure 5. In this apparatus, the crossed Bragg cells have a double function : they split the laser beams into four beams and they determine the velocity sign for each component according to the principle mentioned above.

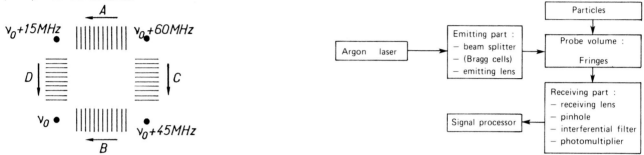

Fig. 5. Beam configuration at the output of the two
AEDC crossed Bragg cells.

A and B : two fringe patterns, moving horizontally
in phase, with a frequency motion of 45 MHz ;
C and D : two fringe patterns, moving vertically
in phase, with a frequency motion of 15 MHz.

Fig. 6. Diagram of a velocimeter.

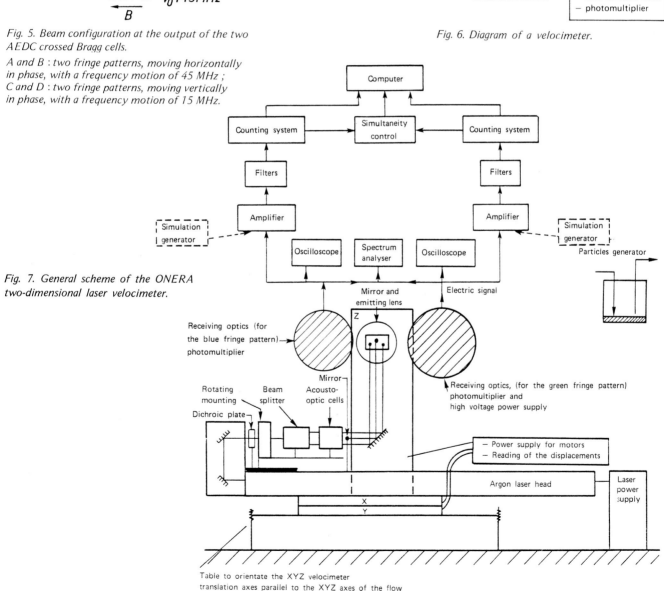

Fig. 7. General scheme of the ONERA
two-dimensional laser velocimeter.

Fig. 8. General view of the ONERA two-dimensional
laser velocimeter.

Fig. 9. Beam dividing system.

A laser velocimeter includes optical, mechanical and electronic elements (see block diagrams on figures 6 and 7). The ONERA velocimeter is shown on photograph of figure 8. It is a two-colour fringe velocimeter, equipped with Bragg cells for velocity sign determination of each component. A description of this velocimeter is given in ref.[8]. Its main components are the following :
- light source : argon laser with an all-lines power of 15 watts,
- separation of blue and green wavelengths by dichroïc plates,
- two beam dividing systems set in parallel, each comprising a beam splitter and acousto-optic modulators (for velocity sign determination) (see fig. 9),
- emitting lens, with a focal length ranging from 300 mm to 2 m,
- receiving optics, with an aperture varying between f/3 and f/10,
- photomultiplier tubes, with S20 photocathode, in front of which are monted 10-nm-bandwidth interferential filters,
- probe volume translations along three orthogonal axes, with 200 mm amplitudes ; optical encoders ensure a 0.01 mm probe volume localisation,
- signal processing by counters[9],
- particle injectors for incense smoke, dioctyl-phatalate vapour[10] and solid particles.

The measurement precision is limited by three factors :
- the particle behaviour (submicron particles must be used)[11],
- the signal processor (less than 1%),
- the measurement of the fringe spacing or the angle between the crossing beams (using a theodolite is a good method) : 0.1 to 0.3%.

4. Various means for shifting the frequency of laser light

There are three types of devices for shifting the frequency of laser light : mechanical, electro-optic and acousto-optic devices. We indicate here the basic principles of these systems, their efficiency and the problems met in the practical application to laser velocimetry.

4.1. Mechanical devices

When a circularly polarized laser beam passes through a half wave plate rotating with a velocity $\omega = 2\pi\nu'$, the laser wave frequency is either increased or decreased by ν' ; with such a rotating half wave plate, the laser beam is not deflected and the efficiency is theoretically 100 %. However the frequency shift does not exceeds a few tens of kHz, and thus is not suitable for laser velocimetry applications, which require a few MHz[12].

The first order beam scattered by a rotating grating is frequency shifted by doppler effect[13] ; the ν' shift is given by laser velocimetry relationship :

$$\nu' = \frac{u_{grating} \cdot \sin\theta}{\lambda} \tag{4}$$

$u_{grating}$ = rotation velocity of the grating,
θ : angle between incident and scattered beam,
λ : wavelength of the incident laser beam.

As in the first order of a grating there is the relation

$$\sin\theta_1 = \lambda/d \tag{5}$$

d being the groove spacing of the grating, we can write :

$$\nu' = \frac{u_{grating}}{d} \tag{6}$$

The values of ν' so obtained extend up to 10 MHz ; however the maximum efficiency is about 70%.

4.2. Electro-optical devices

The transverse Pockels effect is used in KDP crystals : a rotating electric field is induced which either accelerates or decelerates a circularly polarized laser beam[14]. In its principle, it is equivalent to a rotating half wave plate, but with the following advantages :
— no moving part,
— accurate frequency control,
— higher frequency shift, but limited to a few MHz.
A 92% efficiency can be obtained with four cells placed in series. The main disadvantage is that such devices require a voltage power supply of several kV.

An electro-optical phase modulator has also been built, which changes the laser beam phase by Pockels electro-optic effect in ADP crystals[15] ; instead of creating a uniform translation of the fringes in space, the phase modulation tends to impose a cyclical movement to the fringes : a sawtooth waveform signal of about 100 V is applied to the crystal in order to achieve a relative phase change of 2π between the laser beams crossing within the probe volume : in this way the spatial displacement of a fringe is equal to one fringe spacing. The fringe pattern movement obtained is equivalent to a frequency shift extending up to 1 MHz. The fly-back period in the sawtooth signal is very short, but cannot be neglected. The transmission of such cells is 60%.

4.3. Acousto-optical devices

Acoustic sinusoïdal waves of frequency ν', issued from a piezo-electric crystal, propagate through a liquid (Bragg cell) or a crystal (acousto-optical modulator). An incident laser beam of optical frequency ν_0 is scattered by Brillouin effect, and the frequency of the first order (+ 1) scattered beam is $\nu_0 + \nu'$ (fig. 10) ; the phenomenon can be explained by the fact that the piezo-electric crystal generates a moving phase grating : thus the frequency shift is similar to that of a rotating grating, as described above. The current values of ν' vary between 40 and 200 MHz. These values are often too high because commercially available signal processors are limited to a few tens of MHz. To overcome this difficulty, two systems are proposed :

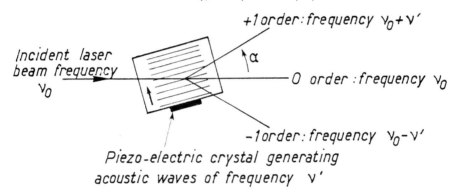

Fig. 10. Acousto-optical modulator (Bragg cell) scheme.

— In the first one, a Bragg cell is placed on one beam after the beam splitter, creating a frequency shift of 40 MHz for instance. The signals issued from the photomultiplier have a frequency $(\nu_D + 40)$ MHz, ν_D being the doppler frequency due to the particle velocity . $(\nu_D + 40)$ is electronically mixed down with a frequency chosen between 31 and 39 MHz[16]. If for instance a 33 MHz frequency is taken, the processor receives signals of frequency $(\nu_D + 40 - 33)$ MHz. This is equivalent to a frequency shift of 7 MHz. In order to avoïd the mixed frequency (ranging between 31 and 39 MHz) entering the processor when no particle is crossing the probe volume, an electronic low-pass filter (with a 20 MHz cut frequency) is set at the processor input. This imposes for instance a higher limit of 11 MHz on the doppler frequency ν_D when the equivalent frequency shift is 9 MHz. Considering the basic relationship of laser velocimetry $u = \nu_D i$, it is pointed out that for high velocities the fringe spacing must be very much increased, which leads to a poorer spatial resolution.
— In the second one, a modulator is set in each of the two laser beams after the beam splitter ; one modulator provides a shift ν' and the other one a shift $\nu' + \nu''$, ν'' being a few MHz. This way, the crossing beams have the following frequencies :

$$\begin{cases} \nu_1 = \nu_0 + \nu' \\ \nu_2 = \nu_0 + \nu' + \nu'' \end{cases}$$

and $\nu_2 - \nu_1 = \nu''$ is a suitable value for the fringe motion. The downshift of ν' is optically achieved here, instead of electronically in the previous system. This second method is advantageous because the higher limit imposed to ν_D is not so drastic : for a usual commercial counter it is 30 MHz (ν'' ranging usually between 2 and 8 MHz), because the highest frequencies processed with a good precision is 38 MHz.

The angle α between the first order beam and the zero order one is given by :

$$\sin \alpha \simeq \alpha = \frac{\lambda}{2\Lambda} \tag{7}$$

λ : incident laser beam wavelength,
Λ : acoustic wavelength in the liquid or the crystal,
$\Lambda = V/\nu'$; V is the sound propagation velocity through the Bragg cell.

In order to limit the stray-ligh amount near the probe volume, it is desirable to let only pass the first order beam through a mask which stops the other diffracted beams and the zero order beam ; for that, the higher α (i.e. the higher ν'), the shorter the distance from the Bragg cells at which this mask can placed.

Prisms are used to compensate α, because in the optical velocimeter geometry it is suitable to make the laser beams at the Bragg cells output parallel to the entering beams. In the first system proposed above, only one prism is necessary, be cause the unique Bragg cell is supplied by a single frequency. It is however necessary to compensate the optical path length with a slab placed in the second beam. In the second system proposed above, it is necessary to place after each Bragg cell a prism compensating α' due to ν' and, moreover, on one beam an angular variable prism to compensate α'' due to ν'' : in fact ν'' varies between 2 and 8 MHz.

The material through which the sound waves are propagating must be chosen with care in order to resist the high power laser beams (about 5 W in each beam, at the beam splitter output, with a nominal 15 W argon laser). In some devices one solution consists in placing the crystal between two cylindrical lenses in order to spread inside the crystal the laser energy along a line parallel to the acoustic wavefronts.

Acousto-optical modulators are widely used in laser velocimetry because they are easy to set up ; they necessitate low voltage power supplies (a few volts) ; a good quality of the emerging beams and an efficiency superior to 85% are obtained (if the laser beam diameter entering the Bragg cell is less than 3 mm). A few MHz frequency shifts, suitable for laser-velocimetry, are available. A good stability of ν' and $(\nu' + \nu'')$ is obtained with quartz oscillators, when ν' is in the 40 MHz range ; if ν' is higher (up to 200 MHz), it is easier to compare $\nu'' = (\nu' + \nu'') - \nu'$ to a quartz oscillator in a frequency loop. Anyway, a 10^{-3} precision of the frequency shift is usually obtained ; this way, the presence of Bragg cells in a laser velocimeter does not affect the overall precision, the main errors coming from the signal processor (about 1%) : cf. §3.

5. Acousto-optical modulator used as a shutter

Another type of velocimeter, based on the optical barrier principle, has been designed for turbomachine applications[17] : the particles pass successively through two bright spots, produced by two highly focussed laser beams.

When measurements are performed inside the channels between the rotor compressor blades, the blades cut the illuminating laser beams and dazzle the photomultipliers. In order to avoid this problem, a trigger optics is placed at the laser output. The laser light can illuminate the probe volume at chosen times, synchronised with the compressor wheel rotation : this allows stroboscopic measurements inside the periodic flow, as well as the photomultiplier protection against dazzling.

The trigger optics is usually an acousto-optic modulator ; at its output, the first order diffracted laser beam is used to illuminate the probe volume : the deflection efficiency is better than 90% and the contrast ratio is about 10^3 : 1. Illumination times of a few microseconds are then possible, with 20 ns risetimes.

6. Results obtained for a typical application in a wind tunnel

In most interesting aerodynamic applications, the velocity sign determination is generally mandatory as soon as the turbulence intensity is higher than 30%. This is the case in studies of boundaries of free jets, recirculation, separation and mixing zones.

To illustrate the laser velocimeter possibilities, when equipped with acousto-optic modulators, we present measurements performed in a typical region of a transsonic wind tunnel (S8A Chalais-Meudon) where the shock-wave —boundary layer interaction is studied. The velocity and its turbulence are measured inside a separated flow, downstream a shock[18]. Figure 11 shows a typical evolution of the velocity u and of its fluctuations $\sigma = \sqrt{\overline{u'^2}}$; $z = 0$ corresponds to the wall. Near the wall the mean velocity becomes negative, but with large fluctuations, as the velocity histogram points out on figure 12. The laser velocimeter, with acousto-optical modulators, is the only apparatus able to provide such results, because it is sensitive to the velocity sign direction ; moreover no material probe is introduced to disturb this rather complicated flow.

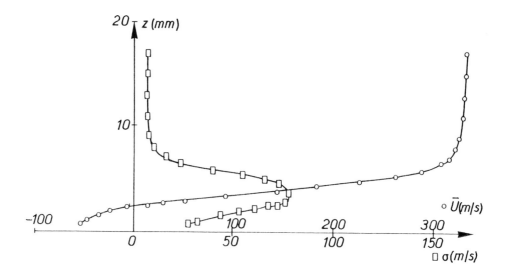

Fig. 11. Vertical exploration of a separation zone downstream of a shock at the S8A Chalais-Meudon wind tunnel.

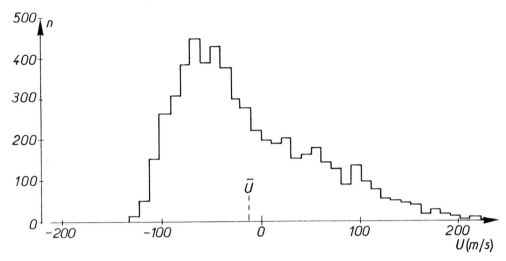

Fig. 12. Typical histogram at S8A, when the mean velocity is near zero (separation zone).

7. Conclusion

Since acousto-optical modulators have been introduced into laser velocimeter equipments, velocity measurements have been performed in very complex flow situations, such as in turbomachines where they are used as fast choppers, and in highly turbulent flows where they permit the determination of the velocity sign by frequency shifting of the laser light.

Acousto-optical modulators are preferred to the other methods listed above because they are easy to use, they necessitate low voltage power supplies and their efficiency is good (better than 85%) ; moreover they do not introduce any extra uncertainty in the measurements.

References

1. Boutier, A., *Vélocimètre compact pour mesures dans des écoulements très turbulents*, N.T. ONERA No 237, 1974.
2. Boutier, A., Data processing in laser anemometry. *Proceedings of the ISL/AGARD Workshop on Laser Anemometry.* May 5-7, 1976, Report ISL 117/76.
3. Iten, P.D., Dändliker, R., Direction sensitive laser doppler velocimeter with polarized beams. *Applied Optics*, Vol.13, No 2, p. 286-290, Feb. 1974.
4. Pfeifer, H.J., vom Stein, H.D., *Un procédé pour déterminer le signe de la vitesse dans la métho¹e différentielle par effet doppler.* Rapport interne ISL (1971).
5. Orloff, K.L., Logan, S.E., Confocal backscatter laser velocimeter with on axis sensitivity. *Applied Optics*, vol. 12, No 10, October 1973.
6. *Proceedings of the Purdue Workshop.* July 11-13 1978. Purdue University, Lafayette, Indiana (USA). To be published.
7. Crosswy, F.L., Hornkohl, J.O., Lennert, A.E., Signal characteristics and signal conditioning for a vector velocity laser velocimeter. The use of doppler velocimeter for flow measurements. *Proceedings of the Purdue Workshop. March 9-10, 1972.*
8. Boutier, A., Lefèvre, J., Perouze, C., Papirnik, O., Operational two-dimensional laser velocimeter for various wind-tunnel measurements. *Proceedings of the Purdue Workshop, July 11-13 1978.* Provisional edition : TP ONERA 1978-74.
9. Pfeifer, H.J., Schafer, H.J., *A single counter technique for data processing in laser anemometry.* ISL Report No 30/74 (1974).
10. Liu, B.Y.H., Whitby, K.T., Yu, H.H.S., A condensation aerosol generator for producing monodispersed aerosols in the size range, 0.036 μm to 1.3 μm. *Journal de Recherches Atmosphériques, (1966)*, p. 397-406.
11. Haertig, J., Particle behaviour, in *ISL Report R 117/76 (1976).* Proceedings of the ISL/AGARD Workshop on Laser Anemometry, May 5-7, 1976, p. 1-39.
12. Durst, F., Zaré, M. Removal of pedestals and directional ambiguity of optical anemometer signals. *Applied Optics*, Vol. 13, No 11, Nov. 1974.
13. Oldengarm, J., van Krieken, A.H., Raterink, H.J., Development of a rotating grating and its use in laser velocimetry, in *ISL Report 117/76 (1976).*
14. Drain, L.E., Moss, B., The frequency shifting of laser light by electro-optic techniques. *Opto-electronics*, vol. 4, Nov. 1972.
15. Foord, R., Harvey, A.F., Jones, R., Pike, E.R., Vaughan, J.M., A solid state electro-optic phase modulator for laser doppler anemometry. *J. Phys. D, Applied Physics*, vol. 7, 1974. Letter to the editor.
16. *DISA LDA frequency shifter type 55N10. TSI frequency shift system model 980.*
17. Schodl, R., A laser dual beam method for flow measurements in turbomachines. *Gas Turbine Conference Products Show.* ASME, Zurich, March 30 - April 4, 1974.
18. Delery, J., Lacharme, J.P., *Interaction onde de choc - couche limite turbulente en écoulement transsonique stationnaire.* Internal ONERA report, April 1978.

QUANTUM EFFICIENCY AND PHASE IN MODULATION TRANSFER FUNCTION (MTF) MEASUREMENT

S. L. Boersma, Consultant*

12 NicBeetslaan 2624 XP Delft, Holland

Abstract

This paper gives design considerations for a new MTF machine (ODETA V).
The new machine measures optics, but most important feature is rapid MTF measurement
of low light level devices like Image Intensifier Tubes. Photon and Detector noise
are of paramount importance. A variety of MTF measuring methods is evaluated for noise.
They include TV pickup tubes and self scanning photo detector arrays, slit scanning
of the line spread function and mask scan methods: Pseudo Noise masks, Hadamard and
Moiré Sine Wave masks.
  A new Moiré pattern generator is shown. New ODETA V is a programmable MTF machine.
It provides a complete MTF read out on CRT (for focussing), X-Y recorder and digital
printer. Scan speed (photon integration) is variable in a 20 to 1 range for optimum noise.
Single frequency read out can be programmed at up to 5 user selectable frequencies.
In this mode very few photons are wasted on the unwanted intervening frequencies.
  A PTF module is available as an optional extra.
Phase is believed to be not really necessary for Image Quality Assessment.
It is shown that phase (PTF) is not independent of amplitude. Constraints exist.
An ideal MTF precludes bad phase response.

Introduction

There is a feeling that micro processors will revolutionize MTF measurement.
But beware of this magical microprocessor myth. The only thing a microprocessor can do
(and does in ODETA V) is printing out the results, if necessary corrected for
Relay Optics MTF and optical magnification.
It can do nothing to improve speed or noise beyond the physical limits set by the
front end of the machine. The very first thing a MTF machine is called upon is to
render the spread function (or some related transform) into an electrical form.
This rendering has to be precise and with the lowest possible noise.
Wrong front ends, like slit scanning of the spread function, produce too much noise.
And once this has happened " All the kings horses , all the kings electronics men,
cannot put together the MTF again".
  Of course a microprocessor memory , or any other integrator for that matter, can
reduce the noise of a bad front end. But only at the expense of a long measuring time.
Much longer a time than needed with a good image scan method.
So first of all lets look at the front end.

Paul Nipkof

The problem of transforming an optical image into an electrical signal is a very old one.
It was first solved in 1884 by Paul Nipkof of Berlin (ref 1).
He devised the famous Nipkof disc with a spiral of pinholes.
Some MTF machines still use his 1884 method. In the one dimensional case of MTF
Nipkof means scanning the spread function by a narrow slit. Just like Nipkofs TV
this system has two serious sources of noise:
(1) <u>Detector Noise</u> : The photocell sees very little light passing through the scanning
slit. So detector dark current is a serious noise.
(2) <u>Photon noise</u> : Nipkof did not bother about photons. He patented his disc 20 years
before the photon was invented. But today most of the photons in the image are wasted,
intercepted by his disc. The few that do pass through the pinhole (slit) show a
bad $\sqrt{N}$ noise. All this was recognized in the early days of television.
It lead to the development of the storage pickup tube (orthicon, vidicon).
Here photons at each picture element are integrated simultaneously <u>before</u> scanning.
This system makes the best possible use of all the image photons.
  So a TV pickup tube looks the best possible front end for a MTF machine.
But alas, TV tubes are no precision instruments. They can be non linear ($\gamma \neq 1$ and
blooming) and do have limited dynamic range: 10^2 – 10^3 against 10^5
for a photo multiplier tube (PM Tube). Also TV tubes do have a meagre MTF and produce
more noise than necessary.
  The latest form of image storing pickup is the self scanning photodetector array.
These arrays can be used in a special purpose MTF machine (Poor Mans ODETA).
But for a general purpose high precision instrument TV pickup devices are still
out of the question.

*) Work performed under contract with OLDELFT NV of DELFT HOLLAND

Mask Scanning

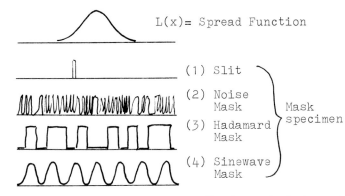

L(x)= Spread Function

(1) Slit

(2) Noise
    Mask   } Mask
                    specimen
(3) Hadamard
    Mask

(4) Sinewave
    Mask

The spread function L(x) can be
scanned by a slit or by extended
masks. Extended masks can be:
(2) Pseudo Noise Pattern
(3) Hadamard Mask
(4) Sine Wave Mask

Full image information is
retained if N mutually orthogonal
consecutive masks are used.

Fig 1   Mask Scanning of Spread Function

Rather surprisingly one found that Nipkof could be improved upon by other means than
image storing. It is not necessary to scan the image by a tiny pinhole or a narrow slit.
To describe an image (spread function) at N picture points one can use a narrow slit
at N different positions. But also possible is the use of N different masks in succession
whereby each masks extends over the entire picture area. See fig 1.
Each mask has a different pattern of transmission. One takes N consecutive measurements.
In each measurement the photocell sees the entire picture, but each time through a
different mask. This yields N values.
If the masks transmission patterns are mutually orthogonal the complete picture information
is fully retained in these N integral vanues. Thus the picture (e.g. spread function)
is transformed into N electrical values. And now electronics can process these N values
into a spread function or MTF curve.
   Mask scanning is an improvement over slit scanning of the spread function:
Through a mask the photocell sees nearly all (about 50%) of the light, therefore the
detector noise is much lower than with slit scanning. A slit is of course a special
case of mask but one of very low efficiency indeed.

Fig 1 shows various possible masks:
(1) Scanning Slit (Nipkof)
(2) Pseudo Noise Pattern , only one noise pattern shown. see ref 2.
(3) Hadamard Mask (ref 3). Here a set of N rectangular waves that are mutually orthogonal.
(4) Sine Wave Pattern. Scanning with sinewaves of varying frequency produces MTF directly.

All these methods are, or could be, used in an MTF machine:
(1) is the old Nipkof scan of the spread function.
(2) a Noise Pattern has been used in a special instrument (ref 2) :
    When scanning with a noise pattern the AC RMS photo current represents directly
    a well known image quality criterium.
(3) Hadamard Patterns are not yet used in MTF machines, but to measure the spread function
    they are far better than slit scanning.
(4) Sine Wave Patterns measure MTF in , among other, Oldelft ODETA.

Noise

The signal to noise ratio (SNR) will be calculated for the various methods, the TV pickup
device, the Nipkof slit and the extended mask scan.
Consider an arbitrary image delivering A usable photons during the measuring time.
We want to resolve N picture elements (pixels). So the average number of photons
per pixel = A/N .

TV storage pickup Tube
All photons contribute to the signal. Signal = A , detector dark current = D.

$$\text{Detector Signal/noise Ratio } SNR_d = \frac{A}{D} \quad \ldots \ldots \ldots \ldots \ldots (1a)$$

Each pixel integrates on the average A/N photons , so the photon signal/noise ratio:

$$\text{Photon } SNR_{ph} = \sqrt{\frac{A}{N}} \quad \ldots \ldots \ldots \ldots \ldots \ldots (1b)$$

These values are the best obtainable by any method. They show the theoretical minimum noise. The other scan methods will be compared to this ideal standard. Practical TV tubes fall short of this ideal .

Nipkof Slit

Only 1/N part of the light in the image reaches the detector. Signal for an arbitrary image = A/N , detector dark current = D so:

$$\text{Detector } SNR_d = \frac{A}{ND} \quad \ldots \ldots \ldots \ldots \ldots \ldots (2a)$$

Each pixel (A/N photons) is only " on the air" 1/N of time, giving A/N^2 photons/pixel:

$$\text{Photon } SNR_{ph} = \frac{1}{N} \sqrt{A} \quad \ldots \ldots \ldots \ldots \ldots \ldots (2b)$$

Thus Nipkovian detector noise is Nx the theoretical minimum and photon noise is $\sqrt{N}$ x the ideal limit.

Mask methods

Mask scanning can improve the signal to noise ratio provided optimum masks are choosen, not low efficiency masks like a slit. When an optimum mask like Hadamards is placed over the image nearky all of the light reaches the PM tube (small factors like 2 ,π etc are neglected here). Therefore:

$$\text{Detector } SNR_d = \frac{A}{D} \quad \ldots \ldots \ldots \ldots \ldots \ldots (3a)$$

This is the best value attainable with any method. It equals the ideal storage TV device. But dark current D of a photomultiplier tube (PM tube) is far lower than for a vidicon. Thus masks scan has a better signal-noise ratio for detector noise than practical TV devices
What about photon noise ? For a complete picture rendering the image is covered by N consecutive masks a_{ij} each integrating 1/N of time. i denotes the mask number, j is the position of pixel L_j .One finds N integrated light values A_i :

$$A_i = \sum_j a_{ij} \cdot L_j \quad \ldots \ldots \ldots \ldots \ldots \ldots (4)$$

These are N linear equations for the N pixels L_j . A pixel L_j at point j is reconstructed (solved) by taking a linear combination of the N light values A_i :

$$L_j = \sum_i b_{ji} \cdot A_i \quad \ldots \ldots \ldots \ldots \ldots \ldots (5)$$

b_{ij} is the appropriate (Hadamard) matrix to solve the N equations for L_j .
The matrix elements b_{ji} are +1 or -1. When these are correctly choosen pixel j emerges full strength and the contributions $b_{ki} \cdot A_i$ of all the other pixels L_k cancel.
This cancelling only holds if there is no photon noise. Foreign pixels k will cancel by a judicious choice of the b_{ki} because the pixel remains unchanged during all the N measurements. But random noise in the N consecutive measurents is not correlated at all. Inverting a later noise and adding to a former one gives no cancellation but a $\sqrt{2}$ increase. The uncorrelated photon noises of all the A_i add up on a $\sqrt{N}$ basis exactly as if the different light values were simply added (that too would give a $\sqrt{N}$ noise addition). So the noise contribution to a single pixel j equals the statistical fluctiation $\sqrt{A}$ in the entire image. Therefore:

$$\text{Photon } SNR_{ph} = \frac{1}{N} \sqrt{A} \quad \ldots \ldots \ldots \ldots \ldots \ldots (3b)$$

This photon noise is as bad as Nipkofs. It is a factor $\sqrt{N}$ worse than ideal TV storage. Mask methods are a good remedy for detector noise (e.g. IR noisy detectors) but their photon noise is essentially Nipkovian.

The Hidden Pinhole

Nipkof performs bad because of the small pinhole in front of the PM tube.
With Hadamard and Sinewave masks full light reaches the PM tube. This is nice for detector noise but photon noise is no better. This is because there is "a Hidden Pinhole" after the PM tube, in the signal channel. It is a tiny pinhole in Hadamard or in Fourier space. Instead of scanning real image space like Nipkof the machine scans through frequency space, looking at only one single frequency at a time, rejecting the others.
Now if a Nipkof pinhole is kept stationary, looking at only one pixel ALL the time we get a good low noise signal of just that single pixel. If only one Hadamard mask is used instead of N consecutive ones we get a good low noise sample in Hadamard space, but who cares ?

If we scan the spread function by a sinewave mask of only one frequency we get a good
low noise sample in Fourier space. And this we can sell for it is the MTF value at the
choosen frequency, now with low noise. So one has for the 3 different mask methods:

Noise scan   : Rather noisy (small wonder)
                  Application: Direct measurement of Image Quality Criterium (ref 2)
Hadamard mask : Detector noise : excellent
                  Photon noise   : Nipkovian
                  Application   : Spread function analysis , Spectroscopy (ref 3 and 4)
                  IR detector noise reduction

Sinewave Scan : Detector noise : excellent
                  Application   : MTF measurement (ODETA)
                  Photon noise   : Excellent for single frequency read out.
                  $\sqrt{N}$ worse for read out of N frequencies.

These considerations were the basis for the design of the new low noise MTF machine.

ODETA V

ODETA V is specially designed for low light level MTF (OTF) measurements on, e.g.
second generation Image Intensifier Tubes.
From the foregoing it will be clear that a mask scan method was choosen here.

Resolving Power

An important parameter in an MTF machine is the product of window diameter W and maximum
spatial frequency F. According to Shannons sampling theorem N=2W.F is the number of
resolvable pixels . It is also the number of discernable frequencies in the MTF curve.
N=2W.F is the machines resolving power in frequency space. Frequency resolution = 1/2W
therefor a small window diameter W "blurrs" the MTF curve. One needs a big window and
a wide frequency range.
  In order to aquire the information contained in 2WF samples one needs at least N=2WF
different masks. For ODETA V N= 400 (1500 optional). If 400 different masks of window
diameter 2 cm are placed on a single plate one gets a slab 8 meter long.
MTF machines with a single target (e.g. on a rotating drum) must have this length of
target in order to have good resolving power. Shorter targets produce a blurred MTF curve.
Shorter targets can only be used if MTF read out is restricted to a few frequencies only.

Composite Masks

High resolving power can be attained by short mask plates if not one but two plates
of masks are used. In the window two masks M1 and M2 are superimposed.
The composite mask has a transmission M1.M2 . The multiplication greatly increases the
number of realizable patterns. If each of the constituent plates contain n masks
n^2 different composite product masks can be realized. As $n^2 \gg n$ two quite small plates
(65 x 30 mm) can give the same resolving power as a single 8 meter slab.

Moiré Pattern

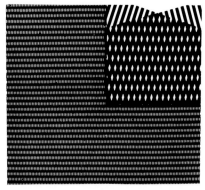

Fig 2
Two counter rotating gratings
form a moiré pattern of variable
spatial frequency. Scan includes
zero spatial frequency, facilitating
100% normalization at 0 C/mm.

The well known moiré pattern, as used in all the
ODETAs since 1963, is an example of the composite
mask technique, see fig 2. Two square wave line
gratings move over the window in opposite directions.
Each individual grating has a small area (65x 30 mm),
but the moiré product pattern forms over 400
different sinewave patterns: 0 - 12 C/mm with a
frequency resolution of 1/2W = 1/36 C/mm (18mm window).
    The gratings move in opposite direction over
the window. Moiré waves move towards the center
of rotation. The number of waves that pass a window
point remains constant if the sine of the grating
angle is made to increase linearly with time.
Then the PM tube behind the window sees a signal
of constant time frequency.
The signal is fed through a bandpass filter of center
frequency ν and bandwidth B.
This filter cuts out pattern harmonics and reduces
noise. It integrates photons over a time 1/2B .

Table 1 on the next page gives the Fourier transformation by a moving Moiré Pattern.
It shows that the electrical bandwidth B is proportional to window width W. Apparently
high resolving power in MTF frequency space requires a big window, necessitating a
wider electrical bandwidth B. So , as always, the penalty for high resolving power is
more noise. Moiré scan gives the lowest noise of any method for one frequency read out,
for N frequencies read out noise is $\sqrt{N}$ worse than an ideal TV storage pickup device.

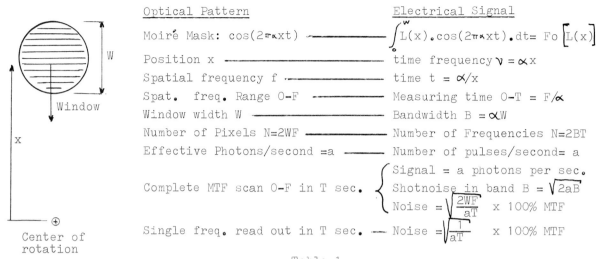

Optical Pattern	Electrical Signal

Moiré Mask: $\cos(2\pi\alpha xt)$ ————— $\int_0^W L(x).\cos(2\pi\alpha xt).dt = F_0 \big[L(x)\big]$

Position x ————————— time frequency $\nu = \alpha x$

Spatial frequency f ————— time $t = \alpha/x$

Spat. freq. Range O–F ————— Measuring time O–T = F/α

Window width W ————————— Bandwidth B = αW

Number of Pixels N=2WF ————— Number of Frequencies N=2BT

Effective Photons/second =a ——— Number of pulses/second= a

Complete MTF scan O–F in T sec. $\begin{cases} \text{Signal = a photons per sec.} \\ \text{Shotnoise in band B} = \sqrt{2aB} \\ \text{Noise} = \sqrt{\dfrac{2WF}{aT}} \times 100\% \text{ MTF} \end{cases}$

Single freq. read out in T sec. —— Noise $= \sqrt{\dfrac{1}{aT}} \times 100\%$ MTF

Table 1

Fourier Transformation by Moving Moiré Pattern.

Triangular moiré waves move over the window towards the center of rotation. Bandpass filter passes only the fundamental (cosine) wave. Noise figures are given directly in % MTF for a complete MTF scan and for a single frequency read out. 2WF , the window-frequency product, is the machine Resolving Power.

The New Pattern

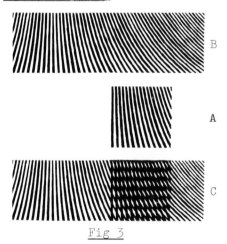

Fig 3

Hyperbolic line gratings form triangular wave Moiré Pattern.

Fig 3 shows a new moiré pattern generator. Instead of two counter rotating gratings one remains stationary and the other is moved rectilinear on top of the first. Small grating A remains stationary in window. Oblong grating B is placed on top of A and moves rectilinear from right to left.(C)
Spatial Moiré frequency increases linear with displacement. Advantages are:
(1) Frequency is linear with displacement, no need for arc sin correction.
(2) Bigger window: window can be full grating width, no arc sector cut off.
(3) Bigger Numerical Aperture: Small stationary grating can act as a dust seal, no extra glass window necessary. Thinner glass means less sferical abberation, allowing a higher aperture of the incident beam.

The Tracking Filter
Table 1 shows noise is considerably ($\sqrt{2WF}$) reduced when the MTF is measured at only one frequency. When measurement is limited to p frequencies the noise is reduced by a factor $\sqrt{2WF/p}$ or rather $\sqrt{2WF/(p+1)}$ as frequency zero has to be measured too.
For 2WF= 400 measuring at p=5 frequencies reduces noise by a factor of 8. Or conversely the light level could be reduced by a factor of 64 for same noise as with full spectrum MTF scan. So appart from full spectrum scan ODETA V should have a MTF read out at a few (5), user selectable, isolated spatial frequencies.
    But when a scan stops at, say, frequency f1 the signal frequency drops to zero and so does the output. Even when scanning is only slowed down at f1 the signal frequency will drop below the passband of the signal bandfilter and the signal gets lost.
    In ODETA V this problem has been overcome by making the band filter tunable. Both the central frequency ν and bandwidth B of this Tracking Filter are automatically scaled up and down in proportion to scan speed.

When the scan speed is lowered the filter frequency goes down too. The filter will
remain tuned to lower signal carrier frequency. Now at any arbitrary frequency f1 the
scan speed can be lowered by a factor of , say , 100 and the filter will keep tracking
the signal . A 100 fold speed reduction means stopping for all practical purposes.
The 100 x lower filter bandwidth gives the required noise integration and thus we have
the wanted spot frequency read out at spatial frequency f1.

Implementing the Tracking Filter

In ODETA V scan speed is controlled by a master clock oscillator driving a stepping motor.
Therefor the filters central frequency ν and bandwidth B should always remain proportional
to this system clock frequency.   Three solution for the Tracking Filter present themselves:
(1) Digital Filter: Filter is seen as a differential operator implemented on a
micro processor and operating at the clock rate. But e.g. a sixth order band pass filter
leads to a lengthy program. That could not be run in real time at the high Nyquist rate
required.
(2) Digital Filter by implementing its weighting function on a multiple tapped digital
delay line: now it is the hardware of which a consideral amount is needed.
(3) Active Analog Filter Digitally tuned. This hybrid solution was choosen for ODETA V
Complete tracking over a tuning range 1000 : 1 has been reached.

Programmable Moiré Scanner

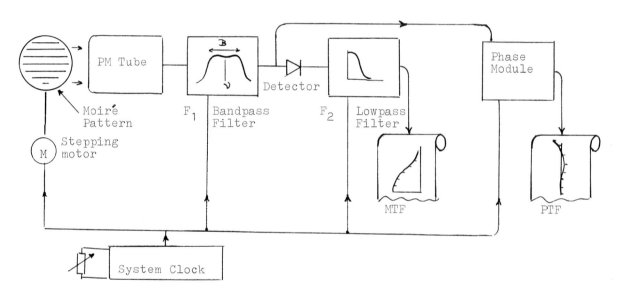

Fig 4   Programmable Moiré Scanner

Entire machine is controlled by a single programmable variable frequency clock.
Clock rate sets scanning speed by means of a stepping motor moiré pattern drive.
Thus signal frequency is strictly proportional to clock frequency.
Maste clock also tunes Bandpass Filter F_1 and Post Detector low pass Filter F_2.

Both filters will always track the signal. They stay in tune with the signal
for optimum noise rejection at all scan speeds. Read out can be a complete
MTF spectrum or MTFs at only a few, user selectable, frequencies.
At these spot frequencies scan goes dead slow for low noise.
System Clock also provides the reference signal for a Phase Measuring Module (PTF).

Fig 4 shows the Programmable Moiré Scanner from ODETA V.   the entire system is controlled
by a single programmable variable frequency clock. This clock is the only time determining
element in the machine. Scan speed, filter frequency, bandwidth and post detector filter
integration time are all tied up to this clock. When scan speed is reduced by the clock
everything stays in unison. Nothing gets out of tune, only the time scale changes.
   By setting the clock rate the user is completely free in choosing his optimum photon
integration time. Both when making a full spectrum MTF scan and when reading out at
5 isolated spatial frequencies.

Scanning Programs

The tracking filter gives complete freedom of scanning mode. ODETA V has four preprogrammed scanning modes:

(1) CRT Display : 1.2 seconds scan time for full range MTF, for focussing the optics.

(2) Single Frequency Read Out : The user selects a single frequency, e.g. for focussing at low light levels. The scanner travels fast to a frequency just below the choosen one, then slows down, lets the filters integrate the noise, makes a reading on a display and repeats this sequence. Read out is updated once every 1 - 13 seconds depending on integration time choosen.

(3) Five Sample Frequencies : On displays and Printer. Scanner scans very slow at frequency 0 (for normalizing) and at up to 5 user selectable spatial frequencies f1 -f5. Intervening frequency ranges are skipped fast. No photons are wasted there. A microprocessor corrects the frequency values f1 - f5 for optical magnification of the relay optics. The 5 MTF values are correcte for the MTFs of the relay optics. So the printed out data are true data for the unit under test.

(4) X-Y Recording of Complete MTF : on X-Y recorder. Scan speed is continuously variable from 60 to 1200 seconds full range. Speed can even be varied during scan (e.g. to reduce noise). Frequency markers are not affected. X-Y Recording of PTF : Optional extra: Phase recording on X-Y recorder.

Is PTF (Phase) really worthwhile ?

For image quality assessment phase recording is only seldom justified:

(a) It requires a considerable effort from the operator: for MTF only the slit can be anywhere inside the analizer window. But for PTF the slit must be adjusted very near the windows center, which is the phase reference point. This takes time.

(b) The PTF is very sensitive to mechanical noise : at 12 c/mm a one micrometer shift of the image represents 4 degrees of phase error.

(c) For image sharpness the value of phase is very marginal: Phase errors produce a skewness of the line spread function (ref 5). But it is width rather than skew that blurrs the image. Quality criteria are always square law functions of OTF, doing away with phase. So MTF only is sufficient for sharpness assessment.

(d) Data reduction: A complete set of OTFs of an optical system represents an enormous amount of data. The urgent need for data reduction is illustrated by the efforts to construct Quality Criteria. A very sensible form of data reduction is to leave out phase. The MTF is the best quality criterium of OTF. All this reminds one of the unimportance of phase in acoustic reproducers.

(e) Phase in OTF is not an independent quantity (see appendix).

To a certain extend PTF is already determined by the MTF , so why measure it?

For checking his design , not for quality, phase can be of interest to the optical designer. Therefore a PTF module is available as an optional extra with ODETA V.

Appendix

Suppose only the MTF =T(f) has been measured, not the phase. What can be deducted now for the phase and the spread function L(x) ?

T(f) is the modulus of the OTF , squaring it yields :

$$P(f) = T^2 (f) \quad \ldots \ldots \ldots \ldots \ldots \ldots \ldots (5)$$

This is the power spectral density function. The spread function L(x) derives from P(f) by:

(1) Take the inverse Fourier transform of P(f) :

$$R(\tau) = {}_f\!\int P(f).\cos(2\pi f\tau).df \quad \ldots \ldots \ldots (6)$$

P(f) is an even function, so there is no need for sines.

(2) Find a L(x) such that its autocorrelation function is identical to R(τ).

$$\int_x L(x).L(x+\tau).dx = R(\tau) \quad \ldots \ldots \ldots \ldots (7)$$

(7) gives no unique solution for L(x). A whole set of functions satify (7), e.g. the two functions L_1 and L_2 in fig 5 have an identical autocorrelation function.

But in fig 5 L_2 contains negative parts, so it cannot be an optical spread function. As light intensity is always positive we require $L(x) \geqslant 0$.

The autocorrelation function R(τ) is always maximum for τ =0 and then drops down. It will reach zero (or a low value) beyond $\tau = \tau_i$:

$$\int_x L(x).L(x+\tau_i).dx = 0 \quad \ldots \ldots \ldots (8)$$

Because both L(x) and L(x+τᵢ) are positive the integral (8) can only vanish if L(x) and L(x+τᵢ) do not overlap (fig 6).

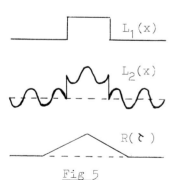

Fig 5

Functions L_1 and L_2 have the same auto-correlation function R(τ). But L_2 cannot be an optical spread function as it contains negative parts.

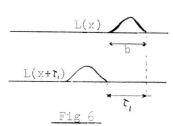

Fig 6

Correlation between L(x) and
L(x+τ_1)vanishes when they do
not overlap, when $\tau_1 >$ b.

Fig 6 shows that if the spread function L(x) is
shifted over $\tau_1 >$ b the correlation disappears.
As τ_1 is determined by MTF via (6) it follows that
measuring MTF , without phase, is sufficient to
find the maximum width b of the spread function L(x).
   The unknown phase distorsion φ can only affect
the shape of the spread inside the width b.
Phase factors that would broaden the spread function
are physically forbidden in optics: They would lead
to negative light intensities.

Fig 7 shows this constraint on phase as a function
of spatial frequency. The PTF can only wander
between narrow limits. Narrower for lower frequencies.
The better the MTF the smaller gets the spread width
b and the narrower becomes the corridor of allowable
phase distorsions φ.
   Phase can only change the inside of the spread.
If the spread function has become of zero width b
there is no inside left over. Any phase would
broaden the spread function , so when MTF = 100 %
there is no phase distorsion at all.

Fig 7

OTF phase distorsion is
limited to narrow range
between shaded areas.
Ranges narrows down to
zero when MTF =100% ,
giving complete
determination of phase
by MTF measurement only.

Spatial and Temporal Filters

The phase in OTF should not be mistaken for the phase
in the optical wave front. The optical wave is a
temporal signal, like a voltage in an electrical
network.
Temporal signals too have their constraints:
Electrical signals can be + and - but the smearing
out of a puls by an electrical network can only be
in the positive time direction, to the future,
the past is inaccessible.
This leads to the causality relations : Real and
Imaginary part of the network response are not
independent but the are each others Hilbert transform.
   An optical wave front cannot leave a medium
before it has entered it.
This leads to the Kramers-Kronig relation between
dispersion and absorbtion in spectroscopy.

But in a spatial filter , like a lens, one has the x coordinate instead of time.
Puls smearing can now be to both + and - coordinate directions (it mostly is).
Therefore the Hilbert Transform relations do not obtain here.
   But OTF phase is not free either. Now phase constraints arise from the fact that here
the signal , a light intensity, is always positive.

References

(1) Paul Nipkof : German Patent 30105 (1884)
(2) Boersma and Kruythof : Optica Acta Vol 10 No 3 p 285 (1963)
(3) Sloane and Harwit : Appl. Optics Vol 15 p 107 (1976)
(4) Ibbet et al :   Appl. Optics Vol 7 p 1089 (1968)
(5) Shack :   Proc SPIE Vol 46 p 39 (1974)

THE SUPPLY OF GERMANIUM FOR FUTURE WORLD DEMANDS

James R. Piedmont and Richard J. Riordan
Night Vision and Electro-Optics Laboratory
Fort Belvoir, Virginia 22060

Abstract

The supply of germanium has been of major concern to industry who utilize it in numerous applications. One application, Electro-Optics Viewing Systems, use the unique optical properties of germanium to operate within a given spectral range. Similarly, germanium is used in such semiconductor devices as radiation detectors, light emitting diodes, and solar cells. The oxide of germanium is used in the textile industries of Europe and Asia as a catalyst. For these commodities, there is a predictable increase in demand for both non-military and military use in the 1980's. The impact of this increase in demand is evaluated and compared with its supply.

Introduction

The US Army's Night Vision and Electro-Optics Laboratory of the U.S.A. and Roskill Information Services in the U.K. have recently carried out independent surveys on the future supplies of germanium.[1,2,3] Both surveys showed that an increased demand for germanium is predicted for the U.S.A., Europe, U.K., U.S.S.R., and Asia. This growth will be due to the increased use of germanium in optics for infrared surveillance devices. These findings indicate the possibility of creating an unprecedented demand for germanium. This paper will briefly describe the supply and demand for germanium then proceed to illustrate the impact that the increasing utilization of infrared devices may have upon the supply of optical grade germanium.

Germanium does not occur as a native element, but is found in low concentrations in certain metallic ores and minerals. There are only a limited number of mines throughout the world which have ores or minerals with a high enough concentration of germanium to make it commercially extractable. The ores from these mines are shipped to smelters where the important base metals, zinc, lead, and copper are separated. A smelter is an establishment where ores are treated to separate crude metallic products. Germanium and other byproducts are found in the fume and residue resulting from the smelter process and are collected and stored in "dumps" at the smelter site. Independent producers of germanium can purchase these residue "byproduct dumps" from smelting companies throughout the world. Table 1 summarizes the world's germanium refiners with their respective smelter and mine sources.

Table 1. Composite of the World's Germanium Producers,
Zinc Smelters and Mine Sources

GERMANIUM PRODUCERS/COUNTRY	SMELTERS/COUNTRY	MINE LOCATION/COUNTRY
HOBOKEN-OVERPELT/BELGIUM	HOBOKEN/OLEN, BELGIUM, VIEILLE MONTAGNE/FRANCE NEW JERSEY ZINC/U.S.A.	SHABA/ZAIRE SAINT SALVY/FRANCE SARDINIA/ITALY TENN. VALLEY/U.S.A.
JAPAN ELECTRONIC METAL/JAPAN TOKYO ELECTRIC COMPANY/JAPAN SUMITOMO/JAPAN HITACHI/JAPAN	U.S.S.R. BELGIUM WEST GERMANY WEST GERMANY	U.S.S.R. ITALY/ZAIRE ITALY ITALY
EAGLE-PICHER/U.S.A.	EAGLE-PICHER/U.S.A.	TRI-STATE/U.S.A. ILLINOIS/U.S.A.
KAWECKI BERYLCO/U.S.A.	KAWECKI BERYLCO/U.S.A.	WEST GERMANY ITALY
PERTUSOLA/ITALY	PENARROYA/FRANCE PERTUSOLA/ITALY	SALATOSSA/ITALY MALINES/FRANCE LARGENTIERE/FRANCE
LENINOGORSKE/U.S.S.R. USTJ-KAMENOGORSK/U.S.S.R. KONSTANTINOVSK/U.S.S.R.	LENINOGORSKE/U.S.S.R. KAMENOGORSK/U.S.S.R. KONSTANTINOVSK/U.S.S.R.	KAZAKHSTAN/U.S.S.R. URALS/U.S.S.R. UKRAINE/U.S.S.R.
PENARROYA/FRANCE	PENARROYA/FRANCE	SAINT SALVY/FRANCE
OTAVI/WEST GERMANY PREUSSAG-WESER/WEST GERMANY	OTAVI/WEST GERMANY PREUSSAG/WEST GERMANY	SARDINIA/ITALY BLEIBERGER/AUSTRIA
MIASTECZKO/POLAND	MIASTECZKO/POLAND	ORZEL/POLAND BOLESLAW/POLAND TRZEBIONKA/POLAND MATYLDA/POLAND
PANTAN/PRC (?)	KWANGTUNG/PRC (?)	KWANGTUNG/PRC (?)

A primary producer of germanium is one who possesses a capability to mine or dig out the germanium bearing ore, smelter to separate the valuable metals from impurities, and equipment to purify the extracted germanium. Germanium is derived by processing fume resulting from the smelting operation. In addition to the processing of fume, some primary producers will often add germanium scrap into the system and reprocess it along with the fume. A secondary producer of germanium is one who has equipment to convert germanium scrap into purified material. Scrap (reclaimed material) or secondary germanium plays a very important role in the supply of pure germanium. A refiner of germanium, on the other hand, has the capability to process the oxide and scrap into purified material. The amount of germanium reclaimed from scrap from semiconductor and optical manufacturing processes is quite high.

The principal producers which process germanium bearing residues are Belgium, Japan, U.S.A., Italy, and U.S.S.R. The largest primary producer of germanium in the world is Metallurgie Hoboken-Overpelt located in Olen, Belgium. This plant has a germanium metal capacity of 50 tons a year.[4] It produces germanium dioxide and all grades of metal from residues obtained from Zaire, Italy, Canada, U.S.A., and other countries. Metallurgie Hoboken-Overpelt does not possess mining operations of its own but obtains its supply on the world market. Japan has three germanium producing companies using mainly secondary material augmented with raw oxide imported from the U.S.S.R. and West Germany. The U.S.A., the world's third largest producer of germanium, uses mainly stockpiled concentrates augmented by imported residues, oxide, and scrap metal. The second largest primary germanium producing company in the world is Eagle Picher Industries located at Quapaw, Oklahoma, U.S.A. Eagle Picher has a metal capacity of 30 tons per year.[5] In addition, Kawecki Berylco and Atomergic Chemical Corporation in the U.S.A. have metal producing capabilities of ten tons, and two tons, respectively.[6,7]

Production

The complex chemical and physical processes used to separate germanium from the ore materials are unique to each producer, although they basically utilize the general flow processes shown in Figure 1. The fume derived from the smelting process is collected, in sag houses, dissolved with sulfuric acid and chemically treated to produce germanium dioxide. The germanium dioxide is hydrogen reduced to metallic germanium which can then be zone refined to produce high purity germanium.

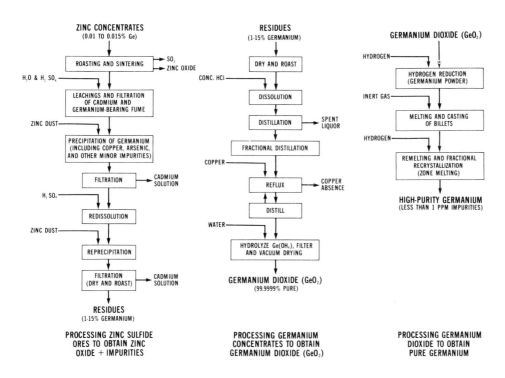

Figure 1. Byproduct Germanium Subsystems

The approximate manufacturing and material costs to produce germanium at various levels of purity are illustrated in Figure 2. The material costs shown are the approximate dealers' prices quoted on the open market. After following the processing steps outlined in Figure 1, the primary germanium concentrate contains approximately 10% by weight of germanium impurities yields pure germanium dioxide. Remember to bear in mind, however, that all zinc deposits do not contain germanium, while some that do may not contain enough germanium to be economically extracted.

Process	MATERIAL COST PER kg	Mfg COST
SMELTER		
PURIFIED GERMANIUM DIOXIDE (CONTAINS 69.405% Ge)	177.50	
HYDROGEN REDUCTION		34.00
FIRST REDUCTION METAL	289.50	
ZONE REFINED		27.00
HIGH-PURITY POLYCRYSTALLINE INGOTS	316.00	
ZONE LEVELED		284.00
SINGLE CRYSTAL	600.00	

Figure 2. Material Cost

Purified germanium dioxide was priced at $167.50 per kilogram in 1970. Periodic price increases since 1970 have resulted in a current price of $177.50. The first hydrogen reduction of germanium dioxide results in germanium powder which is then melted and cast to form an ingot of germanium metal. The $316.00 per kilogram price of intrinsic or ultrahigh purity germanium metal includes zone refining cost of $27.00. Single crystal germanium prices are likely to be double the prices of the ultrahigh purity polycrystalline form. The price of optical grade germanium is somewhat higher than that of zone refined metal.

Companies producing germanium offer it in a number of forms having a variety of specifications. The usual commercial forms are germanium tetrachloride, germanium dioxide, germanium metal, and zone refined germanium metal. Germanium dioxide, which is the most common form sold, is usually sold to consumers who are equipped to produce pure germanium or use it as a catalyst. The commerical metal is usually n-type polycrystalline germanium with a resistivity of from 5-50 ohm-cm. The technique of processing germanium metal to achieve a high purity of 99.9999% is known as zone refining. Some companies offer optical grade germanium that has a similar high purity.

Table 2. World's Production Capacity for Germanium

Company/Country	Tons/Year
Hoboken-Overpelt/Belgium	50
Eagle Picher/U.S.A.	30
Elec Metal - Tokyo Elec/Japan	35
Kawecki Berylco/U.S.A.	10
Atomergic Chemical/U.S.A.	2
Penarroya/Italy	17
Otavi Minen/West Germany	20 (estimated)
Preussag/West Germany	4
Penarroya/France	2
Pertusola/Italy	15 (estimated)
Bor/Yogaslavia	2
Birsod/Hungry	15
Johnson Matthey/UK	15
Leninogorske & Ustjkamenogorsk/U.S.S.R	20 (estimated)
Tsumeb Corp/Namibia	40
Miasteczko/Poland	1
Pantan/PRC	2 (estimated)
Monteponi/Italy	13 (estimated)
Vieille Montagne/France	2 (estimated)

Approximate Total   295

Table 2 lists the current estimate for germanium production capacities for companies throughout the world. The United States Bureau of Mines (USBM) estimated the 1974 world capacity to be 222 tons per year which was about double the output for that year. This estimated capacity is well above any annual output figures yet attained and it is doubtful that any larger capacity will be required in the foreseeable future. In 1976, the world's output of germanium decreased to about 83 tons compared with the 112 tons in 1973. This decline is primarily attributed to the rapid substitution of silicon in place of germanium for certain electronic applications. It is presently estimated that most of the larger germanium producing companies are operating at forty percent capacity.

The estimated capacities are the authors' projections based on available export-import information, literature, and current trends in the germanium market. Capacities listed in Table 2 for the Tsumeb Corporation, Johnson Matthey Corporation and Monteponi interest are given, but those companies are no longer producing germanium; however, these facilities and raw materials from their parent companies are still available for the production of germanium should the need arise. The Tsumeb plant, which opened with a capacity of 40 tons a year in 1954, had an output of only 3 tons in 1966. Operations at this plant were suspended about 1970. The Johnson Matthey facility was established to extract germanium from coal flue dusts. The flue dusts with the highest germanium content were those derived from the manufacture of producer gas. The consumption of coal producers' gas declined and in 1974 when natural gas became the major energy source, Johnson Matthey ceased operations.

Economic Considerations

The world's germanium industry in 1978 registered gains in almost all measurable aspects of its activities except for the continued decline in use of germanium in the semiconductor industry. The gains exhibit a rather surprising recovery from the low demands of the 1970-1976 era. There is a definite indication, at least among developed industrial nations, that a very bright outlook for industry wide growth will exist as 1979 begins. This outlook is primarily due to the increasingly heavy demands for germanium attributable to use in infrared optics. It is anticipated that increased market prices will result from higher costs of raw materials, production costs and labor.

The economic impact of some of the final stages of germanium purification is indicated in the pricing structure of the metalloid as illustrated in Figure 2. Because of the variable content of germanium in the residue, the ultimate initial processing costs for primary germanium are difficult to assess. Another factor is that the value of germanium in the major metal concentrate is small or even negligible in relation to the value of the principle metals with which it is associated.

Smelter production of germanium-bearing residue bears little relation to and is not necessarily always dependent on mine production. Most of the residues produced by the smelting process are stockpiled at the smelter, where they may be purchased by a primary germanium producer. The purchase of the residue is solely dependent upon the amount of germanium or other valuable materials contained in the residues. There are several indications that incremental processing costs are somewhat irrelevant just as they are for many of the byproduct metals. The complex technology necessary for the recovery of germanium is economically feasible only if, at some stage in the processing of the major metal, the germanium or other metals become concentrated in a separable phase. The costs of their separation and purification are a significant consideration, so recovery is limited to the amount of germanium or other metals that can be marketed. The present industry recovers and refines only enough of each byproduct metal to satisfy existing markets. Consumption of germanium, on the other hand, is more a function of technical performance among competing materials rather than a function of relative price.

Most nations are today experiencing high inflation rates and the industrial prices seem to be headed for substantial increases. Labor costs will continue in an upward spiral and the associated costs per man-hour will similarly continue its uphill trend. Profits which traditionally have been reinvested to provide additional jobs and expanded capacity have been increasingly diverted to meet nonproductive environmental and safety regulations.

Demand

In 1973, a general rise in demand for germanium was experienced in all major consuming countries. This was the first increase since 1969. Unfortunately, the 1973 increase in consumption was short-lived and demand continued its downward trend established during 1969 when silicon gained acceptance as a substitute for germanium in semiconductor devices. It is probably that the decline in the demand for germanium has leveled out as a result of new end uses for germanium. Table 3 lists the consumption of germanium by countries.

Table 3.   Consumption of Germanium by Country

	1973	1976
U.S.A.	19.5	20.0
Japan	27.1	16.0
France/Italy	13.7	16.0
West Germany	14.7	12.0
U.K.	4.0	3.5
Others	12.5	4.5
TOTAL	91.5	72.0

The principal consuming countries of the world are the U.S.A. and Japan. Texas Instruments, Inc. is the largest industrial consumer of germanium in the world today. This company produces germanium semiconductors, light emitting diodes (LEDs), infrared and radiation detectors, and Thermal Inaging Systems (TIS). It is probably the largest manufacturer of finished germanium lenses in the world today.

As a result of rapid advances in silicon technology, the use of germanium in transistors, diodes and rectifiers is still declining. This decline will probably continue as companies who are presently committed to using germanium modify their operations toward silicon use. During 1975, consumption in the electronics category increased due to the use of germanium substrates for the LED industry. It is estimated that the LEDs will be partially edged out by LCDs by 1980 after which time the LED market will decline slightly; germanium should, however, continue to be consumed by that market. If the decline in germanium semiconductors continues and semiconductor plants change over to using silicon at the rate seen in the mid-1970's, then in the 1980's the demand for germanium used for electronics should level off to about 1-2 tons per year. The use of germanium in optics, particularly infrared surveillance systems, is one of the most important future use of germanium.[9] The demand for germanium infrared optical devices is forecast to increase to about 55-70 tons per year during the 1980's.

The major end uses for germanium have so far been in transistors, diodes, rectifiers, infrared optics and detectors.[10] Germanium is also used in various alloys, brazing or soldering materials, and as a catalyst for various petroleum refining operations and processes. Other uses for germanium are in X-ray equipment, fluorescent lamps, strain gages, batteries, superconductor materials, biological organo-germanium compounds, thermoelectric devices, optical fibers, several types of phosphors, and in solar energy devices. It is also used in Europe as a catalyst in the textile industry. The fiber optics industry is expected to grow in the 1980's and become quite large during the 1990's as fiber optic-based communication systems become established.

Summary of the World Supply and Demand

The world's production capacity for germanium is shown in Figure 3 and Table 2. The major producers are presently operating at around forty percent of capacity and have indicated that a full capacity could be obtained in less than six months. In addition, a dormant annual capacity of 55 tons could be activated if the political and economic conditions are right. The supply of germanium for satisfying the demand is available in "byproduct piles" which are stockpiled at smelter sites throughout the world. The USBM has assessed the world germanium resources contained in base metal reserves to be 2,400 tons. Table 4 shows this assessment by country. Supplies for all countries are believed to be adequate for the foreseeable future because of the large stockpiles located at smelter sites around the world. Roskill estimates an annual consumption of 70 tons between 1984-1986. It is believed that research into germanium lenses is currently in more advanced stages in the U.S.A. than in Europe. Therefore, the U.S.A. IR optics demand pattern is expected to increase and peak in the early 1980's, while the European and Asian pattern should increase and peak in the late 1980's.

Table 4.   Assesment of World Germanium Reserves

Country	Reserves (tons)
United States	450
Europe	800
Africa	450
Asia	300
Canada	75
South America	150
Other	175
Total	2,400

The potential resources could be expanded even more if all the germanium were to be recovered from coal flue dust.

Figure 3 illustrates the world's production capacity and demand for 1973-1978 and projection forecasted for 1985 and 2000. The statistics are based on information supplied from the USBM.[11] A particular emphasis is made on projected consumption of germanium in infrared optics. The illustrated demand patterns are estimates by the USBM, with the IR optics demand estimated by Roskill, which are in agreement with Night Vision & Electro-Optics Laboratory's predictions.

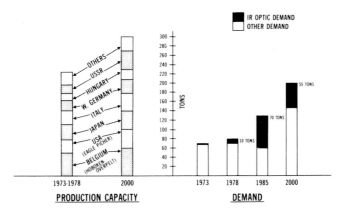

Figure 3. (World's) Germanium Production Capacity VS Demand

The development of optical grade germanium took place in America and Europe independently of each other; although the goals were identical, the approaches were very different. In Europe, the major developer and producer of optical quality germanium, Hoboken-Overpelt, utilizes the CZ (Czochoralski) technique to produce single crystal lens blanks up to 26cm in diameter. In the U.S.A., Eagle Picher and Exotic Materials consider polycrystalline to be adequate and produce large diameter ingots by a modified Bridgeman zone refining method. A recent study carried out in England substantiates that reliability of polycrystalline lenses. Table 5 lists international companies capable of optical grade germanium lens production.

Table 5. A Brief List of Optical Grade Germanium Lens
Producers, Worldwide

Hoboken Overpelt/Belgium
Eagle Picher/U.S.A.
Exotic Materials/U.S.A.
Zeiss-Electro/West Germany
Pantan/PRG
Nippon Electric Co/Japan
A.E.G. Telefunken/West Germany
Texas Instruments/U.S.A.
Barr & Stroud/UK
Thomson-CSF/France
Sopelem/France
Nerviano/Italy
Phillips/Netherlands
U.S.S.R.

Conclusions

There appears to be sufficient reserves of germanium in the world today to meet future demands which will continue to grow for several more decades. The pattern of demand has slowly changed from semiconductors in the early 1970's to the predicted end use in optics for the 1980's. The growth will occur despite the replacement of germanium by silicon in the semiconductor industry. Germanium's future use is expected to be in the field of infrared optics, radiation detectors, solar, fiber optics, etc.

There are basically 16 primary and secondary producers of germanium throughout the world employing less than a thousand people in the industry. Because the number of producers of germanium has remained constant in recent years, it would not benefit any one producer to initiate a price war. If the demand for germanium increases, the dealers' prices would correspondingly increase as a result of a number of complex factors discussed previously.

Since germanium is usually a minor byproduct of ores mined for zinc, the supply may therefore be a function of zinc production. The primary production of germanium under current technology is somewhat dependent upon the availability of germanium bearing residues stockpiled at mines and smelters. The U.S.A. could rely on its own internal resources and secondary supplies for the specialized applications and, when necessary, import germanium from Belgium, United Kingdom, West Germany, and Japan. Ultimately, the economics of supply and demand may encourage the recovery of germanium from other sources, such as coal ash or petroleum residues.

Attempts to estimate the amount of international trade in germanium from available world trade figures is difficult because of two factors: Individual transactions often consist of only a few kilograms, which is less than the unit of weight in customs returns. Secondly, material which is classified under the heading "germanium metal" and "waste or scrap" often includes residues, waste and scrap and other materials containing a very small percentage of germanium.

International trade consists largely of exports of germanium dioxides from Belgium, West Germany and the U.S.S.R. to the consuming countries. The U.S.A. and Japan import relatively large amounts of raw germanium dioxide from the U.S.S.R. Belgium is the most important source of germanium oxide, providing well over half of the total exports to over six countries each year. Trade statistics for germanium oxides are obscured because most European countries record exports and imports of germanium and zirconium oxides as one figure. Zirconium oxides are traded in far greater amounts than germanium oxides (hundreds of thousands of tons compared to less than 50 tons). For these reasons, it is not possible to correlate import and export statistics and only estimates of total world trade are available.

Acknowledgements

The authors wish to express their gratitude to Mr. John Lucas of the Bureau of Mines and Mr. Jack Adams of Eagle-Picher Industries Inc. for their technical assistance in the assemblage of this paper. In addition, the authors gratefully appreciate the assembled data of Figure 1 provided by student Gage Alexander. We are also grateful to Mr. Carlton L. Creech of NV&EOL for reviewing this paper.

References

1. Piedmont, J. R., Riordan, R. J., "The Supply of Germanium for Future United States Demands" SPIE Vol 131, Practical Infrared Optics (1978) U.S.A.
2. "Germanium, World Survey of Production, Comsumption and Prices," November 1977, Roskill Info. Services Ltd., London.
3. The Royal Radar Establishment of the Ministry of Defence, South Site St., Andrews Road, Malvern, Worcs.
4. Private Communication, Hoboken-Overpelt.
5. Private Communication, Eagle Picher Inc.
6. Private Communication, Kawecki Berylco.
7. Private Communication, Atomergic Chemical Corporation.
8. Petrick, A., Bennett, H. J., Starch, K. E., Weisner, R. C., "The Economics of By-Product Metals," 1973, Bulletin IC8570, U.S. Bureau of Mines.
9. Adams, J. H., "Germanium New Markets Show Good Growth," March 1977, E/MJ.
10. Babitzke, H. R., "Germanium," 1975, Bulletin 667, U.S. Bureau of Mines.
11. Staff of Nonferrous Metals, "Minor Metals," 1975, U.S. Bureau of Mines.

SESSION 6

FOCUS ON INDUSTRY

Session Chairman
J. Ragot
SORO Electro-Optics SA
France

Session Co-Chairman
Dr. J. P. Auton
Cambridge Consultants Ltd.
United Kingdom

Session 6 manuscripts were not available for publication.

FOURTH EUROPEAN ELECTRO-OPTICS CONFERENCE

Volume 164

SESSION 7

HIGH POWER LASER METALWORKING

Session Chairman
Dr. Helmut Walther
Fiat Research Center
Torino, Italy

Session Co-Chairman
Dr. J. Wright
J K Lasers Ltd.
United Kingdom

LASER BEAM MACHINING

A. Schachrai

Dept. of Mech. Eng., Technion
Haifa, Israel

Abstract

High power CO_2 Lasers are currently being used for various engineering applications such as welding, drilling, cutting, transformation hardening, surface alloying, cllading and glazing . It is the purpose of the presented paper to describe some aspects of the cutting application. The Laser as a cutting tool is described from both the system and the process points of view. The characteristics and performances of some typical Laser machining systems are described and a survey on the state of the art of Laser machining systems offered by various manufacturers is presented. Recent results of a study on metals cutting by a CO_2 Laser are presented. The workpieces materials are Low Carbon Steel (AISI 1045), Stainless Steel (AISI 314L), Ti6Al4V and Inconnel 718. The most significant metallurical transformations have observed in the Low Carbon Steel, thus, in comparison to the Stailess Steel, wherin the only observed damage phenomena are crakes and intergranular etching. In the H.A.Z. of the Titanium alloy the microstructure changes from α and β phases to a martensitic phase (α') accompanied by the disappearance of the β phase. In the H.A.Z. of the Inconnel an increase of the hardness and microcracks have been found although no microstructure change can be observed.

Introduction

The Laser as a cutting tool has been successfully applied to a large number of materials. In the literature one can find large varieties of reports which detail cutting data (power, speed, thickness etc.) of materials such as: metals, wood, cardboard, fabrics, plastics, composites (Boron/aluminum, Fiber-glass/epoxy, Borom/epoxy, Graphite/epoxy and Kevlar) ceramics, glasses and quartz [1-4]. The main advantages of the Laser beam machining which are very well exploited in the applications are:

(a)   Non dependency on the material's mechanical properties such as hardness and abrasiviness.
(b)   Free of mechanical contact.
(c)   Act in the ambient atmosphere.
(d)   Reduces thermal damage in the workpiece.
(e)   Performs in a clean manner (No noise, fumes, dust, chips, lubriating fluids and vibration).
(f)   No difficulties in performing any complicated patterns starting indifferently from edges or the inside.

The above mentioned capabilities and advantages can be easily transfered to an economical justification required to apply Laser systems in production line. The most common industries where Lasers are applied as a cutting tools are [5]: The Electric and electronic industries (cutting and drilling of the ceramic substrates for integrated circuits and for the microcircuits components trimming). Plastic and rubber industries (cutting of complicated pattern in P.M.M. boards, drilling polyethylene irrigation tubes and aerosol containers). Wood and cardboard industries (plywood cuttings for the preparation of the steel rules dies and for the furniture and shipping cases). Apparel industries (cutting fabrics in a single layer at a very high speed). Aero-space industries (cutting of advanced composite materials and thin boards of metals and high exotic alloys). Light mechanical industries (cutting   thin sheet of metals for various applications). The basic design and properties of some of the cutting system will be described later in the paper.

In the cutting process one can define five sets of parameters which are involved [Fig. 1]: The beam (power, wavelength, modes etc.). The focusing lens (focal length and aperture). The nozzle and gas jet(Nozzle diameter, the gap between the nozzle and workpiece, the gas and the gas pressure). The cutting speed. The material dimensions, and properties (Reflectivity, melting point, density, thermal conductivity, diffusivity etc.). Several types of gas-jet are used for Laser machining: Reactive gas (oxygen) for metal cutting, inert (gas (Argon, Helium or Nitrogen) for flammable materials or when the material has to be protected from undersirable oxidizing effects. In cutting of wood or plastics dry Air is usualy employed.

Briefly, the gas-jet serves three functions: (a) Performs an exothermic reaction from which most of the energy required for the cutting process is obtained. (b) Expels the molten material from the cut. (c) Protects the lens from vapour and ejected material.

Large variety of analytical models describing the cutting process have been proposed in the literature [6-9]. One aspect of the process which have not been treated enough is, the so called, "surface Integrity" of the cut workpiece. The best definitions for the Surface Integrity, which are schematicaly shown in Fig.2, have been given by Field and Khales [10] which proposed some experimental procedures in order to verify the phenomena. It was the purpose of the research, which its results are described later in the paper, to investigate the Surface Integrity phenomena of some common used metals cut by CO_2 Laser beam.

Laser Cutting Systems

Laser System Elements

A typical Laser cutting system consists of the following sub-systems: Laser head, beam handling unit, workpiece handling unit, control system and the safety hardware.

The vast majority of Lasers used for industrial cutting systems are CO_2 Lasers in the power level between 250-2000 W.

The beam handling unit includes plane mirrors and focusing elements which may be of transmission-type lenses or non-planar mirrors.

The type of the workpiece handling unit is determined by various considerations such as the workpiece nature, production rate, level of automation, accuracy, etc. Depending on the above mentioned point, this subsystem may consist of some simple elements for clambing and for driving the part relative to the beam, or more complex high-production facilities such as vibratory parts feeders, indexing dial table and automatically loading mechanism.

The control sub-system which controls the cutting patterns and the service functions (speed, power, nozzle gas flow, shutter position, etc.) may be effected by means of a photo-electric tracing device, simple NC or more sophisticated C.N.C.

For some applications the control of the distance between the workpiece surface and the focusing lens is required. This control is necessary in order to get always the same spot-size on the workpiece surface inspite of the waveness or distortions. The admissible tolerances are generaly some tenth millimeters. This control may be achieved by means of a capacitive feeler (which is usful for metalic workpieces) or mechanical feelers.

The safety hardware comes to protect personnel from electrical, mechanical and Laser radiation hazards.

Types of Laser Cutting Systems

There are three types of Laser cutting systems:

(a) Systems where the workpiece is in motion while the Laser head and the beam handling unit are stationarey. This method which places severe restrictions on the size and weight of the workpiece is generaly used with relatively small workpieces in processes calling for high precision (e.g., scribing of ceramica substrates for integrated circuits).

(b) Systems where the Laser head and the beam handling unit are in motion while the workpiece is stationary . This system carried relatively small and light heads (Lasers up to 500 W.) and cutting speeds are limited. The main advantege of this method is the minimum restriction on the workpiece size and weight.

(c) Systems where the beam handling unit is in motion while the Laser head and the workpiece are stationary. This type is the most flexible of the three as the only moving part is the optical unit with its small and light components. Hence, the following advantages of the method: are very high cutting speeds and economical space requirements. At the same time this method has two marked disadvantages: (1) slight variability of the beam spot size due to the variable distance of the lens from the head and the divergence of the Laser beam. (2) Sensitivity to misalignment and vibrations.

Industrial Laser Systems

Systems of all the above mentioned types are manufactured and offered by various manufacturers. Some of these systems and their characteristics are detailed in Table 1.

Table 1. Commercial Laser Cutting Systems

MANUFACTURERS	POWER [W]	WORKING AREA [m]	SPEED [m/min]	APPLICATIONS
Messer Griesheim (Gr)	300; 500	3.6 x 4	6	Metals and plastics
Held (Gr)	400; 1000; 2000	1.2 x 1.5	5	Wood
United Technologies (U.S.A.)	6000	5x2x0.75x ± 115°x180°		Underbody welding
Hughes-LP1 (U.S.A.)	250	1.5 x 2.4	30	Fabrics
Hughes-LC1C (U.S.A.)	500	1.8 x 3.6	45⁻	Composites
Hughes-LPM (U.S.A.)	1200	1.8 x 3.6	18	Metal
Culham (G.B.)	400	2 x 2	80	Fabrics
Isralaser (IL)	500	1.5 x 2 x o.7	9	Metals and plastics

The Messer Griesheim's system is a type A system (Laser head in motion) while all other systems are of type C (beam handling in motion). One exception is the Culham system wherin the beam motion is achieved by tilting a mirror which reflects the focused beam to the working area, method which enables very high cutting speed. Although the United Technologies system is a welding one it is brought in order to demostrat the capability of the Laser cutting systems to operat in five axis (x, y, z, α and θ).

The position accuracy of the systems is within 0.1 mm which is a typical spot-size of a CO_2 Laser beam. The cut part accuracy depends on the workpiece material, cutting speed, power etc. Typical values are in the range 0.2 - 0.8 mm.

Surface Integrity Of Metals Cut By Laser

Experimental Set Up

Four common metals used in various industries have been selected as workpieces: Low Carbon Steel (AISI 1045), Austenitic Stainless Steel (AISI 316L) Titanium alloy (Ti6Al4V) and a Nickel based high temperature superalloy (Inconnel 718). The Laser was a CO_2/G.T.L model 971 manufactured by G.T.E. Sylvania. All the experiments were performed with a 3.75" lens and the x table enabled variation of cutting speed between 0,1 to 7 m/min.

The Analytical Approach

In order to enable better understanding of the process, the relationship between the results and the relevant parameters have been calculated. The calculations are based on the analytical models developed by Roslenthal [6], Gonsalves and Duley [9] and Bunting and Cornfield [9].

According to the above mentioned references the temperature distribution over a two dimensional plate irradiated by an heat source q is:

$$T(x,y) = \frac{q}{2\pi hK} \frac{K_o \ Vr}{2\alpha} \ \exp\left(-\frac{Vx}{2\alpha}\right) \qquad (1)$$

where x and y are measured from the heat source [see fig. 1]. V is the speed, h is the plate thickness, K is the thermal conductivity, α is the thermal difusivity, K_o is a Bessel function of zero order and $r = \sqrt{x^2 + y^2}$.

An expression for the kerf width (2) is obtained by the derevation of equation (1) for the isotherme $T = Tm$ (Tm - melting point):

where
$$A = r \sin \theta$$
$$\theta = \cos^{-1} \frac{K_o(Vr/2\alpha)}{K_1(Vr/2\alpha)} \qquad (2)$$

K_1 is Bessel function of the first order. It should be mentioned that the above expression do not include the Oxygen influence on the process.

Temperature distribution and kerf width for Low Carbon Steel (AISI 1045) are shown in figs 3 and 4. The calculated results represent the influence of the material and process parameters on the temperature and cut shape.

Experimental Results of The Cut Quality

Cutting Speed

The dependency of the kerf width on the cutting speed is shown in Fig.7. It can be observed that the kerf width decreases when the speed increases, but above certain points the cut is spoiled. This speed is called the crytical speed, (1.5 m/min in Fig.5).

The best cut is achieved in a speed which is lower than the crytical speed. In this particular case the optimal speed is about 1.2 m/min.

Power

The influence of beam power on the kerf width is shown in Fig. 6. The linear dependency between the power and the width is followed by changes in the cut quality. It should be noted that the influence of the power on the cut quality is more significant in stainless steel.

Metallurgical Effects

Low Carbon Steel (AISI 1045). The original microstructure of the Low Carbon Steel consists of pearlite and pro-eutectuid Ferrite. In conventional austenization process, there is enough time for homogenization of the carbon content within the Austenite. As a result of the very high heating rate and the very high cooling rate which immediatly follows the process, no time is available for homogenization of the carbon in the Austenite and the result is martensite which is the outcome of the perletic phase and pseudo martensite which is the outcome of the feritic phase (Fig.7). The measured martensite hardness is above 66 Rc and that of the pseudo-martensite is 46 Rc. The H.A.Z. width is about 0.1 - 0.2 mm in oxygen assisted gas and 0.2 - 0.4 mm in air assisted gas.

Stailess Steel (AISI 316L). No change in microstructure or microhardness has been observed in this metal. This result is explained by the fact that the AISI 316L is Austenitic over all the temperatures range up to the melting point. Due to the very high heating and cooling rates no time is available for recrystalization. Cracks oxydation and inurgranular etching were found near the cutting edge (Fig.8).

Titanium Alloy (Ti6A14V). The microstructure change which was observed in cutting the Titanium alloy is a disapearing of the β phase and transformation from α phase to α' phase. (Fig. 9).

Nickel Alloy (Inconnel 718). No change in microstructure was found in this metal. (Not: the experimentals made before ageing). A change in hardness from 20Rc to 37Rc has been measured.

As a result of the high thermal stresses microcracks were created and some exyde layers have been observed as well (Fig.10).

Acknowledgements

The author wishes to thank Mr. D. Guez for carrying out the experimentals, and Prof. S. Nadiv for helping discussions and evaluation of the metallurgical results.

References

1. Horton, M., CO_2 Applications Coherent Radiation Memo, April 1970.
2. Engel, S.L., Laser Cutting of Thin Materials, SME paper No. Mr74-960.
3. Willis, J.B., A2 KW Laser and its Application, B.O.C. Industrial Power Beam, Daventry, England.
4. Wick, D.W., Application for Industrial Laser Cutter Systems, SME paper.
5. La Rocca, A.V., Laser Applications in Machining and Material Processing, European conference on optical systems and applications, Brighton, U.K., April 1972.
6. Rosenthal, D., The Theory of Moving Sources of Heat and its Application to Metal Treatment, ASME Trans., November 1946, pp. 849-866.
7. Gonsalves, J.N. and Duley, W.W., Cutting Thin Metals with the CW CO_2 Laser, Journal of Appl. Phys., Vol. 43, No. 11, November 1972, pp. 4684-4687.
8. Babenko, V.P., Gas Jet Laser Cutting, Soviet Journal of Quantum Elec., Vol. 2, No. 5, April 1973, pp. 399-410.
9. Bunting, K.A. Cornfield, G., Toward a General Theory of Cutting - a relationship between the incidident Power Density and the Cut Speed, Journal of Heat Transfer, Feb. 1975, pp. 111-121.
10. Field, M., Kahles, J.F. and Cammet, T.I., A Review of Measuring Methods for Surface Integrity, presented at teh C.I.R.P. General Assembly, September 1972.

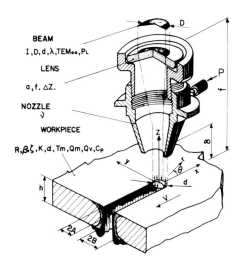

Fig. 1. Parameters of Laser cutting process.

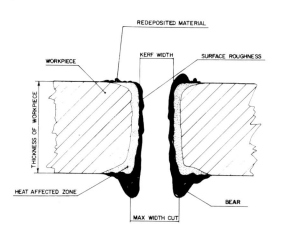

Fig. 2. Schema of the cut area.

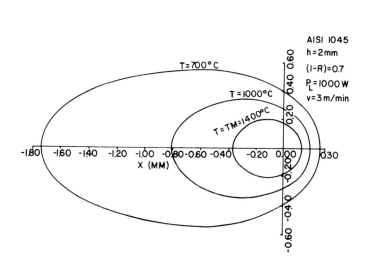

Fig. 3. Temperature distribution for Low Carbon Steel.

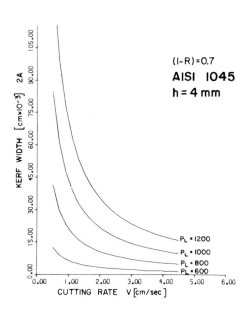

Fig. 4. Cutting rate and power dependence of kerf width for Low Carbon Steel.

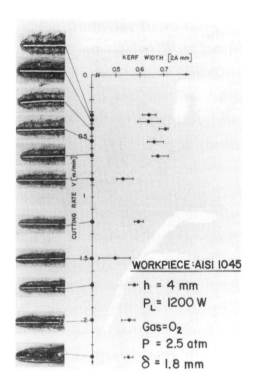

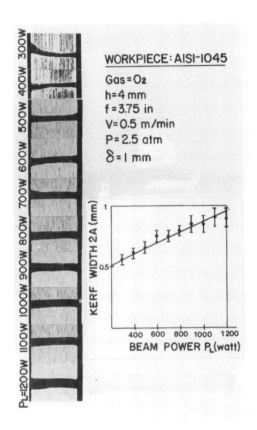

Fig. 5. Cutting rate dependence of kerf width as measured for Low Carbon Steel.

Fig. 6. Power dependence of kerf width as measured for Low Carbon Steel.

Fig. 7. Microstructure of Low Carbon Steel H.A.Z.

Fig. 8. Intergranular etching in the Stainless Steel H.A.Z.

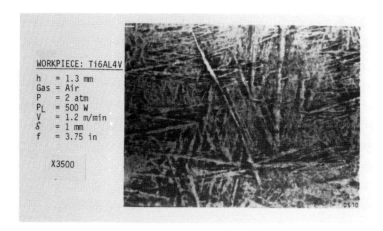

Fig. 9. Microstructure of Ti6 Al4V H.A.Z.

Fig. 10. Cracks in the Inconnel 718 H.A.Z.

LASER ASSISTED MACHINING*

Michael Bass, David Beck and Stephen M. Copley
Center for Laser Studies and Department of Materials Science
University of Southern California
Los Angeles, California 90007
(213) 741-7994

Introduction

The availability of high power cw carbon dioxide lasers with sufficient ruggedness, reliability and simplicity of operation for use in manufacturing facilities has led to the development of new machining methods. These methods currently involve the localized vaporization or melting of the material due to beam heating.[1] In the case of metallic materials beam heating is supplemented with burning enhanced by a flow of oxidizing gases at the point of impingement of the laser beam. In our research, we have developed a new and different method of cutting with a laser, laser assisted hot spot machining (LAM), in which the laser is used to heat the volume of material directly in front of a single point cutting tool to a temperature less than its melting point.[2,3] The application of gas torch and induction heating to assist in the turning of metals was first studied in the United States by Tour and Fletcher (1949).[3] Concurrently, Schmidt investigated the use of gas torch heating in milling.[4] Although many advantages were reported such as reduction in power consumption tool life improvement and improvement in surface finish, hot-machining has not been perceived as a practical and economically viable method by industry. Recently, however, with the development of more intense heat sources such as the plasma-arc[5] and the laser, hot machining has become more attractive in specific applications as a metal removal technique.

The concept of LAM is illustrated schematically in Figure 1. As the tool moves along the workpiece, energy is consumed by three processes: (i) plastic deformation in region ABC; and (iii) friction between the chip and the tool along BC. At constant cutting speed, the power supplied equals the energy consumed by these processes per unit time. The power supplied is given by the equation

$$P = F_C V \tag{1}$$

where F_C is the cutting component of the resultant tool force (cutting force) and V is the cutting speed. It is evident that factors that decrease the energy consumed per unit time during cutting decrease the cutting force generated at a constant cutting speed. Alternatively, if the cutting force is held constant, factors that decrease the energy consumed per unit time permit higher cutting speeds, and therefore, higher rates of material removal.

If the temperature is increased in region ABC of Figure 1, through the action of an external heat source such as a laser beam, then each of the energy consumption processes may be affected. In the case of plastic deformation, increasing the temperature normally decreases the yield stress and strain hardening rate of a material. Thus, increasing the temperature should decrease the energy consumed per unit time in cutting by plastic deformation. In the processes of fracture and friction associated with cutting, the effects of increasing the temperature are more complex. Nevertheless, in many materials, increasing the temperature of the volume of material directly in front of a single point cutting tool is likely to decrease the net energy consumed per unit time.

Several benefits may result from laser assisted hot spot machining in materials where heating with a laser decreases the net energy consumed in cutting per unit time. If the cutting speed is held constant, then heating with a laser should decrease the cutting force. This should increase the accuracy of cutting bt decreasing the amount of workpiece and machine distortion. Also, if ceramic cutting tools are employed that can resist the high temperatures generated, then tool life may be increased by decreasing the cutting forces. This effect may be particularly important in applications involving interrupted cuts. Alternatively, if the cutting force is held constant, then heating with a laser beam should permit higher cutting speeds and thus higher rates of materials removal. An additional benefit may result from LAM; namely, an increase in the smoothness of the machined surface. This effect does not depend, however, on a decrease in energy consumed per unit time during cutting. Several of these benefits have been demonstrated in our research on LAM in steels and nickel-base superalloys. Our results to date are described in this paper.

* This research is supported by the Advanced Research Projects Agency under Contract No. 00014-77-C-0478, ARPA Order No. 3421.

Experimental

Laser assisted hot spot machining (LAM) was demonstrated on several steels, including 420 and 316 stainless steels and the nickel based superalloy, Udimet 700. These materials were machined in the configuration sketched in Figure 2 and Figure 3 give more detail of the tool-light beam-workpiece configuration. Figure 4 shows the coordinate system, directions of motion, angles of incidence and tool forces needed in the discussion of the results.

A three pen chart recorder used to provide a record of the two tool forces and the laser power. The rotation speed of the lathe was determined with a crystal controlled stroboscopic lamp. A 125 mm focal length ZnSe lens was used to focus the TEM_{00} mode laser beam to a spot of ~ 0.016 cm in diameter (full width at $1/e^2$ in intensity). Since 460 W was available in a mode which was measured to be greater than 95% TEM_{00}, the maximum average power density on the target was 2.3 MW/cm^2. This is a maximum because the angle of incidence on the shoulder of the cut (see Figures 2 and 3) was not normal. In these experiments $\theta = 120^O$ and $\phi = 70^O$. In addition, due to the finite radius of curvature of the tool insert (either 0.4 mm for the steel insert or 0.8 mm for the ceramic insert) the incident surface was not flat. This property of the shoulder of the cut made an attempt to do LAM at $\theta = 90^O$ inconclusive. Future efforts include studies of LAM for various values of θ and ϕ. Geometrical constraints in the current experiment allowed the laser beam to strike the shoulder no less than 0.4 cm ahead of the tool.

Pyroelectric strain gauges were epoxied in pairs to the two horizontal and the two vertical surfaces of the insert holder. These were electrically connected in a bridge circuit and the voltage imbalance produced by bending the holder was calibrated in terms of the force applied to the tip of the tool insert. During a laser assisted machining test each force, F_z and F_y was recorded on one channel of the chart recorder. One of the major experimental problems encountered with this method of measuring the tool force was that, in some cases, reflected laser light heated the strain gauges and altered the balance of the bridge circuit. This was for the most severe for the gauge measuring F_z and resulted in a time varying baseline on the chart recorder.

Results

Steels

Micrographs of the grooves cut in 420 stainless steel are shown in Figure 5 along with the machining parameters. The chart recording shown in Figure 6 was obtained while machining this material. The micrographs and the tool force measurements show that the machining process is much smoother when using laser assist. This is consistent with the long, smooth chips obtained with the laser assist on, as compared to the small, ragged chips obtained otherwise. Though 420 stainless steel, when machined under the indicated conditions, shows no reduction in the tool force required, the improved smoothness of the cut is noteworthy.

In laser assisted hot spot machining of several other steels (1018, 1040 and 320 stainless) force reductions of 25-50% were noted. However, in these first experiments the samples were about 1.25 cm diameter x 7.5 cm long rods. As a result the tool caused the sample to bend even though a center support was used at the free end of the sample. Such bending caused significant irregularities in the finishes obtained on these materials.

An interesting example of just how much there is to be learned in this type of materials processing is demonstrated by the micrographs in Figure 7. The laser assisted machining produced extremely smooth, uniform grooves. However, as shown in Figure 8, the average tool force, F_z, though more free of variations, was nearly two times higher with the laser on than with it off. F_y was nearly three times higher with the laser on. Figure 9 shows cross sectional views of the machining grooves in the sample of 1090 steel. The maximum height to minimum depth of the grooves made by LAM is uniformly 0.0043 mm, as compared to an average of 0.0086 mm in the conventionally machined material. Some of the conventionally machined grooves are almost 0.025 mm deep. This is clear evidence for the superior uniformity of the laser assisted surface. The micrograph in the lower left hand part of Figure 9 shows the 0.018 mm thick layer of hardened steel remaining on the surface of the metal machined with a laser assist. For 1090 steel then, the laser irradiation caused the steel to become hardened before it arrived at the tool and though not exactly what one would expect intuitively, it is clear that machining hardened high carbon steel produces excellent quality surfaces.

Figure 10 shows evidence for laser hardening as well as improved machining during LAM. This is a recording of the tool force, F_y, in a sample of 1018 steel which had been previously machined in several places with a laser assist. The reduction in tool force due to application of the laser is clear. However, it is equally clear that previously laser treated metal is harder than conventionally machined metal. Thus, two results can be achieved simultaneously with LAM and proper process design.

Udimet 700

This nickel based superalloy is very difficult to machine. In turning a surface about 19 mm long by 23 mm diameter and taking a 0.25 mm deep cut without a laser assist two ceramic tool tips were broken. With the laser assist the same surface was machined with no tool wear evident to the naked eye. Figure 11 shows the quality of cuts obtained and the machining conditions used. Again smoother, more uniform cuts are obtained with the laser assist. In machining this sample the strain gauge heating problem was severe and little meaningful data could be obtained from the tool force measurements.

Comments

Figure 12 shows a conceptualization of the relationship between tool force and machining speed for conventional and LAM. Here we consider a constant depth of cut and a laser-optical system capable of a single power density. The power density is such that for slow machining speeds the metal could be melted. The tool force in a conventional machining process can be considered in this simple model to be a monotonically decreasing function of speed. In LAM the effect of the laser is negligible at very high machining speeds because at such speeds the laser dwell time on an irradiated spot is too short to heat the metal. At very low speeds the metal can cool between the time it was irradiated and the time the material arrives at the tool. This means that for a material that can be laser hardened the tool force will be higher at low speeds with LAM than without it. For a material that is hardened prior to laser machining the tool force with LAM will always be less than without it. For both types of material the softening at elevated temperatures means that the tool force will decrease to a minimum at some speed which correponds to the maximum amount of tempering of the material arriving at the tool.

Since LAM results so far show several of the possibilities suggested in this simple model we are using it for guidance in interpreting LAM. Additional factors must be taken into account in choosing optimum conditions for laser assisted machining such as tool wear and surface finish. Surface finish improvements obtained through LAM may be very important because they should provide higher fatigue, wear and corrosion resistance than conventionally machined surfaces.

Although early investigations[3,4] clearly demonstrated that metals become easier to machine when heated, no practical and economical hot machining technique based on gas flame or induction heating became established in the production machining industry. One problem was a large heat affected zone due to low heat fluxes, which resulted in distortion of the machine and the workpiece. Another problem was unavailability of adequate temperature resistant cutting tools.

Recently, there has been considerable interest in hot machining where a plasma jet is employed as the heat source.[5] Because the heat flux of a plasma jet is greater than that obtained by gas flame or induction heating, highly localized heating can be obtained by rapid translation of the surface of the workpiece. This minimizes metallurgical damage to the workpiece after machining, avoids workpiece distortion and avoids a build-up of heat that might damage the tool. Temperature resistant ceramic tools are used in this process, which, of course, is limited to the machining of electrical conductors.

In comparison to the plasma torch, the heat flux of the laser is more intense and can be focused on a smaller region. This suggests that the plasma torch may be appropriate for applications involving simple shapes, large diameters and high rates of material removal, while the laser may be appropriate for applications involving complex shapes, small diameters and high accuracy. The laser can be used as a heat source in the machining of non-conductors. Some difficulty is experienced, however, in coupling the laser beam to metal surfaces with a high reflectivity.

Acknowledgements

The participation of O. Esquivel in the experiments is appreciated.

References

1. W.W. Duley, "CO_2 Lasers: Effects and Applications," (Academic Press, N.Y., 1976).

2. H.V. Winsor, private communication. In March of 1977 H. Winsor suggested this application of laser material interactions.
3. S. Tour and L.S. Fletcher, "Hot Spot Machining" Iron Age, July 21, 1949.
4. A.O. Schmidt, "Hot Milling" Iron Age, April 28, 1949.
5. A.E.W. Moore, "Hot Machining for Single-Point Turning--A Breakthrough," Tooling and Production Magazine, November, 1977.

LAM Concept

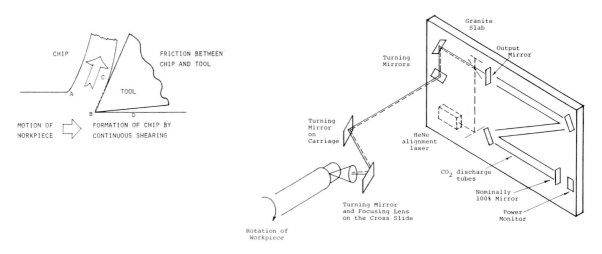

Figure 1
Conceptuallization of Single Point Cutting

Figure 2
General Configuration of the First LAM Experiments

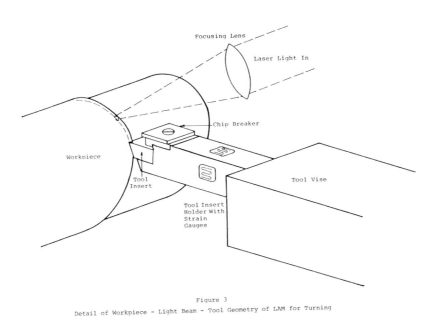

Figure 3
Detail of Workpiece - Light Beam - Tool Geometry of LAM for Turning

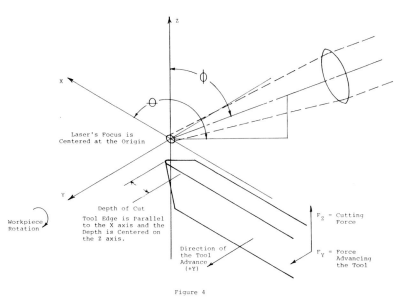

Figure 4

Coordinate Frame Defining Directions of Motion for LAM

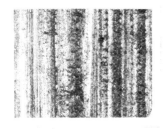

Surface of 420 Stainless
Steel as Conventionally
Machined.

0.0175" (0.445 mm) cut

0.005" (0.127 mm) feed

132 RPM

0.918" (23.3 mm) final
diameter

0.005"
0.127 mm

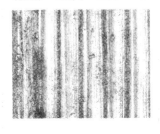

Same Machining Conditions,
but with 3.5 MW/cm^2 CW CO_2
Laser Power Incident ≈ 4 mm
Ahead of the Tool on the
Shoulder of the Cut

Figure 5

Micrographs of 420 Stainless Steel after Conventional Machining and LAM

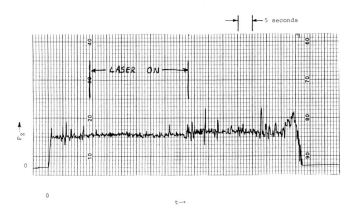

CUTTING FORCE VS. TIME FOR 420 STAINLESS STEEL

Figure 6

Record of the Cutting Force for 420 Stainless Steel both with and without LAM

Surface of 1090 Steel As
Conventionally Machined
0.0154 mm cut
0.127 mm feed
280 RPM
11.68 mm final diameter

0.005"
0.127 mm

Same Machining Conditions,
but with 3.5 MW/cm^2 CW CO_2
Laser Power Incident = 4 mm
Ahead of the Tool on the
Shoulder of the Cut

Figure 7

Micrographs of 1090 Steel after Conventional Machining and LAM

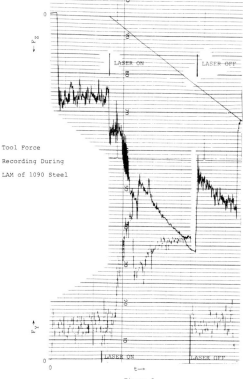

Tool Force
Recording During
LAM of 1090 Steel

Base Line
Drift Due
to Laser
Heating of
the Strain
Gauge

Figure 8

Tool Force Recording During LAM of 1090 Steel

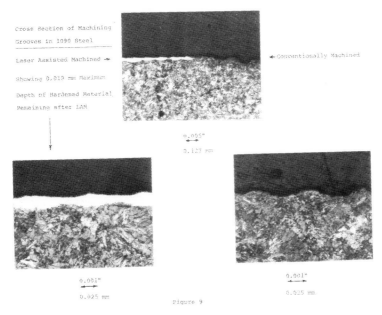

Cross Section of Machining
Grooves in 1090 Steel

Laser Assisted Machined →

Showing 0.013 mm Maximum

Depth of Hardened Material

Remaining after LAM

← Conventionally Machined

0.005"

0.127 mm

0.001"

0.025 mm

0.001"

0.025 mm

Figure 9

Cross Section of Machining Grooves in 1090 Steel with and without LAM

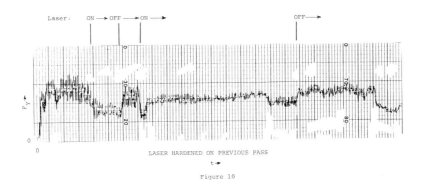

Laser: ON → OFF → ON →   OFF →

F_Y →

LASER HARDENED ON PREVIOUS PASS

t →

Figure 10

Force Advancing the Tool Versus Time During LAM of 1018 Steel
Following Previous Pass which Hardened the Surface

Surface of Udimet 700 as
Machined with a Ceramic
Tool Insert.

0.254 mm cut
0.127 mm feed
244 RPM
22.07 mm final diameter

0.005"
0.127 mm

Same Machining Conditions
but with 1.5 MW/cm^2 CW CO_2
Laser Power Incident ~ 4 mm
Ahead of the Tool on the
Shoulder of the Cut

Figure 11

Micrographs of Udimet 700 after Conventional Machining and LAM

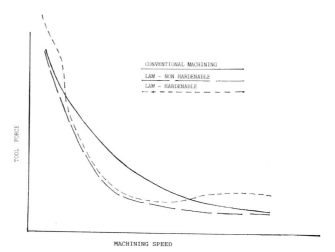

Figure 12. Conceptuallization of the Tool Force versus Machining Speed
Relationship for Conventional Machining and LAM

HIGH POWER LASERS AND LASER METALWORKING

J. H. P. C. Megaw and A. S. Kaye

Laser Applications Group, UKAEA Culham Laboratory
Abingdon, Oxfordshire OX14 3DB

Abstract

The multikilowatt continuous carbon dioxide laser is an efficient, compact and robust system which meets many metalworking requirements. A brief account is given of important principles and characteristics of a range of lasers of this type developed at the UKAEA Culham Laboratory, and techniques are described for their use in welding and surface treatment operations. The underlying principles of these processes are summarised and results are presented which highlight the areas of manufacturing technology for which they are particularly suited.

Introduction

The electrically excited carbon dioxide laser is characterised by the ease with which it can be robustly constructed to deliver large average output powers with high efficiency. At power levels up to approximately 1 kW, it is now routinely used as a cost effective production tool for a wide range of material processing applications. At higher powers, acceptance is growing of its ability to fulfil a similar role, and commercial laser development has now reached a stage where at least four manufacturers offer models at the 1-2 kW level and at least five manufacturers offer models in the 2-15 kW range. The aim of this paper is to outline briefly the principles and features of a particular range of multikilowatt CO_2 lasers developed at UKAEA Culham Laboratory and to describe some of their metalworking capabilities.

The Lasers

The work of the Laser Applications Group at Culham Laboratory is primarily aimed at stimulating the introduction of laser processing technology into British industry. Trials and contract research, often leading to prototype and production systems development, are carried out for customers. Much of the work utilises commercially available lasers, but Culham's own programme of laser development has led to the availability of a range of lasers of powers up to around 17 kW. This development has also fulfilled the requirements of the Atomic Energy Authority to keep abreast of a technology which offers considerable potential for applications in the nuclear industry, for example remote processing in radioactive environments. Indeed the development and application of the highest power laser of the range (designated LTF3 and shown in Fig.2) is at present restricted to AEA needs. However, a scaled-down version of this unit has been built in ruggedised form to operate at 5 kW, since the associated welding capability (at present about 8 mm in steels) and surface treatment rates appear well matched to a sizeable potential industrial market. This laser, designated CL5, is shown in operation in Fig.1. Much of the metalworking to be described in this paper was carried out on this laser, and on an earlier prototype designated LTF2 of output up to approximately 8 kW.

The same general arrangement is employed on all the lasers and is typified by the particular layout of CL5 shown schematically in Fig.3. The directions of the optical cavity, the gas discharge current and the gas flow are mutually orthogonal; the particular merit of this transverse configuration is that short residence times in the discharge can be achieved for modest gas speeds. The laser uses an unstable cavity, which gives an annular output beam of outer diameter typically twice the inner and, to reduce the overall length, the optical cavity is folded into two parallel arms. The water-cooled metal cavity mirrors are mounted on an optical bench which is independently supported on air cushions to isolate it from the effects of thermal distortion and vibration in the rest of the laser head. The beam emerges from the laser head through a zinc selenide output window. Good mode control is achieved. The far field beam pattern has approximately the expected Fraunhofer distribution, the power enclosed within a given diameter being within 20% of the predicted value. The gas discharge is a high pressure ($\sim$ 50 Torr) glow discharge between water-cooled cathodes and segmented anodes, which are modular in design to facilitate servicing. The electrode design and the dynamics of the gas flow have been optimised to enable the running of a stable glow discharge with no auxiliary source of ionisation (eg no electron beam ionisation). The gas is circulated inside the laser head vacuum vessel by two fans mounted in series, and the gas is cooled by heat exchangers in the top and bottom vacuum domes.

The prototype CL5 laser has now been in routine operation at Culham for well over a year and some 1500 hours of operation have been completed with no failure of any of the major components. Two more CL5's are under construction and are due to be commissioned shortly.

Welding

Introduction

Multikilowatt laser welding is a process of joining metal by the fusion of a deep, narrow, parallel-sided seam; it is thus energy efficient and capable of fabrication with minimum distortion. When subjected to focused beams of intensity around $10^6 W\ cm^{-2}$, the workpiece surface will experience a rapid temperature

Fig.1   5 kW laser CL5, photograph courtesy Ferranti Ltd.

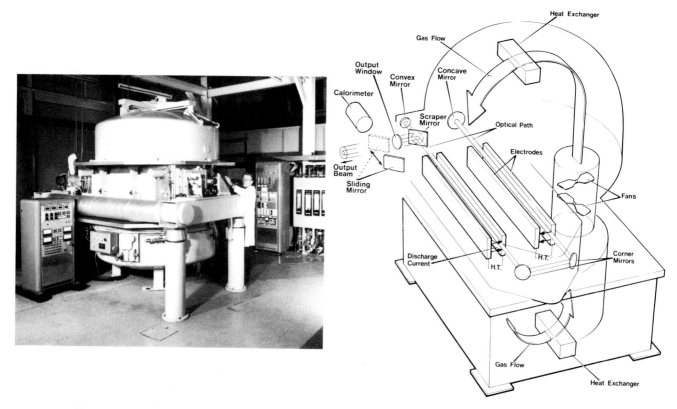

Fig.2   17 kW laser LTF3

Fig.3   Laser schematic

rise leading to increased absorption, possible oxidisation, and ultimately melting, vaporisation and disruption of the surface. The result is very efficient (up to 90%) coupling of the beam into the workpiece, the energy being deposited in a thin layer which is intensely heated and disrupted to form a hole which traps the beam. At beam powers of a few kilowatts, this 'keyhole', which may penetrate through the workpiece, can be several millimetres deep and is kept open mainly by the vapour pressure. In welding, relative movement of the beam and workpiece results in the keyhole being translated along the joint line, metal being melted ahead and flowing around to solidify behind it. Some of the materials exhibiting good laser weldability are steels, titanium and its alloys, and nickel and its alloys. Copper and aluminium are much less readily weldable, mainly due to their high reflectivity at the laser wavelength.

Process

The welding process is being used and investigated by a large number of workers using a variety of lasers. We will discuss it primarily by reference to our own equipment. One of our welding heads (which may be moving or fixed) is shown schematically in Fig.4, where it is seen that the hollow beam permits use of a focusing mirror without off-axis aberrations. Mirrors have the advantage over lenses that they are robust and can be relatively cheaply refurbished if damaged by welding spatter. Mirror height adjustment allows the position of the focus to be altered with respect to the workpiece surface. Using a switching mirror further up the beam line, an expanded, low power (visible) beam from a helium-neon laser can be substituted for the CO_2 beam to simplify seam tracking.

Also shown in the figure is a typical workpiece gas shroud arrangement, the upper and lower 'shoes' being fed with a slow flow of inert gases to prevent oxidisation of the bead during cooling. The inert gas jet J excludes from the interaction point air which could support vigorous plasma-chemical reactions leading to oxidisation and considerable mechanical disruption of the upper bead. In addition, it minimises, by the convective removal of energy, the plasma which tends to form at the interaction point and which would otherwise absorb and redistribute an excessive amount of the laser energy into a widened top bead. The processes in the plasma plume are influenced by the control gas, and by the metal ions and atoms emitted from the workpiece; helium is preferable to argon because of its higher ionisation potential. Plume control is more difficult at lower welding speeds where the workpiece emitting area tends to be large. The momentum in the plume control jet can also be made to play an appreciable role in the formation of the keyhole (a situation analagous to plasma welding) and ultimately it can cause a transition to cutting, when the molten metal is physically removed.

The scaling of the depth of weld penetration with beam power and welding speed is a matter of considerable experimental and theoretical interest. Preliminary conclusions indicate that for very deep welds at slow speeds, the beam intensity distribution and the plasma are important factors (both of these compare unfavourably with the case of a vacuum electron beam, which is inherently a much lower divergence source and suffers less attentuation in the metal vapour). Results from a number of lasers, including our own, indicate that in approximate terms the present maximum practicable depth of penetration d (mm) scales with laser power P (kW) as:

$$d \approx 3.7 \, P^{0.6} \tag{1}$$

Taking into account the commercially available powers of multikilowatt lasers and the power law of the scaling, it is seen that they can be used for single-pass penetrations in steels up to about 15 mm. For deep penetration mode welding above this thickness, electron beam and non-vacuum electron beam may be attractive provided the vacuum requirements and/or X-ray hazards are acceptable.

Studies of the variation of weld penetration with speed indicate that at medium to fast speeds, laser penetration may be comparable with vacuum eb penetration. Results obtained with the CL5 laser are shown in Fig.5, which is a plot of the speeds required to give continuous full penetration of type AISI 321 stainless steel plates of the indicated thicknesses, at powers of 3.6 kW and 5.0 kW. The dotted lines are interpolations to 3.6 kW and 5.0 kW of results of Breinan and Banas[1] at 2.0, 4.0 and 6.0 kW; they used a stable-resonator, oscillator-amplifier configuration and penetrations were measured to the root tip in a blind weld. At speeds of around 15 to 20 mm s^{-1}, the CL5 penetration is comparable with the latter results; at higher speeds the CL5 penetration is greater, but at lower speeds it does not show the same gain in penetration. Hence, if depths greater than 8 mm are essential, modifying the CL5 cavity to stable configuration could be considered. However present indications are that the medium to high speed regime has the greatest potential for exploitation of laser welding and indeed in Fig.5 the chain dotted lines of constant joining rate (area of longitudinal weld section produced per kJ of incident laser energy) show that this is the regime of most efficient beam utilisation. For practical welds, operation below the maximum speed for full penetration improves the geometry of the fusion zone and eases alignment tolerances. Sample welds with high depth/width aspect are shown in Figs.6 and 7. They are respectively sections of 2.5 mm superalloy welded at 3 kW and 60 mm s^{-1}, and of 8 mm (effective) thickness stainless steel welded at 5 kW and 18 mm s^{-1}. The latter is an example of welding in the 'slow speed' regime and illustrates a method of producing a high-integrity root-sealing weld in a very heavy section prior to conventional welding.

Characteristics and Applications

Detailed questions about laser welding performance in alloys of interest to a potential user can only be answered after a careful assessment programme. Nevertheless some guidance can be given based on the

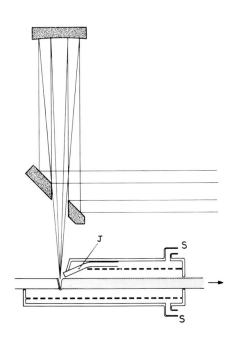

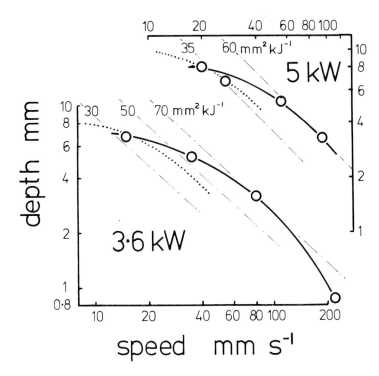

Fig.4   Schematic welding head

Fig.5   Weld speed against penetration

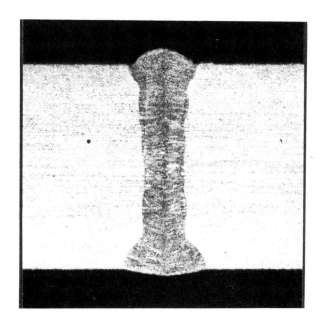

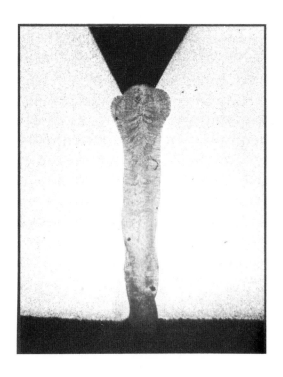

Fig.6   Laser weld in 2.5 mm
superalloy (3 kW, 60 mm s^{-1})

Fig.7 Laser weld in 8 mm
(effective) stainless
steel (5 kW, 18 mm s^{-1})

characteristics of the process and the accumulating body of data. We have emphasised that a significant feature of the process is the minimum distortion; this follows from both the small melt volume and the parallel sided nature of the weld (so that expansions and contractions are in the plane of the material). As an important corollary, the reduced stresses will limit the incidence of cracking during and after welding. Furthermore, the process is particularly amenable to the welding of dissimilar metals and can cope with a reasonable mismatch of thermal properties.

A related feature is that heating rates, temperature gradients and cooling rates are high. Although high cooling rates may be undesirable in alloys of high hardenability, in some materials the resulting fine microstructure will exhibit excellent fatigue resistance properties. If hardening is unacceptable, it can be alleviated by pre- or post-weld heat treatment. We have successfully used lasers to carry out post-weld heat treatment in specific geometries (thereby reducing hardness from 450 HV to 250 HV), and in principle a welding laser beam could be defocused to carry out the pre- or post-heat. Alternatively, undesirable metallurgical changes may be controlled and modified by the use of an appropriate filler. Our investigations in this direction are at a preliminary stage and at present primarily concern the frequently encountered need to weld ascropped edges. Fig.8 shows such a preparation before and after welding with filler. The parent material is 2.8 mm mild steel welded at 3.5 kW and 40 mm s^{-1}, and choice of a dissimilar filler emphasises the melt geometry and convection. In the absence of filler this preparation would have resulted in, at best, a weld with severe undercut and, at worst, the beam passing through with minimal interaction; oscillation between these two modes leads to 'stuttering'.

Finally, the keyhole nature of the process leads to two characteristic effects. First, we observe, in common with workers elsewhere, that non-fully-penetrating (blind) welds in a number of materials exhibit root porosity. This is almost certainly associated with collapse of the keyhole trapping some vapour; it is dependent on the power-speed regime, but tends to disappear when fully penetrating welds are made. This effect is also common to eb welding, where it has been found amenable to control by spinning and oscillation of the beam; these techniques will be applied to laser beams in due course. The second, desirable, characteristic is that the laser can lead to a much reduced inclusion content in the weld zone, with mechanical properties correspondingly superior to the parent plate[2]; it seems likely that the keyhole provides a route to vent the more vaporisable impurities from the melt.

The existence of one or more of the above favourable characteristics has led to the present use of tens of lasers of powers $\geq$ 1 kW in production welding. Although some of the more well established of these have been extensively described in the literature, commercial security frequently precludes detailed discussion of the most recent and economically viable applications; it is probably most helpful to refer a potential user to the following table which lists existing and potential applications together with the salient characteristic(s) being exploited:

Welding Application	Characteristic Exploited
Dissimilar metal joining e.g. saw blade manufacture, fixing of valve seat inserts	Minimum thermal disturbance
Fabrication of stainless steel components e.g. washing machine drums, food bowls and trays	Fast, cosmetically acceptable
Fabrication of electrical cabinets	Hermetically sealed enclosures required
Lead battery fabrication	Fast, no drop out
Automotive components[3] and body parts	Fast, minimum distortion
Joining stock length for strip mills	High integrity, clean bead profile
Fabrication of special section structural members	Cheaper than forming
Pressure vessel fabrication[4]	
Tube to tubeplate joining	Potentially high integrity and reproducible
Tube to tube joining	

Fig.9 shows some sample welds relevant to the above applications: a butt joint in flat plate high yield steel (10.7 mm thick, 7.2 kW at 10 mm s^{-1}), a tube to tube joint in the same material (6.7 mm thick, 5.8 kW at 18 mm s^{-1}), a rectangular section fabricated from stainless steel (6.7 mm thick, 5 kW at 16 mm s^{-1}) and a simulated tube to tubeplate joint also in stainless steel (weld depth approximately 3.5 mm, 3 kW at 50 mm s^{-1}).

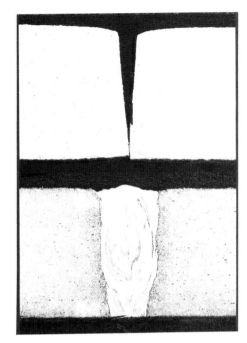

Fig.9   Sample welds, see text.

Fig.8   Laser weld in 2.8 mm mild steel (3.5 kW, 40 mm s^{-1}); top, cropped edge preparation; bottom, final weld with use of filler.

Surface Hardening by Martensitic Transformation

Introduction

The successful operation of a wide range of engineering components like gears, blades and shafts relies on surface hardening; a crucial property in these components is the combination of a core having high strength and toughness and a surface having high hardness and wear resistance. The hardness can be localised to the surface either by diffusing in to a controlled depth suitable additives (for example as in the nitriding and carburising processes), or by starting with an appropriate alloy which undergoes a phase transformation at the surface when it is suitably heated and quenched. We confine ourselves here to the latter category and discuss the role of lasers as the heat source.

The phase transformation involved in a large number of these components is that of martensite formation in the iron-carbon system. We recall that if the alloy is heated to a temperature at which an austenitic structure exists, i.e. where it consists of a solid solution of carbon in gamma iron, and is then cooled above a critical rate, the normal lower temperature equilibrium components of pearlite with ferrite or cementite do not have time to form, but instead a very hard metastable solution of carbon in alpha iron, known as martensite, is formed.

When lasers are used as the heating source, they are normally operated in a power density regime very much higher than flames, but possibly comparable with eb, arc plasma and induction heating sources. However, the laser can be uniquely distinguished from these last three by one or more of the following attributes:

(a)   it characteristically operates as a rapidly scanning source so that overall heat input (and therefore distortion) is minimised, and adequate quenching rates are obtained solely by conduction into the substrate;

(b)   the rapid scanning can take place in atmosphere at long working distances from the source, and the beam can be manipulated and directed into bores and conventionally inaccessible regions without the hindrance of supply cables and pipes;

(c)   its heating pattern may be rapidly altered to suit the application.

Process

Below their melting point metals are poor absorbers of the infrared energy. An absorbing coating is therefore a prerequisite for efficient heating; fortunately suitable coatings (e.g. colloidal graphite, manganese and zinc phosphates) can be easily applied, with the advantage that selective application results in selective hardening. Fig.10 shows the calorimetrically determined absorption of ground stock, 0.4% C steel having four different coatings, as a function of scan speed. A rectangular beam, of length about 3 mm in the scan direction, and of power density about 4.5×10^3 W cm^{-2} was used; in general, the approach to surface melting leads to destruction of the coating and a fall in absorptivity. Although colloidal graphite gives coupling similar to phosphates, we note that it does produce more debris during processing, thereby aggravating attempts to continuously monitor surface temperature by optical pyrometry.

The most difficult technological aspect of the hardening process is to arrange for a suitable beam intensity distribution at the workpiece. In general, flat topped distributions (preferably of variable extent) are required, although the effects of 2 and 3 dimensional heat flow in particular workpieces will require lobes in the distribution if the depth hardened is to have the highest uniformity. In some applications a suitable heat pattern at the workpiece can be obtained by operating at an appropriate distance from the focal plane of a simple lens or mirror (and perhaps additionally arranging for laser multimode operation). In other cases the requirement of uniform intensity over larger areas indicates use of alternative techniques; one that we have used involves use of a vibrating mirror to generate a pattern at the workpiece, but it is difficult to avoid the effects of prolonged dwell times at the limits of the pattern, and masking must be employed. Another technique which we have used successfully involves allowing the annular output beam (in this case 100 mm outer diameter) from the unstable cavity of the laser to undergo reflection at small angle from a mirror consisting of eight individually-aligned, micromachined[5] segments, each behaving as a cylindrical focus element (focal length 800 mm). The images are suitably overlayed to yield a rectangular pattern of length determined by the segment length and width by the proximity to focus. These parameters result in acceptable uniformity of the pattern; Fig.11 shows hardened depths of 0.4 mm and 1.1 mm produced by this technique in quenched and tempered 0.4% C steel. We find that doubling the number of segments leads to excessive diffraction structure in the intensity distribution and the appearance of associated melt stripes in the workpiece. This effect may be amenable to correction by suitable blending of mirror segment edges and/or vibration of the complete mirror.

The hardening cycle may be considered in three stages:

(a)   heating a layer above austenising temperature (A_3);

(b)   holding there for a sufficient time for the carbon to go into solution;

(c)   quenching rapidly.

The timescale in laser hardening is very significantly shorter than that in conventional techniques and much detailed investigation of the metallurgy remains to be done. However, useful guidance on the practical application of the technique is afforded by the use of heat flow calculations. The case of a heat source with gaussian intensity distribution moving over a substrate has been treated[6] but since we have been concerned mainly with sources uniform over areas much larger than the skin depths of interest, we have used one-dimensional analyses, and furthermore have approximated to the moving source by postulating a heating pulse of duration equal to the beam dwell time.

The time dependent solution for the temperature θ along the z axis perpendicular to the surface of a semi-infinite solid exposed to a heat flux F is[7]:

$$\theta_z = \frac{2F \sqrt{\kappa t}}{K} \; \text{ierfc} \; \frac{z}{2 \sqrt{\kappa t}} \tag{2}$$

where K is thermal conductivity; κ is thermal diffusivity; t is time of application of heat flux.

At shallow depth, for heat inputs of interest here, the peak temperatures reached are given by θ_t when t = the beam dwell time.

Additionally we have used a computer programme[8] based on a numerical solution of the one dimensional problem of a plate of finite thickness subject to a heat flux for a specified time on the front, and insulated at the rear. The programme takes account of melting and heat loss by radiation, and plots time resolved profiles in the plate.

To gain some insight into the process, it is convenient to consider predictions and results for a carbon steel, say K = 0.35 W cm^{-1} $^{\circ}$C^{-1}, κ = 7.3 x 10^{-2} cm2 s^{-1}, melting point θ_m = 1500°C, A_3 = 800°C. The amenability of such a steel to heat treatment is conventionally expressed in the isothermal transformation diagram on the right hand side of Fig.12; when a sample, previously heated above A_3, is quenched into a bath of fixed temperature (indicated on the vertical axis), the progress of the transformation is shown by following along an abscissa through that temperature. For example, when quenched into a bath held constant at 500°C, Ferrite starts to form after about 0.8s, cementite after 1 to 2s and the transformation is complete after 5s. Martensite will result if temperatures below 350°C can be achieved in times less than about 1 second.

The left hand side of Fig.12 shows the computed temperature-time history at (a) the surface and (b) depth 0.5 mm for a 10 mm thick sample subject to a power density of 3 x 10^3 W/cm^{-2} for a dwell time of 0.2s. It is seen that quenching rates strongly favour martensite formation, and that holding times above A_3 are short – even on the surface they are only of order 0.15s. In the resulting microstructure the surface, where the carbon has been taken into homogeneous solution, will be fully martensitic while further into the layer considerable hardening will still be observed although the short time at elevated temperature results in much less homogeneity. The upper part of Fig.13 plots a microhardness scan for such a layer. Fig.14 shows a micrograph of similarly treated pearlitic cast iron, and the lower part of Fig.13 shows the corresponding microhardness scan.

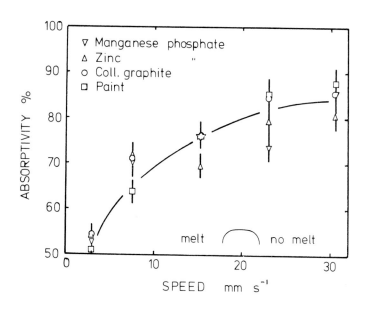

Fig.10   Absorptivity v scan speed for four coatings

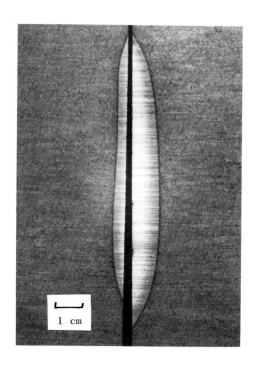

Fig.11   Sections showing laser hardened zones in 2 samples of 0.4% carbon steel (mounted with treated surfaces face-to-face)

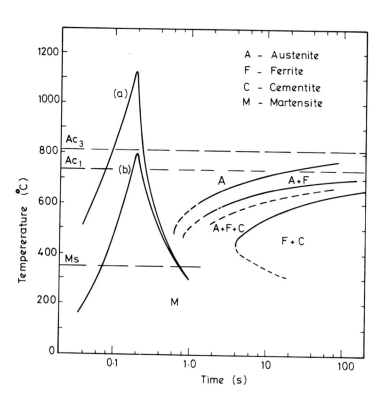

Fig.12   Isothermal transformation diagram for carbon steel with (superimposed) calculated temperature cycle at (a) surface and (b) depth 0.5 mm for 10 mm thick steel sample subject to a power density of 3×10^3 W cm^{-2} for dwell time of 0.2s.

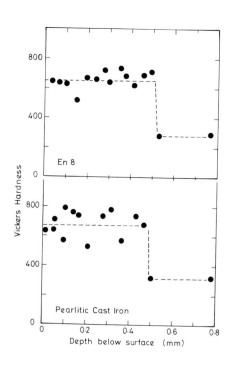

Fig.13   Typical microhardness scans in layers similar to those of Figs.11 and 14.   Upper part 0.4% carbon steel, lower part pearlitic cast iron.

In Fig.15 we have plotted the computed depth raised above A_3 as a function of the reciprocal of the dwell time (which is proportional to the beam scan speed in our approximation), a form which highlights some features of the process. The solid lines show computed depth values for a 0.4% C steel; the broken line (Ft$^{\frac{1}{2}}$ constant) indicates the onset of surface melting at 1500°C. Also shown are experimentally observed depths of hardening for samples having an absorbing coating which were traversed under a line focus of approximately 3 x 10^3 (▲) and 6 x 10^3 (▼) W cm^{-2} respectively, where it is seen that agreement between prediction and observation is fair. The dotted lines show corresponding predicted depth values for a cast iron (K = 0.42 W cm^{-1} °C^{-1}, k = 7.2 x 10^{-2} cm^2 s^{-1}), and the chain dotted line shows the onset of melting, in this case determined by a eutectic, at 930°C. The figure illustrates the strong dependence of depth on the scan speed; this is particularly significant in the case of cast iron close to the melting point. The sensitivity of the process to beam uniformity can be gauged by considering for example the treatment, close to surface melting, of carbon steel using a perfectly uniform beam of intensity 3 x 10^3 W cm^{-2}, when a dwell time of 0.33s will give a hardened depth of 1 mm. With a hot spot in the beam of intensity 20% higher than average, the dwell time must be reduced by 30% to avoid local surface melting, and the resulting depth hardened by the main beam then drops to 0.63 mm.

Discussion and Applications

Those materials most readily hardened by conventional techniques tend to be the most amendable to laser hardening. Thus, alloy and tool steels are particularly easily treated whilst structures with widely dispersed carbide or graphite are less so. Nonetheless, ferritic malleable iron cylinders are successfully hardened in one of the largest production line uses (15 lasers installed) of lasers.[9] Automotive component hardening, featuring high volume production and/or requirements of minimum distortion, is a prime candidate for laser processing, and the field is being actively investigated by a number of groups including our own.

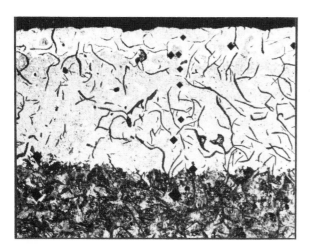

Fig.14 Micrograph of laser hardened layer (~ 0.5 mm thick) on pearlitic cast iron.

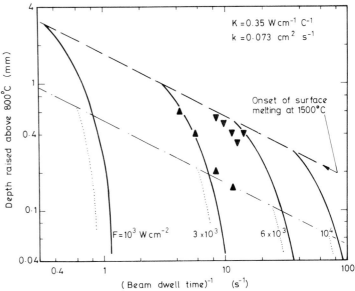

Fig.15 Calculated depth raised above 800°C as a function of reciprocal of beam dwell time. See text.

Other Surface Treatments

Cladding

Surface cladding or coating is employed when localised corrosion or wear resistance is required on a component having a composition not inherently capable of producing those properties. Conventional heat sources, for example flames or plasmas, may be used to fuse a layer of alloy on to a substrate; a laser may be used in an analogous manner but with the following potential advantages associated with the greater controllability of the heating:

(a) optimisation of coating geometry so that additive usage and distortion is reduced;

(b) minimised melting of substrate so that dilution is reduced;

(c) fast cooling, offering possible control of microstructure.

The laser beam is operated in a high power density regime and is rastered or focused to create the required width of cladding band. The cladding alloy is preferably added in powder form since the supply can be more readily matched to the geometry of the beam. Alternatively the laser may be considered as a means of improving the integrity (homogeneity, adhesion) of existing overlay coatings applied, for example, by spray bonding or electro-chemical techniques.

Our preliminary investigations have involved the cladding of elements (in powder form) such as aluminium, chromium, molybdenum and nickle, and of Stellite alloy on to steel substrates. Fig.16 shows a car engine exhaust valve with a hard facing alloy on the seat deposited by a 4.5 kW laser beam; microhardness measurements at the working surface give a value of 600 HV; detailed technical and economic comparisons with other methods have yet to be carried out.

Diffusion of Alloying Constituents into Substrates

Above, the aim was to deposit a working surface of composition undiluted by the substrate; here the aim is to develop a surface of composition determined by both substrate and additive. By this means very economical use is made of additives, but much more exact metallurgical guidance in their selection is required. Our work in this area is at a very early stage, but we have observed some encouraging results when using the laser in a similar manner to an electron beam in experiments by other workers[10] on the diffusion of elements into aluminium alloys to increase surface hardness. In particular, silicon levels can be very readily enhanced.

Rapidly Quenched Surface Layers

When focused to power densities in excess of $\sim 10^6$ W cm^{-2}, lasers can produce on a substrate a thin melt layer which may experience quenching rates of order 10^6 °C s^{-1}. This is a regime of particular relevance to metallurgists working in the field of rapidly quenched metals[11] where similar cooling rates are routinely achieved by the sudden, intimate contact of atoms, ions, vapour or molten droplets or ribbons with a heat sink. The rapid quenching can result in supersaturated and metastable phases, such as supersaturated solid solutions and homogeneous micro-crystalline structures; for example, tool steel undergoing such quenching may be shown to possess extremely high hardness. With careful choice of composition, particular substrates become amorphous on quenching, to yield metallic glasses. These may exhibit useful magnetic properties, high corrosion resistance, and in addition a unique combination of hardness and ductility.

Early experiments were carried out to produce such structures using pulsed lasers, but a recent innovation[12] has involved use of a cw multikilowatt laser. Rapid scanning of the focused beam resulted in the creation on substrates of fine melt tracks exhibiting some of the above properties. Although the ability to effect microstructural modification is in itself of far reaching importance, there is particular interest in the prospect of creating on a component an amorphous corrosion resistant skin. Laser alloying may be used to set up first of all an appropriate composition at the surface, but subsequent glazing using a large number of adjacent narrow laser scans is unattractive because the heat pattern from one will tend to recrystallise the structure of a preceding scan. However it may be possible with sufficiently high power and a line focus to create wider glazed tracks in one pass.

Conclusions

We have described a particular range of multikilowatt lasers, and discussed some of their existing and potential applications in metalworking, where the feature exploited is the laser's unique ability to remotely deliver through atmosphere a controllable, high power density beam. The high process rates of the laser require that it be matched to high volume production lines; when the laser output exceeds the demand of one line it can be readily switched to carry out a different process on another line. Lasers of powers up to 1 or 2 kW are now routinely used as production tools, where the upward trend of the powers of the lasers employed has been determined by the availability of reliable, cost effective, higher power machines and the growth of confidence in them by users. There seems no reason to doubt that conditions are right for this trend to continue.

Acknowledgements

This work has benefitted from the efforts of all our colleagues and support staff of the Laser Applications Group under the direction of D.F. Jephcott. We specially acknowledge contributions in the general area of high power laser development and application of P.J. Andrews, A.S. Bransden, W.J. Brewerton, A.G. Delph, E. Hanley, K. Harries, M. Hill, P. Millward, C.J. Nicholson, S.J. Osbourn, L. Reynard, B.A. Ward and drawing office and workshop staff. We acknowledge also the interest of T. Bell and D.N.H. Trafford of Liverpool University.

References

1. Breinan, E.M., Banas, C.M., Evaluation of Basic Laser Welding Capabilities, United Technologies Report R75-911989-4, 1975.

2. Breinan, E.M., Banas, C.M., Fusion Zone Purification During Welding with High Power CO_2 Lasers, United Aircraft Research Labs. Report R111087-2, 1975.

3. Megaw, J.H.P.C., Hill, M., An Investigation of the Welding of Automotive Components by Laser, Culham Laboratory Report CLM/RR/F7/1, 1978, unpublished confidential.

4. Willgoss, R.A., Megaw, J.H.P.C., Clark, J.N., An Assessment of Laser Welding in BS1501 Types 316 and 310 Stainless Steels and Ducol W30 Carbon Steel, CEGB Marchwood Engineering Laboratories Report R/M/R264, 1978.

5. Ward, B.A., Diamond Machining Metal Mirrors Using Fly-cutting Geometry, Proc.Soc.Photo-Optical Instrumentation Engineers, Vol.93, pp.62-68, 1976.

6. Pittaway, L.G., The Temperature Distributions in Thin Foil and Semi-infinite Targets Bombarded by an Electron Beam, Brit.J.Appl.Phys., Vol.15, pp.967-982, 1964.

7. Carslaw, H.S., Jaeger, J.C., Conduction of Heat in Solids, Clarendon Press, Oxford, 1959.

8. Langosz, G.E., Stefan Problem Solution Programme, unpublished.

9. Miller, J.E., Wineman, J.A., Laser Hardening at Saginaw Steering Gear, Metal Progress, May, pp.38-43, 1977.

10. Schweisstechnik (Switzerland) Vol.65, p.265, 1975.

11. Jones, H., Splat Cooling and Metastable Phases, Rep.Progress Physics, Vol.36, pp.1425-1497, 1973.

12. Breinan, E.M., Kear, B.H., Banas, C.M., Greenwald, L.E., Laser Glazing - A New Process for Production and Control of Rapidly Chilled Metallurgical Microstructures, Soc.Manf.Engs., Western Laser Conf., Los Angeles, Nov., 1976.

Fig.16 Car engine exhaust valve with stellite facing on seat deposited by 4.5 kW laser beam.

A HIGH POWER LASER METALWORKING FACILITY AND ITS APPLICATIONS

H. Walther, L. Pera
Fiat Research Center
10043-Orbassano, Turin, Italy

Abstract

Scope of this paper is to give an overview of the activities developed in the CRF laser laboratory equipped with a 15 kW CO_2 laser source. These activities are focused on the development of processing opportunities regarding transport vehicles as well as turbine and diesel engines. Four examples are discussed: surface transformation hardening, surface alloying, wear resistant coating and hot corrosion resistance coating.

Introduction

Large metalworking industries were forced, during the last five years more than at any time before, to identify opportunities to compensate for the steadily increasing cost of manpower, energy and raw materials. As a consequence they had to look for lower cost- or for higher quality-processes, or better, for new processes that allow to achieve both of these advantages. There is no doubt, that most R & D institutions have considered the processing opportunities offered by high power laser systems now commercially available.

As it can easily be recognized, the most important advantages of laser processing derive from the fact that short interaction times between the photons and the metal target are normally sufficient for obtaining, in well definable regions of the component, the desired metallurgical effects. Useless heating of not interested zones and of the bulk can in general be avoided and evident advantages consequently achieved mainly concerning: limitations in thermal distorsion, high rates of phase transition, energy savings, elimination of finishing operations and increase of production rates.

It may easily be summarized that laser processing offers rather unique opportunities for cost reduction and quality improvements at the same time, and that the intrinsic potential for these opportunities evidently tends to increase with increasing laser power.

Such a tendency can of course be recognized from the well known /1/ display of the various applications (fig. 1) in terms of power density and interaction time. In fact, the total field in which the laser can be applied, as well as single applications (trasformation hardening, melt quenching) are not displayed parallel to the lines of constant specific energy, but appear with a lower slope, indicating that for the higher power density applications less total energy is required.

The present state of the art in laser processing with CW-CO_2 sources can be summarized as follows: Laser sources with proven industrial reliability are now available up to power levels of a few kilowatts (say 5 KW). From a total of 90 units over 1 KW sold, almost half seem to be already in production (table I). For systems with higher power levels (say up to 15 KW), industrial reliability will presumably be reached within a very few years.

Whilst high power laser (HPL) manufacturers have already performed extremely useful process development work with experimental prototype facilities in order to demonstrate laser processing potentialities up to the highest power levels (15 KW and more), much more work must still be done by the users for tailoring this potential towards practical applications.

The Fiat Research Center (C.R.F.) has collaborated with the AVCO Everett Research Laboratories (A.E.R.L.) in demonstrating HPL application potentials by considering various examples such as synchrogear welding, transformation hardening of crankshafts and camshafts etc., and alloying of exhaust valves. Successively Fiat's laser laboratory has been set up and furnished with a 15 KW A.E.R.L. Laser source.

This laboratory is fully operative since April 1978 and it is presently engaged in developing the technological fundamentals for a series of practical applications of high power laser processing. Scope of this contribution is to give an overview of this activity.

The Laser Laboratory

The main activity of the laboratory is obviously focused on the development of laser processing opportunities regarding transport vehicles as well as turbine and diesel engines.

Two laser systems are available, one is based on a 15 KW CO_2 AVCO Laser source with two working stations and displayed over a surface of 180 m^2 as shown in fig. 2. The first station is designated for cutting and welding. An overhead horizontal optical bench serves for supporting welding telescopes (f/7 or f/18) and a high flux, downhand deflection mirror. The telescopes focus the beam on a horizontal X-Y working table to spot sizes of 0.7 mm and 2.5 mm respectively. The working table can be moved in both horizontal directions with maximum speeds of 30 m/min and a positioning accuracy of 0.1 mm. Devices for additional biaxial sample rotation (up to 500 RPM) are available. Total movements can be programmed and numerically controlled.

The second working station for the 15 KW Laser is used for surface treatments (hardening, alloying, coating). The optical system is based on a mirror system (f/150 or f/100) which produces a beam that scans the workpiece by means of two vibrating mirrors (125 and 600 Hz) over a rectangular area of variable size (up to 25 mm, each side). Numerically controlled linear and rotational sample movements are also provided for this station.

The second laser system has been supplied by Valfivre (Italy) and is suitable either for continuous or for pulsed operation; its beam can be focused to spot sizes from 0.1 mm to 1 mm. Maximum power output in continuous operation is 500 W whilst pulsed operation mode is possible at 2 KW and pulse frequencies of 1000 Hz. Bidimensional linear and rotational movements and numerical control are also provided.

Main task of the laser laboratory staff is to operate the facility, to improve its performance and to assist research and development work with specialized tasks such as heat balance calculations, laser optics, numerical control of sample movements etc. The proper R & D in laser processing is performed by process engineers and materials scientists of conventional specialization.

Applications

Prior to setting up the laser laboratory and acquiring the laser source, a series of feasibility studies were performed in collaboration with the laser supplier. Some of these studies concerned:
- precision welding of synchro gears,
- penetration welding of fork lift guide,
- transformation hardening of diesel engine crankshafts and petrol engine camshafts,
- transformation hardening of valve seat inserts,
- alloying of exhaust valves.

Preliminary results on these arguments have been published earlier /2,3/ and will not be discussed here; most of the work is continuing and is focused on metallurgical aspects and corresponding properties.

The four examples discussed here are typical of the metallurgical problems of laser processing.

Surface Transformation Hardening

Surface hardening of Fe - C alloys by phase transformation can in general be achieved in two ways:
- Thermochemical treatment (for materials with low carbon content) in atmospheres rich in Carbon (or other suitable elements). After inward diffusion of C, normal quenching is required.
- Induction treatment (for materials with higher C content), where high alternating currents are induced into the sample and concentrated by the skin effect near the surface. Successive quenching is normally necessary and case thicknesses up to a few mm can be achieved.

Laser hardening of Fe-C materials /4,5,6/ can be seen as an alternative to induction hardening with additional advantages which might enlarge the field of surface hardening.

Amongst these advantages (see table II) one should especially consider: selfquenching, minimal distorsion, and the possibility of treating zones with difficult accessibility for induction coils.

The problem of achieving, by laser transformation hardening, a martensitic surface layer of a certain thickness can be theoretically described by means of a heat balance calculation /7/.

The results of such a calculation are shown in fig. 3, where the specific laser power, necessary for producing at a certain depth of the irradiated sample the transition temperature from ferrite to austenite (say 900°C), is plotted for different interaction times. Experimental points confirm in general the reported diagram /8/.

Also shown is a curve, along which the surface melting temperature (1250°C) of the material is reached. As microcracks have been observed under surface melting conditions, operation above this curve should be avoided. It is recognized, that the problem of avoiding surface melting for greater case depths (say > 1 mm) becomes more and more the problem of exact power control; small power variations can in fact generate great variations in surface temperature. Fig. 4 shows a typical result for a low alloyed steel (AISI 1045) with satisfactory metallurgical features.

Not so easy is the case of hardening nodular cast iron, where the inhomogeneity causes and additional problem. From fig. 5 it is seen in fact, that melting can be induced not only on the surface, but also in the bulk around the nodular graphite particles where the temperature is certainly lower than near the surface. One possible explanation is given by the higher C concentration around the nodules which lowers the melting temperature so that fusion may be obtained even when the surface does not show any melting. A result with metallurgically acceptable aspect is given in fig. 6, where fusion phenomena (with their intrinsic danger of microcrack formation) are completely absent and where the graphite particles appear totally embedded in martensite. Before transformation, this matrix had a ferritic-pearlitic structure.

Surface Alloying

Mechanical properties at the sample surface (e.g. hardness and toughness), different from bulk properties, can be achieved conventionally with compositional changes introduced by:
- diffusion processes (e.g. pack cementation, slurry deposition, etc),
- ion implantation
in addition of coating techniques.

These methods produce limited results because of the long processing times required, and the compositional limits connected with the process time and/or solubility conditions.

Laser alloying largely extends these limits because the high power density allows a rapid mixing of the components and a sufficiently rapid quenching to prevent segregation.

Strengthening of aluminum and aluminum alloys (e.g. by Si) is based on finely dispersed particles which limit the movement of dislocations. Volume fraction and dimensions of these particles determine a suitable combination of hardness and toughness of the material. The effect of "dispersed particle strengthening" on mechanical properties can be evaluated by the usual methods of physical metallurgy /9/.

As properties, it is thereby convenient to consider the resulting shear strenght τ and the elongation to rupture γ (instead of hardness and toughness). The hardness can be calculated from these properties /10/. For the present example of dispersion hardening of duralumin by Si, the model of Ashby /11/ has been modified /12/ by considering, instead of a constant particle radius r, the effective size distribution of r.

The result is:

$$\frac{\tau - \tau_o}{G} = C \left[\frac{1}{2} \ b \ \gamma \right]^{\frac{1}{2}} \cdot (\frac{f}{r_o})^{\frac{1}{2}} \tag{1}$$

$$\gamma = \frac{8\pi(1-\nu)\sigma_c}{G} \left[(\frac{\pi}{6f})^{1/3} - 1 \right] \tag{2}$$

where

r_o = most frequently occuring radius of

$$N(r) = \frac{f}{32\pi r_o^4} \frac{r}{r_o} e^{-r/r_o} \tag{3}$$

f = volume fraction
G = shear modulus
σ_c = rupture strength of Si
ν = Poisson ratio of Al
C = 0,18
b = Burgers vector
τ_o = yield shear stress ($\sim$ 4,5 kg/mm^2)

(1) and (2) are plotted in figg. 7a, b, for a typical case.

It is seen from (1), that the shear strength τ increases with increasing f/r_o ratio.

Conventional Al-Si alloys have maximum volume fractions of about 20% with $r_o \sim$ 6μm, but many Si particles are considerably greater.

With the laser, one can achieve structures having f values even higher than 50%, with $r_o \sim$ 1 μm. Fig. 8 shows structures obtained by laser alloying; the upper one has a rather conventional aspect: Al dentrites in an Al-Si matrix; the total Si content is 12%. The lower picture shows a structure not achievable by conventional techniques: a homogeneous distributon of Si in Al with f = 50% and r_o = 1 μm.

Wear Resistant Coatings

The evolution of the wear process of mechanical components depends, as it is well known, on mechanical properties, compositional details and structural features /13,14,15,16/.

In general, wear resistant coatings require:
- Perfect integrity between the layer and the substrate, such as obtainable by diffusion bonding on one hand and by compatibility criteria for the thermomechanical properties (e.g. differential thermal expansion) on the other.
- Compositional optimization of the coupled materials; it is known that, besides of hardness, the solubility of one material in the other should be low (say < 0,1%) in order to achieve low wear and friction.
- Adequate thicknesses in order to achieve sufficiently long lifetimes to failure (e.g. by delamination); this under increasingly severe conditions.
- Lubricant retainment as achievable not only by adequate machining (e.g. roughness R_a > 3) but also by material porosity in the bulk.

Conventional coating techniques such as electrolytic procedures, chemical or physical vapor deposition, as well as the various flame or plasma spray methods, although widely applied, do not seem to have satisfied simultaneously the listed requirements. Laser processing offers here opportunities which no other technique can offer so far. As an example, fig. 9 shows a lamellar cast iron substrate coated with a wear resistant alloy of Mo, Ni, Cr_3C_2, which was processed by the laser. It is seen, from the concentration and hardness profiles, that the coating is sufficiently thick and perfectly bonded to the substrate. For comparison, plasma sprayed layers of the same material presently being used, have notoriously imperfect bondings and a tendency to delaminate from the substrate. The porosity of the plasma spray coating may however be considered as an advantage for retaining a certain quantity of lubricant.

Without sacrificing the bonding, one can provide by means of the laser a useful porosity by adding a quantity of Si powder to the coating powder material before spreading it on the components surface. Using a surface treatment temperature according to fig. 10 of about 2800°C, it is recognised that the components Mo, Cr-carbide and Ni are brought into the liquid state, whilst Si (and Cr) already evaporate.

This evaporation forms rather spherical pores, which are distributed uniformly over all the zone ranging from the surface to the interface with the base material as shown in fig. 11.

Hot Corrosion Resistant Coatings

The well known problem connected with hot corrosion resistance of materials required the development of a large number of coating processes to allow both an increase of the operating temperature and the use of fuels with higher impurity content.

The usual two categories of coating processes are the overlay coatings like hot spraying, chemical vapor deposition, physical vapor deposition and the surface modification coatings like pack- and slurry cementation, metalliding and hot dipping.

All of these coatings present various limits due to bonding between coating and original surface, presence of micropores and microstructural instability, etc. Some of our work is oriented to produce "overlay coatings" by re-treatment of coatings obtained by plasma spray processes. A preliminary result is shown in fig. 12. It indicates that the original coating (upper figure) even having a very irregular surface with the presence of micropores, becomes, after laser treatment, very regular and assumes a lower porosity content. The micrograph concerns a Ni-Cr-Al coating on a IN 738 LC substrate. First evaluation shows in fig. 13 a reduction of oxidation rates.

Comments on laser operational problems

After one year of operation with the 15 KW laser some comments concerning the realiability of the system can be made. Our system was in the average available for applications 63% of the time; 23% of the operators time was dedicated to system repair and maintenance. During the last few month the situation improved further and we expect for the second year a reliability of $\geq$ 80%.

Most of the operational problems were related to electric devices and electro-pneumatic components, (not to the laser itself); in addition, alignment of beam was quite time consuming. Some other comments can be made on the transfer of energy. Separated work stations dictate a somewhat complex layout of optical devices. The resulting set-up requires a large number of mirrors: this means low total transmission coefficients for the beam. In our system five mirrors are required for one station and eleven for the other. The overall transmission coefficients are respectively 0.85 and 0.7.

To reduce dissipation of energy, the quality of mirror surfaces must be constantly inspected for infrared absorbtion. Atmospheric effects were also found to be of importance for beam transmission to guarantee the quality of focused beam. It was found that in order to avoid beam deflections due to temperature induced zones of local convection, it is important to maintain a certain turbulence of the air in the beam path. The effect of thermal blooming was found to be of much less importance on the quality of the focussed beam.

References

1. E.M. Breinan, B.H. Kear, C.M. Banas : "Processing Materials with Lasers", Physics Today" - Nov. 1976.

2. A. La Rocca, G.P. Cappelli, H. Walther : "Development Activities for the Industrial Application of H.P. Laser to Metalworking", Int. Seminar on Laser Processing of Engineering Materials, Inst. of Physics, London, June 1977.

3. A. La Rocca: "Laser Application in Machining and Material Processing", Europ. Conf. on Optical Systems and Applications, Brighton, U.K., April 1978.

4. F.D. Seaman, E.V. Locke: "Metalworking Capability of a High Power Laser", SAE 740864 - Oct. 1974.

5. C. Wick: "Laser Hardening", Manufacturing Engineering - June 1976.

6. M. Yessik, R.P. Scherrer : "Practical Guidelines for Laser Surface Hardening", SME Technical Paper MR75-570 (1975).

7. G. Alessandretti : Int. Fiat Report - 1978.

8. A. Blarasin, G. Boda : Int. Fiat Reports - 1978.

9. A. Kelly, R.B. Nicholson: "Strengthening Methods in Crystals", Elsevier Publ. Comp. - 1971.

10. D. Bacci : Int. Fiat Report.

11. M.F. Ashby : "The deformation of Plastically Non-homogeneous Materials", Phil. Mag. Oct. 1969.

12. D. Bacci : Int. Fiat Report.

13. N.P. Suh : "An Overview of the Delamination Theory of Wear", Wear 44 - 1977.

14. S. Iahanmir, N.P. Suh : "Surface Topography and Integrity Effects on Sliding Wear", Wear 44 - 1977.

15. B. Peterson, K. Gabel, I. Devine : "Understanding Wear", ASTM Standardization News - Sept. 1974.

16. E. Rabinowicz: "Friction and Wear of Materials", John Wiley and Sons Inc. - 1965.

CW – CO2 LASER SOURCES FOR PRODUCTION (ESTIMATED)

MANUFACTURER	POWER (kW)	NUMBER SOLD	FOR PRODUCTION
PHOTON SOURCES	1	50	20
SYLVANIA 971	1.2	20	11
BRITISH OXYGEN	2	8	3
UTRL	3	2	2
SYLVANIA 975	5	10	7
UTRL	6	1	0
UTRL	15	1	0
AERL	15	6	0
TOTAL		98	43

Table I

BASIC LASER PROCESSES WITH THEIR POSSIBLE ADVANTAGES

PROCESSES / ADVANTAGES	CUTTING	WELDING	HARDENING	ALLOYING	CLADDING
OF QUALITY	• NEGLIGIBLE THERMAL EFFECTS • SMOOTH BORDERS • HIGH PRECISION • MINIMUM KERF WIDTH	• LITTLE DISTORSION • DEEP PENETRATION • HIGH PRECISION	• HIGH HARDNESS • LITTLE DISTORSION • EXACT DEPTH CONTROL • POSSIBILITY OF TREATING LIMITED ZONES • TREATING ZONES WITH DIFFICULT ACCESS	• INFINITE POSSIBILITIES OF COMBINATIONS • COMPATIBILITY OF PROPERTIES BETWEEN SURFACE ZONES AND BULK	• SUFFICIENT THICKNESSES • COMPACT LAYERS • HIGH FUSION TEMPERATURE OF THE LAYER ACHIEVABLE • COMPATIBILITY OF PROPERTIES OF THE LAYERS AND THE BULK
OF COST	• HIGH PRODUCTIVITY BY SPEED • A SINGLE LASER SOURCE CAN PERFORM VARIOUS PROCESSES • ELIMINATION OF FINISHING PROCESSES • TIME AND POWER SHARING ON THE SAME OR ON DIFFERENT SOURCES AND WORKING STATIONS ($\sim$ 100% TIME EFFICIENCY) • POSSIBILITY TO CREATE A LASER ENERGY NETWORK WITH DIFFERENT SOURCES AND MANY WORKING STATIONS • GREAT DISTANCE BETWEEN SOURCE AND SAMPLE WITHOUT ANY MECHANICAL CONTACT • FOLLOWS COMPLEX WORKING PATH'S • MATERIAL SAVINGS • ENERGY SAVINGS				

Table II

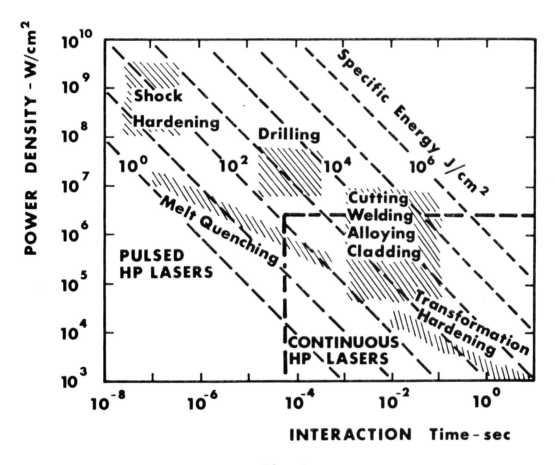

Fig. 1

DISPLAY OF LASER PROCESSING TECHNIQUES VESSUS POWER DENSITY AND
INTERACTION TIME

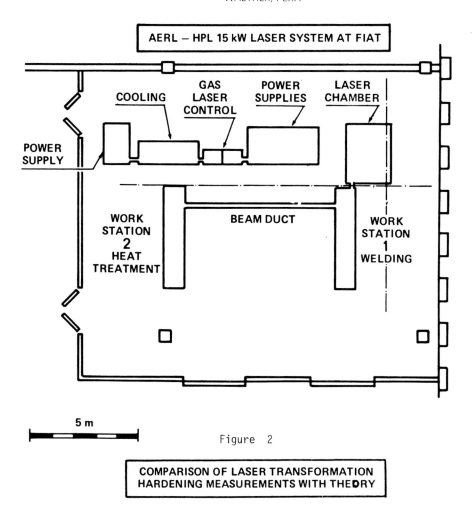

Figure 2

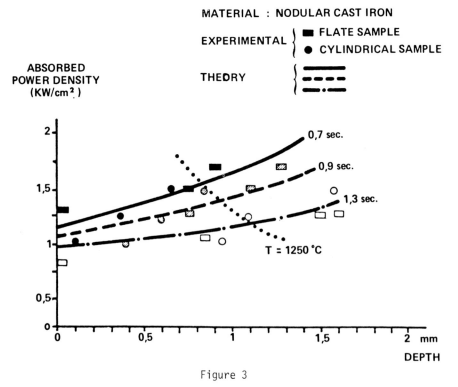

Figure 3

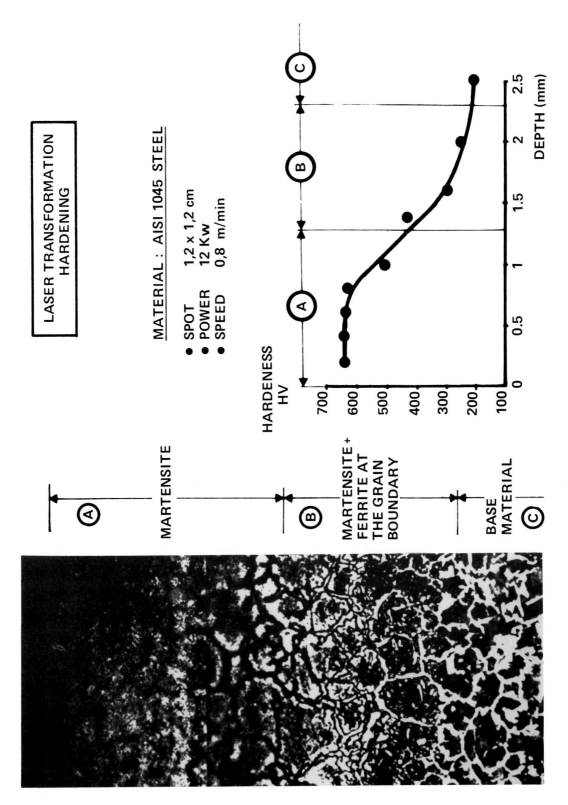

Fig. 4

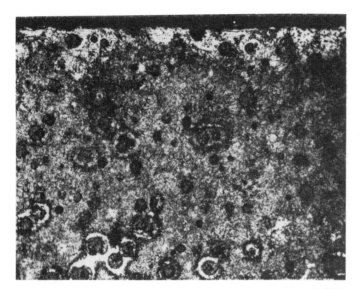

x 120

SURFACE MELTING
OF LASER TREATED
NODULAR CAST IRON

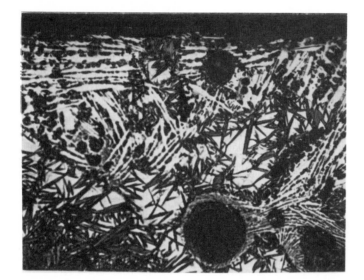

x 800

Fig. 5

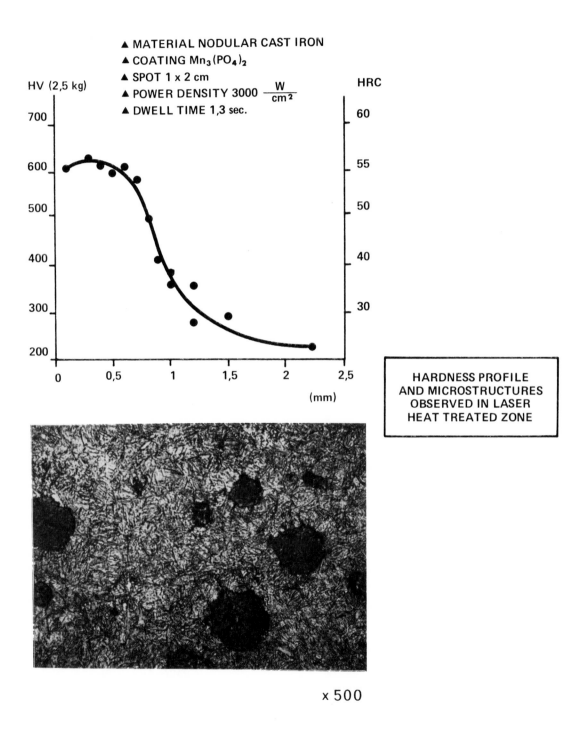

▲ MATERIAL NODULAR CAST IRON
▲ COATING $Mn_3(PO_4)_2$
▲ SPOT 1 x 2 cm
▲ POWER DENSITY 3000 $\frac{W}{cm^2}$
▲ DWELL TIME 1,3 sec.

HARDNESS PROFILE
AND MICROSTRUCTURES
OBSERVED IN LASER
HEAT TREATED ZONE

x 500

Fig. 6

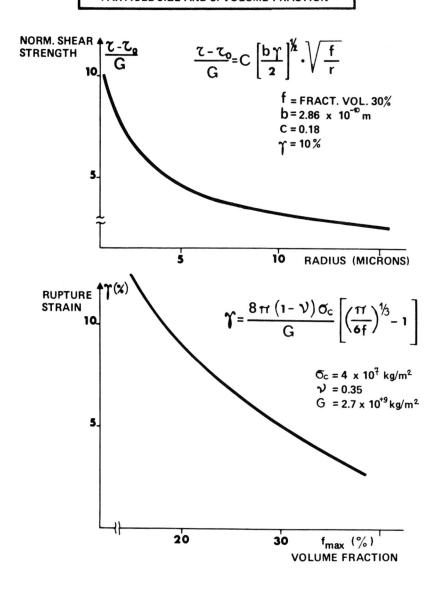

SHEAR STRENGTH τ AND RUPTURE STRAIN IN Si – ALLOYED ALLUMINIUM IN FUNCTION OF Si PARTICLE SIZE AND Si VOLUME FRACTION

NORM. SHEAR STRENGTH

$$\frac{\tau - \tau_0}{G} = C \left[\frac{b\gamma}{2}\right]^{1/2} \cdot \sqrt{\frac{f}{r}}$$

f = FRACT. VOL. 30%
$b = 2.86 \times 10^{-10}$ m
$C = 0.18$
$\gamma = 10\%$

RADIUS (MICRONS)

RUPTURE STRAIN

$$\gamma = \frac{8\pi(1-\nu)\sigma_c}{G} \left[\left(\frac{\pi}{6f}\right)^{1/3} - 1\right]$$

$\sigma_c = 4 \times 10^7$ kg/m^2
$\nu = 0.35$
$G = 2.7 \times 10^{+9}$ kg/m^2

f_{max} (%)
VOLUME FRACTION

Fig. 7

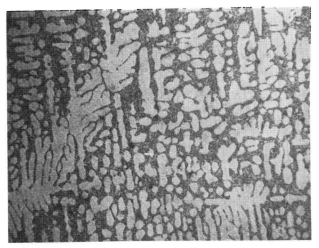

**DENDRITES OF Al
MATRIX WITH Si
PARTICLES**
(12% Si in Al)

60 x

**HIGH Si
CONCENTRATION**
(~50%) in Al

300 x

Fig. 8

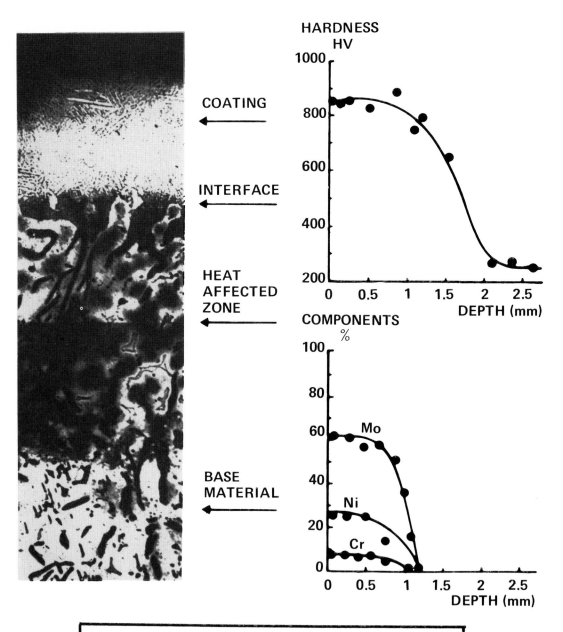

LASER COATING

HARDNESS HV

COMPONENTS %

HIGH DENSITY Mo / Ni / Cr COATING ON LAMELLAR CAST IRON OBTAINED BY LASER PROCESSING

Fig. 9

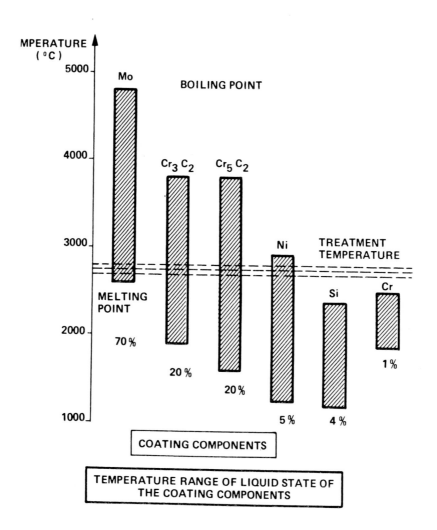

Fig. 10

POROUS LASER COATING

POROSITY VARIATION AT SEVERAL DEPTHS

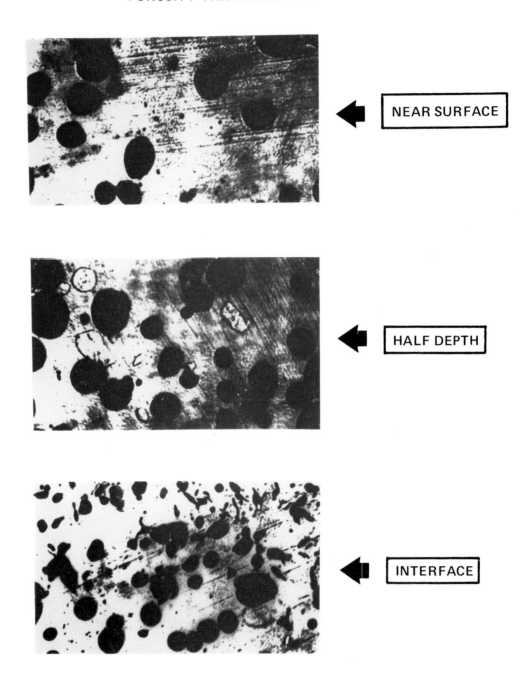

NEAR SURFACE

HALF DEPTH

INTERFACE

Mo / Ni / Cr COATING WITH LUBRICANT RETAINING
PORES ON LAMELLAR CAST IRON OBTAINED
BY LASER PROCESSING

Fig. 11

STRUCTURAL ASPECTS OF A Ni / Cr / Al
LAYER BEFOR AND AFTER LASER TREATMENT

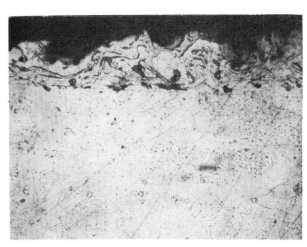

**AS PLASMA
SPRAYED WITH Ni Cr Al**

**BASE MATERIAL
In 738 LC**

300 x

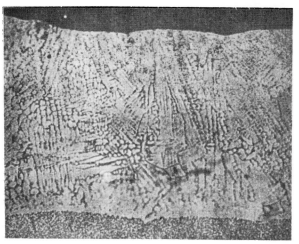

**THE SAME AS
ABOVE BUT
LASER TREATED**

500 x

Fig. 12

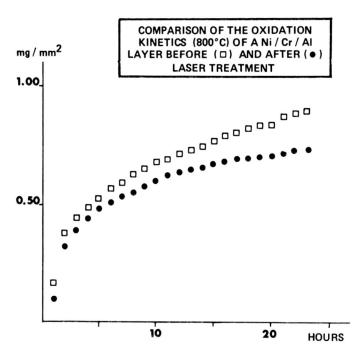

Fig. 13

LASER WELDING & DRILLING

C. M. Sharp
BOC Industrial Power Beams
Daventry, England

Abstract

This paper describes the absorption of 'laser light' by a workpiece and shows that the physical, chemical and thermal changes that result, can be controlled, to cause either drilling or welding of the workpiece. The subject is treated from a practical engineering/applications viewpoint and includes case studies of established industrial applications. The unique features of laser processing are highlighted and where appropriate, comparisons are made with other industrial processes.

Introduction

In the early 1960's the power of the pulsed ruby laser was often measured and expressed in terms of the number of laser pulses required to penetrate a razor blade. This was probably the first example of laser material processing, which showed that a focussed laser beam could melt and if necessary, vapourise material, thus establishing the feasibility of laser welding and drilling. The step from feasibility to becoming a reliable and economically viable production technique has necessitated the development of both the laser and laser processes. Today after over a decade of development, the laser is no longer a solution looking for a problem but a machine tool looking for applications which will exploit its unique advantages. These applications can be where the laser is replacing a conventional method or highly innovative and only possible by means of the laser.

Absorption of laser energy by a workpiece

Laser welding and drilling are both thermal processes which rely on efficient absorption of the laser energy incident on the workpiece. The factors involved in the absorption process are :-

Reflectivity

This is a measure of how much of the incident energy is absorbed by the workpiece to affect the welding or drilling operation and is dependent upon :-

    Workpiece material
    Wavelength of incident laser beam
    Surface roughness of workpiece
    Surface contaminants e.g. oxide, oil, etc.
    Surface temperature

Taking all the above factors into account, at room temperature the reflectivity for a metallic workpiece will lie in the range 60-90% Non-metallic workpieces will have lower reflectivities typically 10% at 10.6 microns. In the case of metals it is fortunate that the reflectivity drops when a focussed beam impinges on the workpiece. This is due to the rapid rise in temperature which lowers the reflectivity and ensures efficient coupling of the incident laser energy into the workpiece.

Thermal/Physical properties

Following absorption at the workpiece surface, the photon energy is converted into heat which travels by conduction into the bulk of the workpiece. The rate at which this occurs is determined by thermal diffusivity which is dependent upon thermal conductivity, specific heat and density. In order to subsequently produce melting and in the case of drilling, vapourisation, phase changes must take place which are dependent on density, specific heat, latent heat of fusion and latent heat of vapourisation.

Plasma effect

This effect is observed when vapourised workpiece material becomes ionised forming a "plasma cloud" between the incident beam and the workpiece surface. This results in absorption of the incident energy and can effectively arrest the drilling or welding operation. The steps that are taken to minimise this effect are discussed later in the text.

Conclusion

The process of absorption of laser energy is complex and involves many parameters which require careful optimisation for a given welding or drilling application.

Laser Drilling

Principles of pulsed laser drilling

The majority of laser drilling applications are carried out using either pulsed solid state lasers e.g. Nd/Glass or pulsed gas lasers e.g. CO_2 Material can be removed during laser drilling by two mechanisms :-

a) Vapourisation
b) Ejection of molten particles.

In the unfocussed laser beam the power density is too low to raise materials to their melting and boiling-points. Focussing is therefore required to increase the power density to the order of $10^5 - 10^7$ Watts/cm^2. The precise power density that is achievable will be determined by the duration and energy of the laser pulse, the mode, wavelength, diameter and divergence of the laser and the focal length of the focussing optic. A typical situation would be a Nd/Glass laser operating at 1.06 microns giving a pulse of 50 joules in 1 millisecond, focussed by a 50mm focal length lens.

In practice the pulse length, pulse energy and focussed spot size are chosen to suit the particular application taking into account the following factors :-

a) The energy density must be sufficient to melt/vapourise material to the required hole depth.

b) For materials with a high thermal diffusivity the power density should be high to ensure rapid vapourisation before heat is lost by conduction.

c) If the power density is too high a plasma may form, blocking off the incident beam, resulting in poor energy utilisation and limited penetration.

d) If the power density is too low, then surface reflectivity will remain high and the workpiece material may not absorb sufficient energy to melt and vapourise.

e) For a given focussed spot size power density and pulse length are not normally mutually exclusive - power density is reduced if the pulse length is increased.

f) Minimum spot size and hence maximum power density is achieved, when using short focal length lenses, but in this case the depth of focus is small and the cone angle of the incident beam is large which both result in limited hole depth. In some cases when multiple shot drilling, refocussing is carried out as the depth of the hole increases.

Many of the above factors are inter-related, therefore the parameters chosen for a particular application are always a compromise.

Features of laser drilled holes

Holes produced from single shot drilling are tapered, being of larger diameter at the "input" and limited in depth to about 4mm in Stainless Steel. Both taper and depth can be improved using multishot drilling and refocussing.

When compared with conventionally drilled holes, from the point of view of geometry and surface finish, laser drilled holes appear rather poor but from a processing point of view, laser drilling has the following advantages :-

a) It is a non-contact method hence there is no contamination of the workpiece material.

b) No cutting forces, therefore tooling is simple and cheap.

c) No tool wear or breakage.

d) Will drill hard, brittle materials e.g. diamond or fired ceramic.

e) Will drill soft, pliable materials e.g. rubber.

f) Can drill in inaccessible or hostile environments using beam steering mirrors e.g. atomic reactors.

g) Material removal rates are high compared with, for example, spark erosion techniques.

h) Will drill fine holes < 0.01 mm dia.

i) Beam switching techniques can be used to direct the laser output to a number of work stations in sequence, thus ensuring high throughput and efficient utilisation of the laser.

j) Using a high pulse repetition rate laser a series of 'blind' overlapping holes can be produced for scribing, cleaving and engraving operations.

k) Material is removed rapidly therefore components can be dynamically balanced using a succession of fast drilling pulses whilst the component is spinning.

Applications

The following are examples of applications either in commercial use or under development.

Drilling of ceramic substrates for thick film circuits
Dynamic balancing of gyroscopes
Drilling of irrigation pipes
Drilling of diamond wire drawing dies
Drilling of gemstones to remove impurities
Drilling of cooling holes in turbine blades
Scribing of silicon and gemstones
Engraving serial numbers on aircraft components.

Case Study - Laser diamond die drilling

This case study is presented to illustrate the technical and economic advantages of laser drilling

Diamond is widely used for wire drawing dies because of its hardness and hence superior wear properties when compared with other die materials such as tungsten carbide. To produce such dies however, requires the diamond to be drilled as shown in Fig 1. Conventionally this is an expensive time consuming and laborious process, consisting essentially of reciprocating a pointed metal needle in a slurry of diamond powder and lubricant. The process can take up to 24 hours for a single hole so that, to get a reasonable throughput, machines with up to twenty needles or spindles are used which take up space and use large quantities of diamond powder.

Although diamond is transparent, it has been found possible to drill and cut this material with a pulsed ruby laser. Unfortunately, owing to the nature of this laser, the pulses tended to be of a fairly high energy and frequently caused the stone to shatter. The discovery of neodymium-doped calcium tungstate and YAG gave a laser with a small power per pulse, but a frequency of 5 or 10 pulses per second, compared to the maximum of 1 per second for ruby. The neodymium laser is used to 'nibble' the diamond away and when the stone is drilled from both sides gives a bell-shaped profile which is ideal for wire drawing, so reducing the amount of additional polishing necessary.

The picture in figure 2 shows a laser diamond-die driller with the laser behind the operator's head. A die is being loaded into the work chamber while the xyz tables are used to position the die at the laser focus. The die is rotated during drilling to make sure the hole is symmetrical and in addition the axis of rotation can be offset from the laser focus so that larger holes can be shaped and trepanned. The whole operation is watched on the CCTV and controlled from the panel on the right which not only enables the operator to set the pulse rate, pulse energy and rotation speed but also controls the rate at which the pulse energy is increased as the hole deepens. This ensures that there is minimum damage at the entrance to the hole, and the maximum drilling rate is used in the later stages.

In order to show the cost advantages of the laser system over the traditional method using a twenty-spindle mechanical system, a comparison is made for two sizes of die (Table 1)

The economic advantages of laser drilling became abundantly clear, in addition to the benefits of cleaner, faster working, with more accurate work owing to the extra control given by the CCTV viewing system.

Fig. 1 Section through diamond wire drawing die

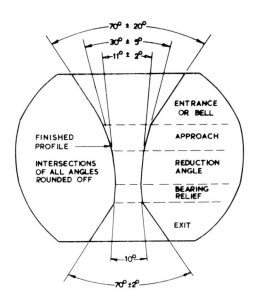

Fig. 2 Nd/YAG laser diamond die drilling machine

Table 1. Comparison of laser and conventional die drilling

Basic costs	Laser £	Mechanical £
Labour plus overheads per hour	4.00	4.00
Spares and servicing per hour	0.70	0.50
0.25 mm diameter dies		
Consumables per hour (Flash tubes, diamond powder etc.)	0.24	4.18
Die production per hour	12.00	5.00
Cost per die	0.41	1.74
Cost saving per die is	1.33	-
Cost saving per machine hour is	15.96	
1.5 mm diameter dies		
Consumables per hour	0.62	17.24
Die production per hour	4.00	1.3
Cost per die	1.33	13.3
Cost saving per die	11.97	-
Cost saving per machine hour	47.88	-

Laser Welding

Principles of laser welding

In laser welding the principles of 'beam coupling' into the workpiece are similar to those discussed in respect to laser drilling, except that the aim is to melt material with minimum loss of material from the workpiece due to vapourisation or ejection of molten material. The power densities employed for welding are therefore somewhat lower, typically 10^5 WATTS/cm^2. In order to achieve this power density focussing optics are employed and in the case of pulsed welding the pulse length is adjustable.

Welding can be carried out with both pulsed and continuous output lasers. Pulsed systems produce either discreet spot welds or a series of overlapping spots forming a seam weld. Pulsed welds are 'conduction limited', that is the penetration and shape of the weld are determined largely by the thermal properties of the workpiece in particular, the thermal diffusivity. Because heat flow takes place radially from the focussed spot, both down into the workpiece and laterally, conduction limited welds tend to have a low depth to width ratio.

Welds produced from continuous lasers are also conduction limited up to about the 1KW power level. Above this level assuming reasonable focussing optics, the welding mechanism changes. The power density at the surface is sufficient to cause localised boiling and since the vapour is highly absorbent to the incident beam, the rate of boiling accelerates, creating a vapour pressure which causes a depression in the melt pool. If the incident power is sufficient then the depression propagates downwards through the workpiece to produce what is referred to as a 'keyhole'. In order to produce a continuous weld the workpiece is moved relative to the keyhole and molten material flows from the leading edge to the trailing edge of the keyhole where it 'freezes' to form a deep penetration weld.

Welds of this type are characterised by their high depth to width ratio and narrow heat affected zone.

Because the incident beam is trapped by the keyhole, energy is transferred directly into the weld cross-section. Compared with conduction limited welds, keyhole welds use the incident energy more efficiently for two reasons.

1) The transfer of energy to the workpiece is more efficient.

2) The weld cross section is narrow therefore for a given absorption of energy, high welding speeds are achievable.

There is however a factor which can limit the performance of laser keyhole welding, that is the production of a plasma at the workpiece surface which has been mentioned previously in the section on absorption of laser light. The vapour within the keyhole cavity is an efficient absorber of the incident beam. However, because of the vapour pressure some of the vapour is ejected from the cavity in the direction of the incident beam, combining with the weld shielding gas to form a dense energy absorbing cloud over the keyhole. The effect of this cloud over the keyhole is twofold, it can prevent a large proportion of the incident energy reaching the keyhole and acts in itself as a secondary, diffuse energy source. This secondary source radiates directly to the workpiece surface surrounding the keyhole and heat is then conducted from the surface into the bulk of the material. The resultant weld is therefore a combination of conduction limited and keyhole welding processes and is of the form shown in Fig. 3.

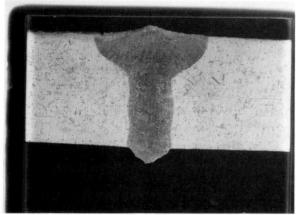

Fig. 3

The plasma thus has the effect of broadening the weld and reducing the welding speed for a given power. The weld shape produced, is often less acceptable from an engineering viewpoint compared with a parallel sided pure keyhole weld. Thus when using multikilowatt lasers on thick sections, techniques are often employed to control or destroy the plasma. On thinner sections where welding speeds are high the plasma effect is less pronounced. Much of the work on plasma control has been pioneered by the British Welding Institute. An example of a pure 'keyhole' laser weld is shown in Fig. 4.

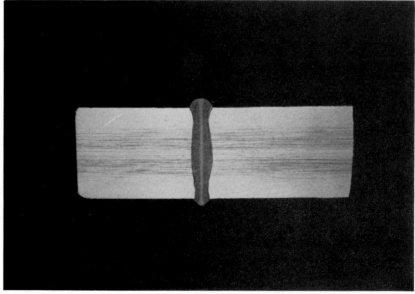

Fig.4.

Features of laser welding

a) Low heat input, therefore distortion is minimised.

b) Small heat affected Zone.

c) No contamination of the weld from electrodes compared with resistance welding or TIG welding.

d) Narrow, deep penetration welds with aspect ratios similar to electron beam welds, can be produced out of vacuum.

e) There is no pump down time hence high throughputs are possible.

f) No filler metal is required - this does mean however that joint fit-up must be good.

g) Time sharing and/or power sharing to a number of workstations is possible ensuring high utilisation of the laser.

h) In the case of 1.06 micron lasers, welding can be carried out through glass envelopes.

i) The technique allows welding to take place into inaccessible or hostile environments.

j) Will weld materials that are difficult to electron beam weld eg mild steel

Applications

The following are examples of applications that are either in production or under investigation.

High speed seam welding of cans.
Securing electric motor lamination stacks
Fabrication of gear clusters for the automotive industry
Continuous welding of bi-metallic strip for hacksaw blade manufacture
Joining of galvanised steel strip in continuously operating strip mills.
Pulsed seam welding of heart pacemaker cans.
High speed seam welding of microelectronic packages avoiding thermal damage to contents.
Fabrication of complex sheet metal components requiring two or three dimensional weld paths.

Case study - Fabrication of sheet metal components for the aircraft industry

In order to achieve stiffness and light weight, many high temperature aircraft components are fabricated from nickel alloy sheet. The welds required are inaccessible, complex in profile and often the application demands that the welds should be 100% gas tight. Components of this type can be joined using a combination of TIG, plasma, resistance seam and electron beam welding, but in practice many problems are encountered with restricted access, component size (in the case of electron beam) excessive heat input (TIG and plasma welding) and electrode problems when resistance welding.

Currently an application of this type is being resistance seam welded using specially shaped electrode wheels but in order to maintain weld quality, constant electrode changing is required. The operation is therefore costly and labourious. The weld integrity is variable often necessitating expensive reworking operations.

In future a 2KW CO_2 gas laser of the type illustrated in Fig. 5 will be used with specially designed high speed moving optics. In order to cope with a range of weld profiles the moving optics drive motors will be controlled by a mini-computer which will translate the weld path into a series of straight lines and arcs of circles. The computer is also used to control absolute position of the weld relative to the workpiece, welding speed and slope-in and slope-out of the laser power when producing 'envelope' welds. The system is designed to operate up to 120mm/sec welding speed with a maximum acceleration of 0.3g and a positional accuracy of $\mp$ 0.2mm. The application requires the production of 3500 metres of gas tight weld per structure which will be achieved in approximately 10 hours welding time.

The use of the laser for this application is justified in terms of :-

a) Elimination of electrode and reworking costs.

b) Increased productivity due to higher welding speed and elimination of electrode changing/dressing operations.

c) Improved product quality

Fig. 5

The future for laser drilling and welding

The number of Companies currently manufacturing lasers is not so much related to current usage but is more an indication of the growth that is expected in the use of lasers by Industry, in the future.

In common with all new technologies the laser is taking time to establish itself as a production tool. To prove technical and economic viability has necessitated the development of the laser, in terms of its output and reliability and equally important, the development of laser process technology.

Laser drilling and associated processes such as engraving and scribing are already well established. They are however normally carried out on 'difficult' materials. The area of application is thus specialised and it is most unlikely that the laser will ever replace the twist drill.

Laser welding has been slower to take off apart from specialised microwelding applications. However, now that multikilowatt industrialised lasers are available capable of producing deep penetration welds up to 15 mm in a variety of materials, it is expected that laser welding will be a major growth area. Many current electron beam applications will, in future, be laser welded without the need for a vacuum chamber. However, the main applications area will be in high volume production taking advantage of the high welding speeds and the ease with which laser welding can be controlled and automated.

PROCESS ASPECTS OF LASER WELDING

R. C. Crafer
The Welding Institute, Abington
Cambridge, U. K.

Abstract

The possibility of producing laser welds by the so called "deep" penetration welding mechanism similar to the well established electron beam welding technique has been known for almost ten years. However, unlike electron beam welding, the beam metal interaction mechanisms can have a serious effect on the weld depths achieved in a single pass. The relevant laser process and material parameters will be discussed in terms of their effect on welding performance and compared with other welding methods.

The Process in Perspective

High power continuous CO_2 lasers have now been employed in welding applications for almost a decade. The Welding Institute's 2-3kW CO_2 laser has itself been in operation for some six years and its 5-6kW laser for several months. Considerable experience has thus been obtained of this novel technique.

Laser welding produces relatively deep narrow welds similar in profile to those produced by electron-beam welding (EBW). A comparison of the two is shown in Figure 1 for an 18/8 stainless steel 2mm thick. As in the case of EBW, the laser process is characterised by a high power density at the workpiece resulting effectively in a low heat input low distortion weld with a narrow heat affected zone (HAZ). The rapid thermal cycling of the narrow fusion zone limits grain growth during heating, and can impart improved joint properties to certain materials.

Butt and lap welding of sheet materials are both possible with current techniques, typical butt welding thicknesses being 4-5mm at 2kW and 8-10mm at 5kW. Butt welding is usually preferred because of its higher joining rate and more efficient use of the laser power. In both methods the narrow profile of the focused laser beam, usually less than 200µm in diameter requires accurate fit up to ensure good coupling of power from the laser beam to the workpiece. A suitable laser application thus appears to be joining precision premachined components. Alternatively, high volume production is possible using the very high welding speeds associated with keyhole penetration. One advantage accruing from the required precision of fit up is that filler material is not normally required; consequently the laser welding process is basically simple. The overall heat input required by the narrow fusion and HAZs, together with the lack of filler material and its attendant heat capacity, makes laser welding a fast process, particularly for materials in the thickness range up to 6 or 7mm.

Laser welding proceeds via the intermediary of a keyhole - a vapour filled cavity surrounded by a wall of molten metal - extending the full depth of the weld. In butt welding applications the keyhole usually extends the full depth of the material, resulting in a guaranteed complete penetration weld. In these respects the laser process bears similarities to EB and microplasma welding used in a keyhole mode. The first stage in the formation of a laser weld is absorption of laser light at the surface of the workpiece. At the CO_2 laser wavelength of 10.6µm (in the infrared region of the spectrum) most cold metals are excellent reflectors, the exact value of reflectivity depending on composition and cleanliness. For example, most common steels, in clean condition, reflect about 90% of the incident light. The initial heating rate is therefore low. As the temperature rises the reflectivity gradually falls so that the heating rate tends to accelerate. By the time the metal has reached its boiling point, which may take only a small fraction of a millisecond, the reflectivity will probably have fallen to around 70 or 80%. Metals with higher reflectivities will approach boiling point by progressively slower rates until, for a given laser beam, a situation will be reached where boiling cannot be achieved. For this reason highly reflecting metals such as gold, silver and copper cannot at present be welded at the 2kW power level, although welding has been reported at higher powers.

When the metal boils the emitted vapour becomes highly absorbent to the laser beam, and the greatly increased boiling rate causes local surface depression and the formation of a keyhole in the metal surface. If the laser power is sufficient the keyhole will penetrate completely through the metal. To produce a weld the joint is moved relative to the beam so that the vapour-filled keyhole is translated along the joint line. Liquid metal moves from the front to the back of the keyhole under the action of surface tension forces and temperature gradients, and solidifies to form the characteristic narrow fusion zone. Figure 2 illustrates this keyhole process diagrammatically.

Comparing Welding Processes

Any comparison of laser welding with other processes is always somewhat artificial because each welding process has its own optimum range of metal thicknesses, power dissipations, joining rates, and joint geometry. However inadequate it may appear, when a comparison is called for it should usually be restricted to similar powers applied to similar metals of similar thickness to make it as meaningful as possible. For example, powers between 1 and 2kW are relevant to the laser welding of an 18/8 stainless steel in the thickness range 1.6 to 3.0mm and also to TIG and EBW welding of this material. As a further refinement, the comparison is limited to those processes not requiring filler material, ie: EBW and TIG. As representative of a more common process, MIG welding data are also included but numerical comparisons should be treated with caution.

1. WELDING SPEED is the linear travel speed of the welding gun relative to the workpiece. In ascending order, for a material thickness of 2mm:

MIG and TIG	10mm/sec
CO_2 laser	45mm/sec
EBW	45mm/sec

2. OVERALL JOINING RATE is the area of butt/lap joined per unit of electricity from the mains:

MIG and TIG	$10,000mm^2/kWhr$
CO_2 laser	$15,000mm^2/kWhr$
EBW	$50,000mm^2/kWhr$

3. PROCESS JOINING RATE is the area of butt/lap joined per unit of energy applied to the metal surface:

MIG	$5mm^2/kJ$
TIG	$7mm^2/kJ$
CO_2 laser	$55mm^2/kJ$
EBW	$55mm^2/kJ$

4. OVERALL ELECTRICAL EFFICIENCY is the power developed by the process compared with that drawn from the mains:

CO_2 laser (overall)	~10%
CO_2 laser (resonator)	~25%
TIG	~25%
MIG	35-40%
EBW	~40%

It will be noted that, although the CO_2 laser efficiency is comparatively low (though not by laser standards), the highly efficient manner in which it uses its energy to produce a weld by means of its high power density gives it an overall joining rate similar to the MIG and TIG processes, but with a welding speed very much greater, and comparable with the EB performance.

The laser produces EB-type welds in thin gauge material without the need for a vacuum chamber and its associated equipment. No limitations are therefore imposed on workpiece size or on the operational duty cycle. In addition, the infrared photons comprising the laser beam are of low energy in themselves and therefore do not produce the dangerous X-rays associated with the EB process. Only low cost acrylic shields are required to confine the laser beam to the working area. The atmosphere itself is transparent to the beam from a CO_2 laser which may be transmitted a distance of several hundred metres with virtually no power loss, though for safety reasons a confining metallic tube would be used as a conduit. This property allows remote laser working, such as welding in a hostile environment through a pressure window, and also permits multiple work station operation either on a time or power division basis. Finally, beam manipulation and focusing is carried out entirely by conventional optical techniques using mirrors and lenses, components which are readily available from optical suppliers.

Laser welding is at present restricted to about 15mm for a single sided weld and around 25mm for a double sided weld. As an approximate guide, 2kW of laser power will weld most steels up to 5mm thick and certain aluminium alloys up to 3mm. 5kW will weld 12mm steels and lesser thicknesses of aluminium or copper. Powers in excess of 10kW will weld up to 15mm thick material and may weld even thicker if the power is sufficient and the focused beam profile adequate.

Process Limitations

There are two particular process limitations having a direct bearing on laser welding. The Threshold effect, already described briefly, signals the change from conduction-limited welding to deep penetration keyhole welding and is associated with vaporisation at the metal surface. At 2kW the effect is only evident with good thermal conductors such as copper and aluminium, and at 5kW it can be avoided almost entirely provided a well focused laser beam is employed.

The more important effect is that of plasma absorption. It occurs particularly at low welding speeds in thick materials when the metal vapour emanating from the keyhole forms a dense plasma above the weld and either wholly or partially prevents the beam from reaching the workpiece. The beam power absorbed by the plasma is reradiated as from a point source and the weld consequently becomes conduction limited. To improve the profile and increase penetration, a plasma control device has been developed. Improvements are specially dramatic with the 2kW laser where a maximum penetration of 5mm was increased to 6.5mm in stainless steel. For moderate thicknesses, say 2 to 4mm, the plasma control device plays a special role. Although not necessary to achieve penetration it has a considerable effect on weld profile. In the example shown in Figure 3, the first section (a) shows an uncontrolled weld in a 3.7mm thick martensitic stainless steel. Note the distortion inducing weld assymmetry and the presence of root porosity commonly associated with non-penetrating or marginally penetrating welds. With partial control (b) the assymmetry is reduced, while with full control (c) the weld is more parallel sided and exhibits no root porosity.

Welding Steels

It is instructive to compare directly the performance of both the 2kW and 5kW lasers on similar thickness and composition workpieces. Figure 4 shows a transverse metallographic section of a laser weld in 6.5mm thick type 304 stainless steel produced at a workpiece power of 2.0kW and at a welding speed of 6mm/s. These conditions are beyond the limitations of a "straight" 2kW laser and require the use of the plasma control device described above, to extend penetration depth beyond 5mm and to maintain weld shape. Note the relatively inconsistent top bead and the poor transverse profile. In contrast the weld section shown in Figure 5 exhibits consistent bead shape and good transverse profile. It is in fact very similar to an electron beam weld. Conditions in use here are 4.5kW laser power and 20mm/s welding speed. Owing to the high welding speeds achieved with this more powerful laser, plasma control was not required.

At a similar thickness, Figure 6 shows an example of a weld in a somewhat different alloy. This is a Ni-Cr-Mo low alloy steel and the weld conditions are 5.7kW and 25mm/sec. Of particular significance is the limited extent of the HAZ which shows up well in this example compared to the stainless steels where the HAZ is not visible. In fact at its widest point, the total weld zone is only 1.6mm wider than the fusion zone.

Figures 7 and 8 are both mild steels, 8.0 and 9.5mm thick respectively, welded at 4.5kW and 12mm/sec and 6kW and 19mm/sec. The deepest full penetration weld made in the single pass mode with the 5kW laser so far is shown in Figure 9. The material is a low alloy steel, the power 6kW and the welding speed 6.4mm/sec. Although the speed in this case is almost identical to that of the 2kW weld in Figure 5, plasma control was not required, and no indication of impending plasma absorption is visible from the transverse section.

Welding Non Ferrous Metals

High conductivity materials such as aluminium and copper are traditionally difficult to weld with long wavelength lasers (eg: carbon dioxide types) owing to their high initial reflectivity and excellent thermal conduction properties.

Recent work has shown that aluminium alloys can be welded at the 2kW power level without the need for absorbing surface coatings provided great care is taken in the selection and maintenance of the mode structure of the laser beam. However the applications is somewhat critical, and limited to materials up to 3.5mm in thickness.

The higher power laser however welds aluminium alloys with few of the criticality problems associated with the 2kW laser. A section of 6mm thick H15 alloy welded at 5.6kW and 44mm/sec is shown in Figure 10. The fusion zone itself is roughly 2mm wide, with a total weld zone width extending to a maximum of 5mm.

In addition to aluminium alloys, some preliminary work has been carried out on titanium alloys. Figure 11 shows a traverse section of a weld in 8.9mm thick alloy made at 5.7kW and 28mm/sec. The feature at the top of the weld is a rolling defect and is not attributable to the welding process.

Conclusions

The industrial carbon dioxide laser has been developed to the stage where it can tackle many of the thin gauge applications currently carried out by electron beam welding and by some of the arc processes. Future development is likely to be concentrated in the fields of work manipulation and process control. With its many unique features, the laser should see a marked increase in industrial applications within the next decade.

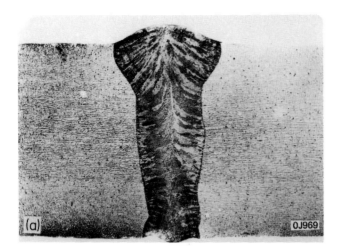 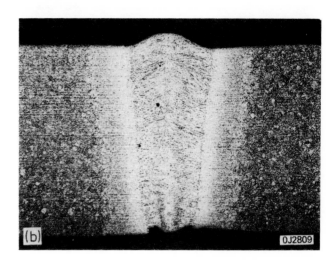

Fig. 1. Weld profiles: (a) laser, in 2mm thick 18/8 stainless steel, power 1.5kW, speed 25mm/sec. (b) electron beam in 2mm thick martensitic stainless steel, power 1.25kW, speed 30mm/sec.

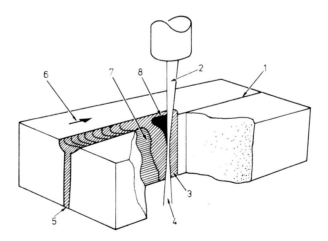

Fig. 2. Perspective drawing of keyhole formation and welding process. 1 - close butt joint; 2 - focused energy source, eg: laser beam; 3 - molten metal; 4 - proportion of energy passes through keyhole; 5 - full penetration weld; 6 - direction of welding; 7 - solid weld bead; 8 - keyhole.

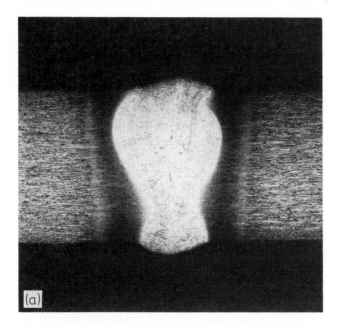

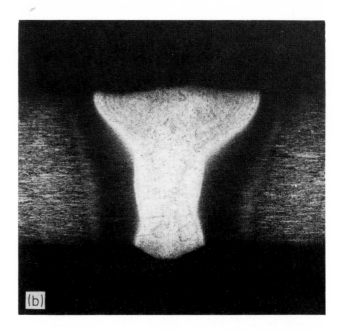

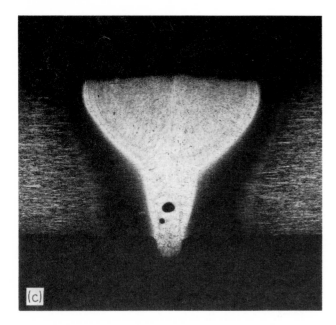

Fig. 3. Effect of plasma control on 3.7mm thick martensitic stainless steel (a) no control, (b) partial control, (c) full control.

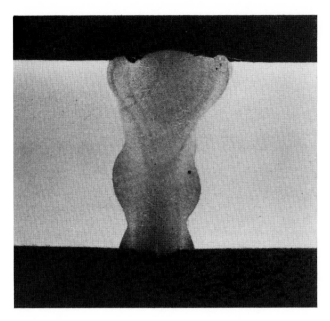

Fig. 4. Weld in 6.5mm 304 stainless steel, power 2kWm speed 6mm/sec

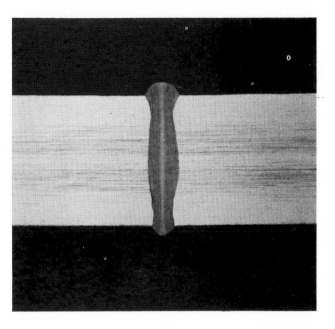

Fig. 5. Weld in 6.5mm 304 stainless steel, power 4.5kW, speed 20mm/sec

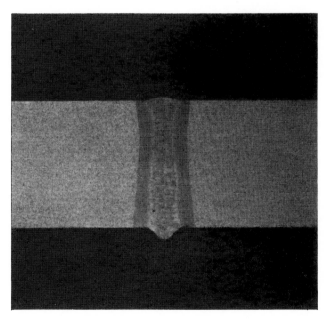

Fig. 6. Weld in 6.5mm Ni-Cr1Mo low alloy steel, power 5.7kW, speed 25mm/sec

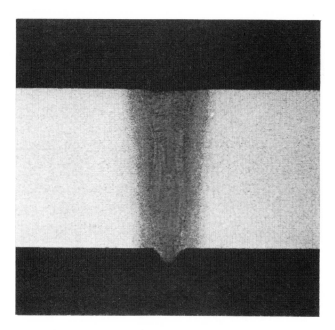

Fig. 7. Weld in 8mm mild steel, power 4.5kW, speed 12mm/sec

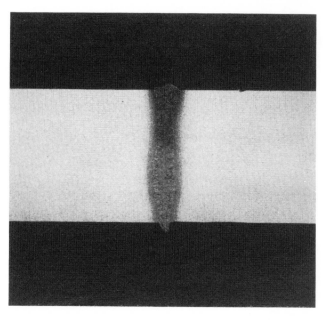

Fig. 8. Weld in 9.5mm mild steel, power 6kW, speed 19mm/sec

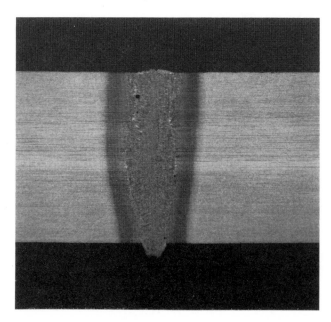

Fig. 9. Weld in 12mm low alloy steel, power 6kW, speed 6.4mm/sec

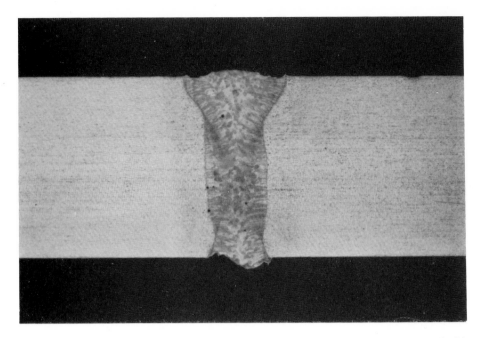

Fig. 10. Weld in 6mm H15 aluminium alloy, power 5.6kW, speed 44mm/sec

Fig. 11. Weld in 8.9mm titanium alloy, power 5.7kW, speed 28mm/sec

AUTHOR INDEX

SUBJECT INDEX